E-Commerce Systems Architecture and Applications

Related Artech House Titles

Desktop Encyclopedia of the Internet, Nathan J. Muller

Electronic Payment Systems, Donal O'Mahony, Michael Peirce, and
   Hitesh Tewari

Internet Commerce Development, Craig Standing

Internet E-mail: Protocols, Standards, and Implementation,
   Lawrence Hughes

Internet and Intranet Security, Rolf Oppliger

Practical Guide for Implementing Secure Intranets and Extranets,
   Kaustubh M. Phaltankar

Secure Electronic Transactions: Introduction and Technical Reference,
   Larry Loeb

Security Technologies for the World Wide Web, Rolf Oppliger

For further information on these and other Artech House titles,
including previously considered out-of-print books now available
through our In-Print-Forever® (IPF®) program, contact:

Artech House
685 Canton Street
Norwood, MA 02062
Phone: 781-769-9750
Fax: 781-769-6334
e-mail: artech@artechhouse.com

Artech House
46 Gillingham Street
London SW1V 1AH UK
Phone: +44 (0)171-973-8077
Fax: +44 (0)171-630-0166
e-mail: artech-uk@artechhouse.com

Find us on the World Wide Web at:
www.artechhouse.com

E-Commerce Systems Architecture and Applications

Wasim E. Rajput

Artech House
Boston • London
www.artechhouse.com

Library of Congress Cataloging-in-Publication Data
Rajput, Wasim E.
   E-Commerce systems architecture and applications / Wasim E. Rajput.
       p. cm. — (Artech House telecommunications library)
   Includes bibliographical references and index.
   ISBN 1-58053-085-0 (alk. paper)
     1. Electronic commerce.   I. Title.   II. Series.

HF5548.32 .R35 2000
658.8'4—dc21

00-035553
CIP

British Library Cataloguing in Publication Data
Rajput, Wasim E.
   E-Commerce systems architecture and applications. —
   (Artech House telecommunications library)
   1. Computer architecture   2. Computer networks   3. Electronic commerce
   I. Title
   004.2'2

   ISBN 1-58053-085-0

Cover design by Igor Valdman

© 2000 Artech House
685 Canton Street
Norwood, MA 02062

International Standard Book Number: 1-58053-085-0
Library of Congress Catalog Card Number: 00-035553

10 9 8 7 6 5 4 3 2 1

The man who can make hard things easy is the educator.
—Ralph Waldo Emerson

To all my mentors . . .
especially the very first ones, my parents.

Contents

Preface

E-commerce initiatives are proliferating in virtually all aspects of the socioeconomic environment. Innovation in computing technologies has extended the reach of e-commerce transactions into business-to-business, business-to-consumer, and intraorganization business processes. Engaging in e-commerce initiatives requires enterprises to embrace intelligent IT-enabled solutions to reengineer their business processes. These systems cohesively integrate an organization's customers, suppliers, distributors, retailers, partners, and employees to further various e-commerce business propositions. E-commerce systems draw customers closer to an enterprise's services and thus strengthen customer relationships, streamline internal processes to deliver cost efficiencies, and enable intelligent collaboration between enterprises.

An e-commerce system is a conglomerate of applications powered by the Internet, the World Wide Web, and other innovations. These applications employ various information appliances, offer various backend transaction processing and information access services, and run on a web of interconnected private and public networks. Complex interfaces between various applications interplay to fulfill a discrete e-commerce business objective. For example, various applications work in conjunction with each other to shape a shopping Web site that lets customers browse through various items, order online, and engage in online customer service interactions. The various applications that materialize this e-commerce business objective collectively constitute an e-commerce system. Similarly, a bank's bill payment e-commerce system on the Internet could enable customers to log on to the bank's Web site using any Internet-enabled device (PCs, handheld devices, etc.) and transfer funds from their checking account to the payee's account. Numerous applications interact to facilitate the execution of this simplistic transaction over a multitude of public and private networks. These include applications that process registering of payees and their respective information, building of an Internet site to enable the customers to process such transactions, settlement applications that settle funds from the bank to the payees' accounts through settlement houses, and so on.

Numerous factors contribute to the effective design and deployment of e-commerce systems. These include managing the technical complexities inherent in the technologies of e-commerce systems, innovative interorganizational synergies, leasing of various e-commerce services, refurbishing one's technology infrastructure to support operations of e-commerce systems, and managing various e-commerce specific risks. This book presents an e-commerce systems architecture that identifies all the major building blocks of an e-commerce system and describes the interaction of these building blocks to shape effective e-commerce systems. For each building block, the book highlights issues that influence the design of e-commerce systems for various e-commerce business domains.

The primary building blocks of e-commerce systems architecture are information appliances, computing networks, e-commerce services and applications, e-commerce core services, organizational technology infrastructure, and networks of electronic payment systems. Information appliances facilitate access to enhanced e-commerce services and access these services through a web of public and private computing networks. E-commerce services and backend applications make up the core component of the e-commerce systems architecture and provide various e-commerce services to various users.

Intelligent integration of e-commerce systems depends upon various e-commerce services, which is the next architectural component shaping e-commerce systems. E-commerce services are usually available from external organizations that provide specialized services such as e-commerce hosting, a certificate authority infrastructure, portal functionality, domain name registration, and others. Finally, an appropriate internal technology infrastructure provides services such as groupware, systems management, and so on, to augment and control the other architectural components discussed earlier.

Plan of the Book

This book provides the readers with a thorough understanding of the various components that shape e-commerce systems. The book delves into the components' inherent technologies, discusses pertinent design and development issues, and presents management strategies that collectively shape e-commerce systems. The various issues presented in this book will provide the readers with insight into the design, development, and operations of e-commerce systems corresponding to all e-commerce business domains (business-to-business, business-to-consumer, and intra-business).

Chapter 1: E-Commerce–Enabled Business Paradigm

This chapter introduces the reader to the various e-commerce value propositions and provides examples of e-commerce systems that fulfill those propositions. The chapter demonstrates how e-commerce initiatives affect an organization's IT strategy. Three e-commerce business domains and respective business requirements are identified. The chapter then highlights characteristics of e-commerce systems that fit each of the identified e-commerce business domains. Finally, the chapter introduces the readers to the primary components of e-commerce systems. Subsequent chapters address these components.

Chapter 2: Information Appliances

Chapter 2 delves into the details of information appliances. Information appliances constitute the architecture's client component. These appliances provide users with access to various local and remote network computing services. Information appliances extend beyond the dumb terminal of the first computing era and the PC of the client/server era to encompass intelligent devices such as pagers, cellular telephones, handheld computers and many others. This chapter describes the various types of information appliances and inherent technologies required for connecting to enterprise computing services. The chapter also analyzes business requirements that drive the choice of information appliances for business processes and analyzes strategies to "e-commerce enable" an organization's technology infrastructure.

Chapter 3: E-Commerce Systems Computing Networks

Computing networks enable users equipped with various information appliances to connect to backend enterprise computing services. Though the Internet and Web are the prominent networks enabling various e-commerce services, other networks such as the PSTN, GSM, and other cellular networks and pager networks play an active role in integrating diverse e-commerce services. This chapter illustrates the structure and architecture of popular computing networks and describes the network services that these networks provide to enable the design of e-commerce systems. The chapter also describes strategies and issues related to bridging an enterprise network to these networking backbones in order to formulate a virtual network.

Chapter 4: E-Commerce Services and Application Repositories

Chapter 4 discusses the technical details of the design and implementation of backend e-commerce applications and services. The chapter presents an e-commerce application architecture that consists of access gateways, middleware, and backend services. It identifies three types of applications (consultative, transactional, and inquiry types of applications) that fit the e-commerce application architecture. The chapter then delves into the application development facets of the various types of applications. This chapter also discusses intricacies inherent in integrating new e-commerce services with legacy applications.

Chapter 5: Establishing E-Commerce Application Access Infrastructure

This chapter introduces the reader to various technologies and strategies that enable enterprises to maximize the value of their IT infrastructures by providing users with intelligent systems access. Extranets and portals are various means that enterprises can exploit to enable intelligent access to information and enterprise services. This chapter discusses those concepts in detail. The chapter also sheds light on the network and systems management issues that enable maximum availability for enterprise IT-enabled services.

Chapter 6: E-Commerce Systems Technology Infrastructure

The design of various types of e-commerce systems requires the rejuvenation of internal enterprise technology infrastructures. Chapter 6 describes the various technologies and services required for the design and operation of e-commerce systems. These services and technologies include middleware technologies, groupware, directory services, domain name systems, and technologies related to application development.

Chapter 7: E-Commerce Payment Infrastructure

E-commerce services have triggered requirements for new forms of electronic payments. Chapter 7 introduces the various types of electronic payment systems and processes. More specifically, the chapter discusses the various process flows for using electronic cash, electronic checks, and credit cards. The chapter discusses the implementation details of the prominent electronic instruments and inherent technologies that enable e-commerce systems with various forms of electronic payments. Finally, this chapter addresses various issues relevant to the integration of various types of electronic payment instruments with e-commerce systems.

Chapter 8: E-Commerce Systems Security

This chapter introduces the reader to information security issues that prevail in the implementation of e-commerce systems. After summarizing the popular technologies relevant to various security controls, the chapter covers security issues relevant to all tiers (client and server) of e-commerce systems. The chapter presents strategies that an organization can use to assess the cost and security-related business risks associated with the implementation of e-commerce systems. Finally, the chapter presents various procedural controls pertinent to the security of e-commerce systems.

Chapter 9: Managing E-Commerce Systems Implementation Risks

The last chapter identifies certain e-commerce systems implementation and operational risks, and proposes appropriate strategies for managing those risks. The strategies include aligning internal processes with the new forces of e-commerce and properly understanding customer requirements before rolling out e-commerce services through diverse information appliances to external customers. Because the integration of e-commerce systems requires the leasing of certain e-commerce services from external service providers, this chapter discusses the need to formulate appropriate Service Level Agreements (SLAs) to manage and share those risks. The chapter also discusses legal and regulatory issues surfacing from engaging in e-commerce services through e-commerce systems. Finally, the chapter highlights the merits of formulating and enforcing various policies to control technology implementation risks.

1

E-Commerce–Enabled Business Paradigm

Welcome to the new business era, where the letter *e* seems to have become a prefix to all business terminology. These terms include *e*-business, *e*-commerce, *e*-marketing, *e*-sales, and many more. Ranging from high-level business models to low-level firm activities, most frameworks, technologies, and business processes are freely using the "*e* word" to signify the new way of doing business and transacting commerce. This phenomenon is indicative of a trend that goes beyond introducing new words into the English language and involves the weaving of a universal and integrated web of processes and activities using *e*lectronic and information technologies.

Much has been said, discussed, and theorized about the power hidden in this *e* word. However, as organizations move toward materializing these theoretical strategies and business models into real systems and processes, they face new challenges. These challenges are embedded both in the intricacies of new technologies that seek to revolutionize and reface past innovations and in the integration of these new technologies with the legacy platforms and systems that power today's systems and business processes. Furthermore, competing in this new era compels an organization to build appropriate business and technological capabilities to respond to the new business frenzy.

This book refers to the new breed of information technology (IT) systems that are revolutionizing such a change as e-commerce systems. The Internet, powerful networking technologies, sophisticated information appliances, and power-packed applications are appealing to enterprises and public audiences alike. The novel characteristics of these systems and their inherent technologies are reengineering the business processes for business-to-business, business-to-consumer, and intrabusiness commerce. To maximize an enterprise's return on investment (ROI) in IT technologies, an enterprise should equally exploit opportunities in all of three business domains.

E-commerce is about selling, buying, and conducting other ancillary activities through electronic channels. These ancillary activities encompass a wide array of functions that include establishing the means to sell, establishing incentives to sell, and collaborating with staff, partners, and customers through electronic channels in an interactive and noninteractive fashion. Together, all these activities both enable e-commerce and provide enterprises with the means to achieve the appropriate ROI on IT initiatives.

E-commerce systems primarily fit into three categories. First, a business-to-business e-commerce system leverages the Internet to streamline an organization's processes by cost-effectively chaining them with other strategic partners' systems. An e-commerce system that enables an organization to transact with other organizations for its business activities (e.g., corporate procurement) is an example of a business-to-business e-commerce system.

An e-commerce system fitting the business-to-consumer domain represents the second type of system, which involves commerce activities between the public domain customers and the organization. An e-commerce system that enables an organization to sell goods and services through the Internet to public domain customers is an example of such a system. Finally, intraenterprise e-commerce systems streamline an organization's internal activities and enable staff members to work effectively and efficiently through remote connectivity technologies, enhanced information sharing, team collaboration, and other similar means.

1.1   E-Commerce Business Drivers

Engaging in e-commerce through investment in e-commerce systems is becoming a competitive necessity while also providing opportunities for seeking competitive advantage. The following paragraphs describe the external market drivers that are pushing enterprises in the realm of e-commerce.

- *New customer/partner/regulatory business requirements:* New business requirements have a profound impact on an enterprise's IT strategy. For example, an organization may require its suppliers to engage in business transactions through special extranets. This drives the supplier to design appropriate e-commerce systems to transact with its customer organization. Similarly, the government requires some of its contractors to build technology capabilities to receive electronic checks. This in turn is driving the contractors to build e-commerce systems to engage in such transactions.

- *Customer empowerment:* A firm needs to recognize its customers' empowerment in order to build required capabilities in its e-commerce systems for offering its products and services. For example, by recognizing that a wide range of its customers are equipped with wireless telephones, an organization can build capabilities into its systems to offer its products and services to customers on wireless telephone displays. Consumers empowered with personal computers and other information appliances have come to expect certain services from their service organizations (through various electronic channels, such as Web sites and WebTV). This, in turn, is driving enterprises to deliver their products and services through various electronic channels. Recognizing customer empowerment compels organizations to further scrutinize and analyze customer behaviors, preferences, and needs in the online world. This requires the deployment of intelligent e-commerce systems that meet customers' requirements.

- *Competitive necessities:* A gap analysis of an organization's systems against its competition may highlight deficiencies that affect an organization's bottom line. This could compel the organization to deliver appropriately competitive products and services. For example, an organization may provide electronic bill presentment capabilities to its customers with no accompanying features for electronic bill payment, which a competitor may be providing to its customers.

- *Competitive advantage opportunities:* An organization can identify sources of competitive advantage by deploying appropriate e-commerce systems. For example, an organization may enable its systems to let its customers contact the organization's customer service department through the Internet using various channels such as Voice over IP (VoIP), and so on.

- *One-to-one marketing:* One-to-one marketing involves targeting an organization's products and services to individual customers as opposed to aggregate markets. This is pushing enterprises to invest in appropriate e-commerce systems. Building customer profiles enables organizations to target individual customers based on their demographic information and offer customized and personalized products and services. Customization enables organizations to shift from the traditional norms of delivering standardized products and services to delivering customer-specific products and services. For example, customization and personalization features enable customers to view customized content. Additionally, when ordering through the Internet, customers can specify their preferences in shipping products. For example, customers can specify the type of gift-wrap to be used, personalized messages, and so on.

- *Global competition:* The global presence of the Internet is facilitating easier access to global markets. Organizations can offer merchandise on their Web sites and expect global penetration overnight. Because the Internet is the only channel that provides an overnight global presence, an organization not venturing into e-commerce can be uprooted by its competition.

- *Branding:* The Internet and cyberspace in general present new opportunities for organizations to brand their products and services. The medium also threatens existing brand names within the traditional business landscape. Establishing an early presence on the Internet therefore is vital to establish a brand. Yahoo.com is an example of a public portal service that established its presence on the Internet early on and has become a popular portal brand among millions of users. Amazon.com, a company with no brick-and-mortar and no pre-Internet presence, branded itself by an early offering of an Internet bookstore, and it outperformed traditional brand names such as Barnes and Noble and Borders.

1.2   E-Commerce Value Propositions

There is a reason for the e-commerce mania. E-commerce systems purport to push enterprises into the new realm of doing business. E-commerce–related technologies provide firms with opportunities to reengineer their value chains' business processes. The term *reengineer* rightly fits this phenomenon, because the gains that a firm achieves as a result of these new business systems are tremendous. The gains come as a result of an exponential increase in customers, streamlining an organization's processes, and an associated increase in a firm's revenues.

Internet-enabled e-commerce has evolved through various phases. The first phase emerged with the dawn of the Internet in the 1970s and continued well into the 1980s. Entities on the Internet in this phase primarily focused on educational and research activities. The second phase started in the 1990s when government and industry recognized the Internet's power to influence the commercial sector. Numerous organizations established an online presence with static pages to build awareness and present content. During the third phase, which started in the middle 1990s and continues today, organizations got on the e-commerce bandwagon and engaged in e-commerce transactions. The final phase will necessitate an organization's online presence and capacity for electronic transactions. The U.S. government's requirement for the use of electronic payments is proof that the dawn of the fourth phase of the e-commerce evolution is upon us.

E-commerce trends are driving enterprises to embrace appropriate business propositions that will further their missions and objectives. This in turn requires deployment of appropriate e-commerce systems. This section highlights the various e-commerce value propositions that e-commerce systems promise to fulfill. An e-commerce systems enabled enterprise provides a mix of these value propositions to further the e-commerce objectives of the enterprise. The following sections elaborate on some of the popular e-commerce value propositions that enterprises can adopt to further their missions.

1.2.1 Customer-Oriented Propositions

The e-commerce paradigm enables enterprises to reach out to a large base of customers covering a wide demographic range. For enterprises, success on the customer frontier implies streamlining all interactions with a customer, including marketing, sales, and customer service. The increasing empowerment of customers with access to the Internet and other e-commerce technologies is pushing enterprises to automate customer interactions with the enterprises for all three facets in innovative ways. The Internet and e-commerce paradigm is shifting organizations away from addressing aggregate market concerns into the realm of addressing individual customers' issues.

On the customer frontier, an organization can deploy e-commerce systems to increase its customer base and retain customers by addressing customer requirements. The following sections elaborate on these value propositions.

1.2.1.1 Increased Customer Activation

Activating customers through marketing (or e-marketing) is the initial step required for establishing a presence in the e-commerce landscape. Although the volume of customers and their associated sales determine a firm's revenue, online enterprises are venturing to activate customers regardless of the direct revenues that they generate from those customers. The firm's initial thrust is to build brand awareness and establish online relationships with their customers. The reasoning behind this strategy is that brand loyalty and strong relationships directly and indirectly translate to higher revenue. Amazon.com is an example of an organization that has established a large customer base and has built tremendous customer loyalty and impressive brand awareness, but has yet to show matching revenues.

Activation in the online world means any of the following:

- Entice customers to bookmark a firm's Web site for future visits.

- Through promotion or other techniques, enable customers to register with the enterprise by prompting for basic demographic information. Organizations can subsequently use this profile information to offer customized products and services to customers.

- Convincing customers to open an online account with the organization (e.g., open a banking or brokerage account or open a shopping account). An enterprise may offer enhanced and full-feature services to its registered customers.

- Content providers (e.g., news organizations) can partner with paging and other wireless service providers to offer value-added content to their subscribers, thus increasing chances for brand awareness.

E-commerce systems enable firms to enroll customers for business transactions in numbers that were simply incomprehensible through traditional channels. Firms can reach customers through a variety of information appliances connected through the Internet and overcome barriers associated with traditional channels. The "physical access" channel, which requires customers to be in the physical presence of enterprise representatives, has continued to be the predominant way to reach customers. For example, in the past, customers would have to make a special trip to a bank to fulfill their banking needs or physically go to a store to buy goods. The emergence of the telephone and the postal channel shifted certain steps to those channels. Customers, for example, started to rely on the postal channel to pay bills and began to use the telephone to register complaints and obtain basic information about their business transactions. These channels (physical access, postal, and telephone) continue to be the primary channels for conducting business.

However, businesses are exhausting all options to reach customers in an innovative fashion through these channels. The primary reason is that inherent weaknesses in each of these channels have inhibited businesses to push their use into new business frontiers. Encouraging physical presence for conducting transactions results in longer queues at business sites. The telephone channel, though appropriate for inquiry types of transactions, has numerous inherent security issues (e.g., authentication schemes are weak) and cannot be relied upon for communicating sensitive information. The mail channel, although very reliable, has tremendous processing costs associated with it (e.g., printing and mail handling).

E-commerce–enabling technologies show enormous potential to overcome such restrictions, as evidenced by early pilots and production deployments. E-commerce systems enabled by the Internet's landscape allow customers to reach business services through various channels, such as TV, personal computers, wireless telephones, handheld computers, and a plethora of other appliances. Together, these appliances enable mobile and virtual means for initiating as well as fulfilling business requests. Businesses, therefore, can enroll customers online, sell products and services, and receive payments online, while overcoming all barriers associated with traditional channels. For example, organizations can use video kiosks at popular locations such as shopping malls and video stores to enroll customers. These video kiosks can present valuable information to customers and, if desired, can facilitate live customer–agent interaction, thus fulfilling real-time customer requests remotely.

Thus, enabling an enterprise's e-commerce systems with diverse information appliances can help enroll more customers. Organizations can also rely on other means to activate customers. The use of Internet advertising channels (banner ads, etc.), e-mail, Web sites, and promotions offered through the Internet can help enterprises activate more customers.

1.2.1.2 Customer Retention

Activating customers is a major step in the online world considering the number of Web sites popping up daily. The next challenge is to retain those customers, because changing businesses no longer means standing in lines to open another account or taking the time to drive to the new business. Customers can change businesses using a few mouse clicks and keystrokes. Retaining customers is therefore a primary attribute for success within the e-commerce world.

A strategy for retaining customers includes the following elements:

- *Offer value-added and differentiating services:* Providing customers with rich content and enhanced products and services at a competitive cost, as well as innovative and differentiating services, can help an enterprise retain customers. The primary challenge for organizations is to identify differentiating niches and offer comparable solutions. In this age of fierce competition, differentiating services quickly becomes necessary. For example, the use of ATMs for withdrawing cash was once considered a differentiating service, but it is now a necessary service that financial institutions offer to customers. Within the Internet world, organizations that established an early online presence by offering differentiating services succeeded in establishing their brand identities. Amazon.com is an example of an organization that established its presence by offering a differentiating service of selling books online.

An organization can offer differentiating services through multiple means. An organization, for example, should exploit multiple access channels to reach customers. By offering services through diverse channels, firms can differentiate their services and retain customers. For example, a financial institution that enables its customers to transact through various electronic channels (e.g., PCs, cellular phones, WebTV) has a better chance of retaining customers than a financial institution that offers limited services. Similarly, a shipping organization can enable its customers to track shipments through a variety of channels. Another method of differentiating services is to offer supplementary services. Organizations can package free services in their Web sites to retain customers. For example, some banks provide financial calculators and other tools that enable customers to analyze their financial needs, assist them in their financial planning, and so on.

- *Offering customers personalized products and services:* Customers are more likely to stay with an online business that is sensitive to individual preferences and offers profile-based products and services. A news site that enables customers to personalize site content will probably attract more customers than one that simply provides noncustomizable Web pages to its visitors and readers.

- *Offering differentiated customer service:* Innovations in Internet and online technologies is enabling organizations to offer customer service through electronic channels. For example, sites can enable customers to interact with a customer service representative as they browse online. Alternately, customers can dial a telephone number given on a vendor's Web site and speak to a customer service representative as they browse online.

- *Provide customers a simplified and pleasing interaction:* It is easy to clutter Web pages with information, thus confusing customers. Good Web page design is vital for enabling customers to find information without a struggle. Intelligent navigational functions built into a Web site will further facilitate browsing and prevent customers from abandoning the Web site.

- *Building stability and reliability functions:* Other factors such as security and performance also play an essential role in retaining customers.

Sites that lack security in their offering of sensitive information and services will observe lower customer retention rates. Similarly, customers are more likely to abandon sites that have poor response times.

- *Up-selling and cross-selling:* Organizations can deploy intelligent customer relationship management (CRM) based systems that up-sell and cross-sell products and services based on the customer's profile. For example, when a customer calls a bank's call center for a specific reason, the bank can use the opportunity to cross-sell and up-sell products and services to that customer. The bank's systems—through the customer's profile—may detect that the caller has two young children, and could thus offer the customer an education investment plan.

- *Soliciting customer feedback:* Organizations have opportunities for installing numerous channels for receiving customer feedback. Organizations should take due care in responding to customer feedback, as the customer is king in the new e-commerce paradigm. To handle the massive volume of e-mail, for example, organizations can deploy intelligent e-mail systems that categorize e-mail based on the received feedback and automatically send that feedback to the appropriate department. For example, the e-mail system can take customer complaints about an ISP's slow bandwidth connections and route that feedback to the appropriate department. Such concerns are serious enough for an ISP to lose customers.

- *End-to-end fulfillment services:* End-to-end fulfillment refers to all the steps that a customer goes through to order a product or service. Offering a complete set of services for customer fulfillment motivates customers to return to the enterprise. For example, the following steps comprise end-to-end fulfillment of services on a shopping Web site:

  — A customer visits a shopping Web site while connected over the Internet by typing in the URL.

  — Innovation of new technologies enables the customer to view enhanced images of products. Similarly, multimedia presentations provide additional information about the products.

  — If a customer desires more information, sophisticated Internet-based groupware tools enable customers to communicate interactively with a sales agent.

  — By getting enough information about the product, the customer can order using a multitude of payment schemes. The strength of various security technologies enables customers to enter credit card information over the Internet.

— The customer receives the order number and can use it to track the shipment.

— The organization provides the customer with a link to the shipping organization's Web site, where the customer can enter the order number to track a shipment.

1.2.2   Organizational Efficiencies

Internet and e-commerce technologies facilitate streamlining of business processes. As subsequent sections and chapters will demonstrate, an organization can use the Internet and e-commerce technologies to not only attract customers and generate revenue but also streamline internal and core business processes. Furthermore, organizations can establish an effective knowledge-management infrastructure that facilitates knowledge sharing within the enterprise.

1.2.2.1   Streamlining Core Business Processes

Generating higher revenue is not the only motivation an e-commerce paradigm and its underlying technologies present to the organization. Numerous organizations are exploring the Internet as a means for reengineering business processes to achieve cost efficiencies. For example, Electronic Data Interchange (EDI) has always been considered to be the primary channel for conducting business-to-business transactions. However, high costs associated with EDI networks and transactions discouraged smaller organizations from reaping the benefits of e-commerce. The Internet, on the other hand, provides organizations with an alternate inexpensive medium for processing transactions. Consequently, most organizations are either shifting away from the traditional EDI networks or exploring usage of EDI-based applications over the Internet. Chapter 4 discusses this in further detail. An organization can use Internet and e-commerce technologies to intelligently chain its business processes. For example, extranets enable organizations and other organizations that are part of the same value system to form virtual business value chains. Linking with the value chains of other organizations that are part of the same value system helps reduce cycle time. For example, a shopping Web site can offer its customers appropriate links to a shipping organization that it uses to ship merchandise to customers. This alleviates the shopping Web site from receiving customer service calls related to shipment status while benefitting the shipping organization in getting a large chunk of the business from the shopping Web site. E-commerce systems can also help organizations build intelligent synergies with suppliers and other strategic business partners. For example, an electronic bill payment provider (EBPP) brings billers, financial institutions, and customers together on one platform and provides advantages to all parties. Billers achieve enormous cost reduction associated with the printing, mailing, and handling of bills, while customers get one-stop service to pay all their bills through the Internet.

1.2.2.2   Streamlining Intraorganization Business Processes

Internet-enabled groupware tools enable an organization's staff to communicate more effectively, collaborate remotely, and exchange information. This helps reduce costs and increase productivity. For example, staff members can use multiple Internet e-mail-enabled appliances to communicate with each other remotely and use videoconferencing to reduce travel costs. Furthermore, the use of Internet-enabled personal information management (PIM) applications enables staff members to organize their work and collaborate on projects remotely, which in turn increases productivity.

By enabling an organization's staff to connect remotely over virtual private networks (VPNs), organizations empower their employees to work remotely and access an enterprise's systems securely. In some cases, this can provide tremendous benefits. For example, a sales person can access an enterprise's database to deliver value-added information to customers on the spot, instead of promising to send it to them later.

1.2.2.3   Knowledge Management Infrastructure

Organizations generate new ideas and knowledge almost daily. However, with the passage of time and staff attrition, this information either disappears from the organization or is unusable. The primary reason is that the inability to intelligently store, process, and access large volumes of data inhibits knowledge-sharing within the organization. Analysts at IBM have estimated that a business barely uses 7% of the information available to it [1]. This implies that 93% of the information the business could use to devise intelligent decisions or save time simply remains untapped.

Internet-enabled technologies facilitate the establishment of a robust knowledge management infrastructure. As will be explained in Chapter 4, intelligent storage, processing, and access to an organization's information assets facilitate the building of a knowledge management infrastructure, thus maximizing knowledge retention, usage, and processing for intelligent and timely decision making. A department, for example, can immediately use available solutions that were devised in the past and in another part of the world by someone who no longer works for the organization to solve current problems.

1.2.3   Revenue Generation

Generating revenue, directly or indirectly, is always the primary consideration when an organization deploys an e-commerce system. An organization can indirectly increase its bottom-line results by streamlining its operations and increasing the number of customers, as explained earlier. On the other hand, an organization offering products and services can directly maximize revenue by increasing exposure of its products and services, providing more information on those products and services, and thus making the entire process of ordering and payments easy, fast, and reliable. The following paragraphs discuss factors that enable an organization to increase its revenues.

1.2.3.1   Advertising

Advertising is the primary means for boosting an organization's revenue. In fact, the notion of advertising on the Internet is strong enough to solely warrant numerous organizations' existence on the Internet. Public portals are popular examples that fit this model. Public portal sites offer numerous free services to their users. The large number of customers visiting these portal sites encourages businesses to advertise on them, thus generating more revenue for the portal sites. Yahoo, AltaVista, and Excite are examples of such portal sites. Similarly, a financial institution that receives many visitors can also open its site for advertising. Recognizing the bank's popularity, other sites can advertise their products and services on the bank's Web site. The bank's Web site thus becomes a vertical portal for the financial industry.

1.2.3.2   Facilitating Electronic Payments

The ability of an organization to display its products and services on the Internet and receive electronic payments through the Internet provides numerous revenue-generating opportunities. Organizations can truly sell products and services 24 hours a day, unlike conventional schemes that involve physically opening the doors for business. The porting of traditional payment instruments, such as credit cards and checks, onto the Web facilitate such commerce.

1.2.3.3   Organizational Alliances

Organizations can form alliances that provide mutual financial benefits. Such strategic partnerships allow each party to benefit from the other's advantages. For example, organizations can partner with public portal sites. This enables the organization to provide its site links on the portal site. A news channel, for example, can partner with a public portal site to provide news updates on the portal site. This enables the news organization to benefit from the portal site's large number of customers, and in return, the news organization may offer some compensation to the portal site.

1.3   E-Commerce–Driven IT Strategy Realignment

The design, development, deployment, and continued maintenance of e-commerce systems requires the same key IT infrastructure processes that enterprises have leveraged to build conventional IT systems. However, certain new elements—such as Internet technologies, access to a wide number of customers, and collaboration with various service organizations for the delivery of appropriate IT services—have necessitated new methods and practices to build these e-commerce systems. Embarking on the initiative to build e-commerce systems therefore requires a renewed IT strategy. An enterprise's IT strategy embodies a number of elements that influence the design, development, deployment, and maintenance of IT systems. These elements include technology acquisitions, system engineering methods and processes, infrastructure development, strategic external relationships, and others. This section examines the impact of e-commerce systems business requirements on the pertinent facets of IT strategy.

IT strategy is a set of short- and long-term actions that an enterprise adopts to support an enterprise's overall business strategy and that maximizes an enterprise's ability to achieve objectives. An effective IT strategy provides a roadmap for implementing IT solutions for the business. An IT strategy aligned with the business strategy helps an enterprise meet four primary objectives. First, it helps in the *delivery of aligned IT solutions* that suit the business requirements and ultimately reap the appropriate ROI. These solutions support a firm's primary business processes and foster the sharing of knowledge within the organization.

Second, an effective IT strategy paves the foundation for the *efficient delivery of systems*, thus responding to the rapid pace of the business. For e-commerce, this requires an enterprise's IT department to meet the challenges of rapidly evolving requirements resulting from the surfacing of new business models and changing customer needs. Third, an IT strategy provides directions for the *ongoing operations and support* of IT systems. In the e-commerce domain, where an enterprise is rolling out Web sites and new e-commerce solutions daily, the IT strategy should stipulate solutions that are mature enough to support the complexities of ongoing operations. Finally, a balanced IT strategy facilitates a *managed injection of new technology initiatives* within the corporation, thus preventing the chaos resulting from unstoppable technology innovations. The following discussion elaborates on these issues.

1.3.1 Formulating E-Commerce System Requirements

Because an e-commerce value proposition (e.g., selling through the Internet) encompasses numerous IT applications (payment applications, content hosting applications, etc.), collecting requirements for building e-commerce systems requires a more holistic approach than for traditional IT applications. For building e-commerce systems, an enterprise should therefore not merely depend upon the requirement analysis of individual constituent IT applications. Rather, the collecting of requirements should encompass all elements required to materialize a functioning e-commerce value proposition. An enterprise's IT strategy should appropriately regroup to address such holistic requirements. The following depicts some of the aspects of the e-commerce system requirements analysis phase:

- *Identification of all IT applications that will comprise the e-commerce system:* It is vital to fully delineate all applications that will interoperate to shape the final e-commerce value proposition. For example, building an e-commerce Web site to sell products includes applications for hosting content, applications for accepting payments, customer service applications, and more.

- *Focus on individual application requirements analysis:* This refers to the normal practice of collecting requirements for designing an IT application. Each of the constituent applications of an e-commerce system should undergo a thorough traditional requirements analysis phase. This includes understanding all functionality (e.g., platform requirements, security requirements, screen design). Requirements analysis at this level should ensure that each application's requirements are aligned with the overall e-commerce system's requirements. For example, if the value proposition calls for performance thresholds of 2 seconds for customer service requests, the IT applications should ensure conformity to these requirements.

- *Customer facing requirements:* This refers to explicitly focusing on all functionality that directly interfaces with the end customer, especially if the e-commerce system involves external customers. Not long ago, enterprises primarily focused on internal customers for the delivery of IT solutions. A bank, for example, developed applications for its tellers, financial controllers, and other business users. However, the emergence of e-commerce systems has shifted the focus more toward the external customer. These external customers are the end users of the e-commerce system, and their satisfaction with the system can make or break a business relationship. E-commerce systems can vary dramatically in their functionality depending on the user demographics. Organizations should assess customer demographics before introducing appropriate functionality into such systems. For example, offering WebTV-based products and services to a user population that does not have such access would not benefit the enterprise or the end users. Similarly, in designing a user interface for a wireless device, the enterprise should focus on factors such as user friendliness (the user should not have to navigate through many screens to trigger a transaction), convenience (the application should not cram lot of text onto the device's limited display), and so on.

- *Systems engineering requirements:* This refers to gathering requirements related to performance, security, scalability, operations, and so on. For example, slow performance of an Internet-based stock trading system may seriously effect the success of the system. Similarly, inherent security issues in Internet-based systems require addressing all vulnerabilities in the system. Operational requirements, on the other hand, refer to a constituent application's requirements related to startup, shutdown, and backup procedures in an operations environment. An e-commerce system may span multiple operation centers internal and external to the organization, so the requirement analysis should explicitly identify requirements for all operations centers (e.g., ISPs, data centers, NSPs). An organization may incorrectly assume that external operations centers will be able to cater to the special requirements of the specific system. For example, in implementing an e-commerce system, an organization may stipulate manual procedures (e.g., startup, shutdown) during nightly hours, whereas the specific external operations center may run a lights-out data center. Delineating such requirements and validating them with the operations service providers ensures a smooth transition to a production environment.

In stipulating systems engineering requirements, an organization should also assess its options in delivering e-commerce systems to its businesses. An organization has four primary avenues of delivering systems to its businesses. The first involves in-house development of e-commerce systems. This entails acquisition of development tools and developing applications internally. The enterprise chooses the appropriate development language paradigms and tools to facilitate development of applications. One disadvantage to this model is that the adoption of nonstandard technologies can considerably degrade development cycle times. For example, enterprises that used Java across most initiatives experienced an initial slow down due of lack of training.

In another model, the organization outsources its development initiatives to external parties. This requires the organization to manage the appropriate vendor interfaces in order to ensure aligned delivery of systems. In the third model, the organization acquires solutions—as opposed to developing them—and integrates them locally. Most enterprises seem to be following this model. The final approach involves renting applications. Application renting through application service providers (ASPs) is a new trend in which an enterprise rents vertical industry applications (e.g., human resources applications) from various service hosting organizations.

- *Infrastructure requirements:* E-commerce system operations usually require a diverse array of technology infrastructure components. These include public key infrastructure (PKI) for security of transactions and payment infrastructure (e.g., ACH, VISA network, high-speed network infrastructure) to cater to multimedia traffic (e.g., voice, video, and others). In planning an e-commerce system, the enterprise should specifically identify the project's assumptions on such dependencies. As will be explained later, certain infrastructure elements require a substantial investment and effort for deployment.

- *Integration requirements:* After soliciting the aforementioned requirements, the organization should assess requirements for the overall functionality of the e-commerce system. The ultimate overall functionality of the e-commerce system may involve various business interfaces, such as continual communication with various departments for updating content.

1.3.2   Instituting IT Processes

Rapidly changing and evolving customer needs and fierce competition necessitates delivery of high-quality IT solutions in reduced cycle times. Delivery of such solutions, similar to any other product that an organization manufactures, requires repeatable and stable processes. These IT processes should be predictable and flexible to adapt to deliver IT systems of varying requirements. Immature and unstable IT processes consume an organization's valuable resources, deliver defective products unsuitable for external customers, and ultimately ruin an organization's image. The need to meet deadlines is driving organizations to ignore the "bureaucracies" of the IT processes. While meeting such deadlines may cause organizations to beat the competition in the short term, the delivered systems may suffer some from these symptoms.

IT processes comprise all the activities and steps that are required for the engineering of IT systems. The lack of or ad hoc implementation of such processes has been blamed for reaping poor ROI on IT initiatives. Missed commitments, systems not appropriately aligned to business requirements, and lack of control of software work products are typical symptoms of the lack of IT processes. Within the e-commerce paradigm, where quality systems and rapid response to market conditions are prerequisites for an enterprise's competition and survival, investing in the IT processes of a firm becomes a vital strategy.

The e-commerce paradigm does not negate the need for well-defined IT processes. On the contrary, new rules of the game, such as interfacing with multiple organizations, integration with multiple components, technologies from numerous vendors, and so on, further strengthens the case for building and assembling e-commerce–related IT systems methodically. An organization that embarks on a formal initiative to streamline its IT processes drives the organization to implement all processes required for the implementation of IT systems that are tailored to an organization's culture and needs.

Well-defined IT processes (e.g., requirements management, configuration management, quality assurance) enable an organization to quickly respond to market requirements in a controlled manner. Control in this context implies that software work products maintain high quality levels and the organization can predict the delivery of such systems to its customers (business) within the committed time frames. Second, controlled IT processes ensure that an organization delivers IT systems that are aligned to business requirements. Furthermore, adherence to methodical processes ensures that the numerous complexities involved in building e-commerce systems, such as coordination with multiple groups (outside vendors, ISPs, NSPs, system integrators, etc.), do not distract the organization from its goals and objectives.

An organization can adopt multiple approaches to implement IT processes within an organization. The Software Engineering Institute's Capability Maturing Model (CMM) is the most popular approach and is further described in Chapter 9. The International Organization for Standardization's ISO 9000-3 is another set of standards that applies ISO 9001 quality practices to the development of high-quality software and systems. The Institute of Electrical and Electronics Engineers (IEEE), on the other hand, maintains its own list of software standards for various software processes (e.g., software testing, verification and validation, configuration management). An enterprise may find any of these standards suitable for the effective and efficient delivery of IT systems within its environment.

Another reason for instituting software processes within an organization is that an organization's IT department, which has traditionally struggled to create intelligent synergies among the design, development, and deployment processes within the organization, must now learn to intelligently deliver solutions in an open environment. This environment includes multiple vendors, value added resellers (VARs), system integrators, external developers, and operations people who deliver a firm's IT systems.

1.3.3 Operational Infrastructure

One of the challenges and necessities of operating in the e-commerce paradigm is establishing a stable and high-performance computing operational infrastructure. The public switched telephone network (PSTN), which is the primary foundation for supporting worldwide telephone-based communications, is an example of this desired stability and performance. People expect and depend upon the PSTN's availability and performance as it rarely frustrates customers due to unavailability issues. Doing business in cyberspace is creating similar expectation levels among customers. Establishing an IT operational infrastructure that is aligned to customers' expectations is therefore an essential element of an organization's IT strategy.

IT's operational infrastructure ensures availability of IT-enabled business services to its customers. The operational environment includes production-level applications, databases, networks, and associated security and technical procedures. A data or operation center's primary role is to ensure the successful operation of the IT systems. This also implies operating IT systems at predetermined service levels (network throughput, system performance, etc.). To effectively establish such levels of service and operations, an enterprise must pursue a two-pronged strategy. First, the deployment of an effective systems management infrastructure that provides effective control of the enterprise's operations and enables it to monitor various quality service levels is needed. Second, the enterprise must decide whether to outsource the operations environment or retain operations in house.

Most enterprises operate their data centers internally. However, the emergence of e-commerce is pushing organizations to outsource data-center operations to external entities. A glass-house data center has traditionally served as the operational environment of the entire enterprise. Traditional data centers house a set of large processors that host an enterprise's application portfolio. Larger enterprises use multiple data centers that serve respective business segments' IT operations. However, increasing number of e-commerce systems, processing requirements, and other requirements are pushing most large enterprises out of their data centers into outsourced data centers.

Enterprises once considered outsourcing their data centers selectively to contain their costs and risks. However, in delivering e-commerce systems, an enterprise could rarely escape the outsourcing phenomenon. Numerous reasons can be attributed to this phenomenon. The primary factor is the inability of enterprises to keep up with upgrading their internal data centers to meet the requirements of e-commerce systems. The increase in complexity associated with hosting large e-commerce systems introduced new elements, including high-speed network connections, firewalls, routers required for Internet connectivity, switches, and so on. Furthermore, a growing number of databases housing sensitive information require special security controls and data handling procedures. Keeping e-commerce systems segregated from the traditional legacy systems (especially if there is no connectivity to legacy systems) enables enterprises to run clean operations and not jeopardize existing operations.

ISPs provided the first entry into the Internet and opened their operations centers to allow enterprises to collocate systems. This trend continued and resulted in enterprises outsourcing their entire data center operations to certain ISPs. Most large ISPs now offer data center operations services. For example, Exodus Communications, Inc. provides Internet data center (IDC) services that enable organizations to outsource their Internet-based operations to Exodus.

Another reason for the increase in the operations outsourcing trend is that existing data centers could not keep up with the increased demand of hosting the larger number of servers and systems. Building new data and operations centers requires a tremendous investment in space, technology infrastructure, physical security controls, and ancillary administrative procedures and processes.

1.3.4   Intraorganization IT and Business Alignment

Traditionally, IT departments had rather loose affiliations with business departments. Businesses interfaced with IT departments for requirements management or testing of IT systems. This has been one of the primary reasons for the misalignment between business and IT. The e-commerce model requires an enterprise to continually roll out new systems driven by market forces and update content on its Web sites. This requires an IT department to strengthen its relationships with research, product, and service groups, as well as advertising, marketing, and sales organizations, to maintain up-to-date content on the organization's Web site.

Enterprises could pursue various avenues to establish tighter relationships with their businesses. For centralized IT departments that deliver IT solutions to various businesses of an enterprise, one solution is to increase business representation in all key IT processes, such as architecture and solution conceptualization phases, requirements management, analysis, design, testing, and operations. The alternate method requires each business to drive its own IT initiatives. This approach, despite its drawbacks from the perspective of establishing the common infrastructure elements of the enterprise, results in more aligned systems, since the businesses control all aspects of IT and can better measure their ROI on IT investments.

1.3.5 Formulating a Technology Architecture

An enterprise's IT architecture includes its application portfolio strategy, data and business models, computing platform standards, and networking strategies. An enterprise's IT architecture requires adoption of new strategies and transformation of some existing ones to respond to the new and continually changing requirements of the e-commerce–enabled business. Technology architecture provides organizations the direction for various IT standards and products that an organization uses to build its IT systems. For example, an organization may standardize on the use of XML for all its data storage requirements. Similarly, an organization may standardize on Microsoft Internet Information Server (IIS) as the Web server for its Web-enabled applications and systems. Formulating an enterprise's technology architecture thus entails adopting the right mix of standards and vendor products, which it can then use to build and deliver systems to its business customers.

In choosing the direction for an enterprise's technology architecture, the organization should be wary of the following factors:

- *Breadth of functionality:* An organization will find numerous products that seem to fit the description of the desired product or service. The organization should institute thorough evaluation processes and procedures to match the desired product or service with an organization's long-term direction. A nonstrategic choice, especially for infrastructure technology components (e.g., operating systems, choice of a network service provider), can cost the enterprise financially as well as in terms of customer satisfaction. The choice of ISPs for hosting e-commerce services is a well-known example in this regard. Numerous ISPs offer e-commerce hosting services, but few may suit an enterprise's requirements for quality of service guarantees, business contingency, and other related attributes. Similarly, an organization's e-commerce requirements may place a special emphasis on scalability. Certain products that fit the description of the desired product may not scale to large numbers of users within the appropriate e-commerce business domain.

- *Openness, acceptance, and maturity of standards:* New requirements within e-commerce are triggering numerous standards. An organization should adopt a controlled approach in embracing such standards as part of their technology architecture due to the lack of maturity of those standards. For example, XML is a popular standard for data formatting on the Web. Numerous initiatives are in the works that call for storage of data in XML format. An organization should assess the implications of accepting such technology before standardizing it and embracing it as part of the technology architecture.

- *Interoperability with legacy systems and infrastructure:* Embracing certain technologies, standards, and products that may not interoperate with an enterprise's legacy systems may be required for tight supply-chain integration. An assessment of vendor products and standards therefore should include such interoperability issues. For example, an organization whose goal is to "Webify" its legacy systems should consider investing in products that provide a Web server for the legacy platform, rather than choosing an alternate platform that may require numerous interfaces.

- *Vendor stability:* In the past few years numerous vendors have surfaced that claim to offer a multitude of e-commerce products, services, and solutions. An organization should consider the stability of such vendors in providing technical support. For new vendors, organizations should assess the various types of support that the vendor enjoys from the industry. This includes assessing successful case implementations, standards support, and other vendor support.

- *Component-based design and integration:* The infancy of the e-commerce paradigm coupled with customer's evolving requirements necessitates that an organization immediately responds with new systems or modify existing ones. This requires adoption of object-oriented and component-based technologies that facilitate plug-and-play functionality. Technologies such as CORBA- and COM-based technology models allow such functionality and are further discussed in Chapter 6.

1.3.6   Intelligently Embracing the E-Commerce Glut

During the early days of information processing, organizations were captivated by the inherent power of information processing but failed to foresee issues associated with the enormous information glut. Similarly, as organizations scurry to embark on various e-commerce initiatives, they experience similar problems and issues. For example, certain organizations offered clients the ability to send e-mail to their customer service representatives. However, organizations were flooded with e-mail from customers and found it extremely challenging to respond to all customer queries, thus resulting in enormous customer dissatisfaction. Newer solutions enable organizations to deploy intelligent e-mail systems that process most mail and automatically generate responses to common questions. Similarly, organizations are facing numerous challenges in linking their e-commerce systems with their backend systems. In some cases, such systems cannot handle the enormous transaction load generated by millions of customers accessing an organization's Web site.

An uncontrolled dash to embrace e-commerce initiatives may hurt an organization from a legal and security perspective as well. Customer privacy on the Internet is a very sensitive issue that can affect an organization's bottom-line and in some cases may even jeopardize its existence. Customer piracy is another related issue. The Internet provides numerous opportunities for distribution of illegal software. Employees of an organization who engage in such piracy can directly affect the organization financially as well as ruin its reputation. Finally, security breaches can jeopardize an organization's information assets, credibility, and bring down an entire business.

1.4   Characteristics of E-Commerce Systems

Before delving into the architectural details of designing and implementing e-commerce systems, it is vital to revisit their bottom line objective. E-commerce systems:

- Facilitate the means to enable *external customers* to conduct various e-commerce services in ways that suit their life styles and are not driven by technology.

- Provide the necessary knowledge to *corporate employees* through innovative means, thus empowering and energizing them to produce products and services effectively and efficiently.

- Provide to *corporate employees* the knowledge base and mechanisms that enables them to service corporate customers in stimulating ways to fulfill customers' intentions and attract them to an organization's other products and services.

- Provide ingenious means to collaborate with *suppliers and partners* that in turn minimize the overhead of delivering products and services to customers.

Underneath the covers, e-commerce systems are essentially IT systems that inherently share the same technological characteristics. Both types of systems involve networks, databases, applications, and so on. The differentiating factors lie in the new functionality needed to match requirements inherent in the three e-commerce business models described earlier and the new infrastructural elements required to support the design, development, and operations of an e-commerce system. These characteristics enable fulfillment of the various domain-specific e-commerce business objectives. As the earlier sections demonstrated, e-commerce imposes new business requirements for the underlying IT systems (e-commerce systems) from the functional and infrastructure perspectives. This section delineates and analyzes those characteristics.

1.4.1 Functional Characteristics

Functional characteristics highlight both a system's functional capabilities along with the user's interaction with the system. Delving deeper, functional characteristics highlight a system's behavior and its constituent components, their functionality, user interfaces, types of users of the system, nature of content that a system processes, and so on. For e-commerce systems, the Internet, new types of customers, and new modes of doing business impose new sets of functional requirements. The following delineates some of the key functional characteristics:

- *Support for diverse information appliances:* An e-commerce system enables access to information and services through a multitude of information appliances such as desktop PCs, notebook computers, Palm PCs, pagers, and cellular telephones. This is unlike traditional IT applications that limited user access through dumb terminals and PC workstations. Accessibility through diverse channels and appliances empowers users to engage in computing activities when they are mobile and remote. For example, by using information appliances such as cellular telephones or handheld computers, consumers can obtain information about bank balances, while an organization's staff can access an enterprise's mission-critical applications when staff members are mobile and away from their offices. Furthermore, a variety of information appliances enables users to be accessible irrespective of their location.

Enabling e-commerce systems with information appliances depends on the type of appliance, the nature of functions supported by the e-commerce system, and the maturity of the underlying standards and technologies. For example, offering users a Web interface on wireless telephones requires installation of additional servers that map regular Web page size and content to the limited display size of the wireless device. Similarly, data connectivity requirements through mobile and wireless appliances are limited due to the data throughput limitations on wireless networks. An organization therefore has to consider such factors before enabling information appliances for various information appliances. Chapter 2 discusses these issues in further detail.

- *Support for varied data and information sources:* E-commerce systems enable information access across local and remote information warehouses. Standardization of database access technologies coupled with the wide use of TCP/IP-based technologies facilitates such a wide access of network resources. An e-commerce system can access data sources distributed over the public Internet, internal mainframe data sources, ERP systems data, servers on an enterprise's intranet, and ultimately provide users a unified view of desired information. Such capabilities have triggered the push toward a knowledge management infrastructure, where users can tap a wide base of information sources for making intelligent business decisions. A consumer Web site, for example, can access a wide repository of information distributed across the TCP/IP-enabled internal and external networks to allow users to compare prices on certain merchandise. Furthermore, accessibility to a wide repository of information has motivated the industry to introduce standards for data mapping and access. For example, XML standardizes data access and empowers for granular information access. Chapter 4 elaborates on these details.

- *Support for rich content:* E-commerce systems indulge in processing new forms of content. While traditional applications handled only text-type data, e-commerce systems employ richer data formats that include the use of animated graphics, video, audio, interactive content, and other forms of rich content. For example, Web sites can use multimedia presentations to advertise products and services. Similarly, Web sites can use graphics to enliven their Web interfaces. Such rich content both embellishes applications with eye-catching presentations and provides value-added services to the end customer. For example, the use of audio and video on the Internet enables users to view presentations, speak to customer service representatives, and take educational instructions.

Incorporating such functions into e-commerce systems, however, requires an intelligent analysis of the customer demographics, the customers' empowerment with specific information appliances, and availability of high-speed networks. Furthermore, security issues have to be addressed for certain e-commerce implementations. For example, the use of voice over IP networks has certain inherent security issues. Chapters 2 and 8 discuss these issues in more detail.

- *Varied user interfaces:* Traditional client/server applications are primarily comprised of standard-sized GUI interfaces for PCs. The newer applications have user interfaces that include PC consoles, user interfaces of small wireless handsets, handheld computers, and TV screens. Vendors marketing various information appliances provide respective APIs for those devices to enable building of graphical interfaces for varying screen sizes and resolutions. Chapter 2 covers the underlying technology elements for some popular information appliances.

- *Human-like interaction:* Through support for collaborative features and functions, e-commerce systems enable users to interact with customers. Such an interaction provides the customer with the feeling of interrelating with a live human-like system, unlike traditional systems that exhibited robot- and machine-like behavior, thus sometimes frustrating users. For example, using the new features, customers can communicate with a customer service representative over the Internet using VoIP (Voice over IP) or engage in a collaborative session exchanging information online using whiteboards. Chapter 6 covers groupware features in more detail.

- *Intelligent learning:* E-commerce systems support automatic and intelligent learning. These systems track customer behavior and preferences as they interact with the enterprise services and information and use this information to customize and personalize user's future interactions with the Web site. For example, a customer who buys toys for infants on a given day can be lured into buying toddler toys through promotions when the customer revisits the Web site in a couple of years. Organizations can incorporate intelligent software that enables such tracking within their Web sites. Chapter 9 discusses such issues.

E-commerce systems also support intelligent software-agent–driven computing (shopping agents, searches, etc.). Numerous vendors are providing software-agent–powered services that alert users to predefined events. For example, specialized applications loaded on a wireless telephone on appropriate activation can alert a roaming user to nearby businesses such as restaurants and hotels. Likewise, intelligent search engines collect users' requests and scour the Web to search for customized information.

- *Managing information glut through portal applications:* Portals provide a unified and organized view to a wealth of information and IT-enabled services. By customizing portals to users preferences, organizations can allow users to rapidly access the desired information and services, thus easing the process. Businesses can offer portal services on a specific subject (e.g., a sports portal that includes sports-related information customized for the user or a portal for an organization's users that displays applications and information links for users' daily job activities). Chapter 5 discusses the advantages of portals within the e-commerce domain.

- *Branded and friendly user interfaces:* Design of user interfaces is of paramount importance for e-commerce systems, especially for business-to-consumer systems. User interfaces portray a business brand, reflecting an image of experience, trust, reliability, and security that customers want. Furthermore, friendly interfaces facilitate easier navigation, which helps retain customers. Customization features, such as customizing the Web interface, personalize users' experience with the online vendor.

- *Scalability/global connectivity:* E-commerce systems, especially those that fall within the business-to-consumer computing environment, expect global exposure. Accordingly, the systems have to be designed to be scalable to react to increasing customer connectivity. Increasing connectivity can result in the need to connect to a large number of customers, register a large number of customers, process their requests, and so on.

- *Richer functionality:* E-commerce systems pack more functionality than their traditional counterparts. For example, it is common for an e-commerce system to pack communication functions, collaboration functions, interactive communication, transaction, and inquiry functions into one e-commerce system. Intelligent design of such functions to incorporate the desired functionality thus poses enormous challenges.

- *Integrate diverse applications:* E-commerce systems are a conglomerate of legacy, Internet-based, and other types of applications. This integration is due to the varied functionality and data access requirements needed for the fulfillment of e-commerce value propositions. For example, an Internet shopping Web site may update its catalogs on a newly designed Internet-based application, while it may process customer payments through its mainframe legacy application. Therefore, instead of porting the entire payment application on to a new platform, the enterprise can continue to leverage the existing functionality on its mainframe systems.

- *Provide rapid user response and fulfillment:* E-commerce systems enable rapid fulfillment times. In the business-to-consumer domain, these times usually do not exceed 3 to 4 seconds. Building systems to deliver rapid response times is necessary to keep users attracted to the site.

- *Support electronic payments:* E-commerce systems allow automatic triggering of payments for products and services, unlike traditional systems that require a customer service representative to interact with a customer either on a one-to-one basis or through a telephone.

- *Flexible and secure application access:* The emergence of security technologies and extranets enables the design of e-commerce systems that internal users, external users, and business partners can access equally, thus streamlining business processes. A business partner, for example, can access an enterprise's stock replenishment system to order stocks. Similarly, a vendor can access an enterprise's catalog system to update its catalog items instead of sending the catalog to the enterprise, which has to perform that process.

1.4.2 Infrastructural Characteristics

The construction, deployment, and operations of an e-commerce system require the existence of an ecosystem that includes relevant technological and nontechnological elements. This is analogous to building a house that depends upon water, electricity, and other utilities, as well as infrastructure, roads, and so on, to make the house livable. An e-commerce system similarly requires the existence of a network infrastructure that extends the users' reach, a payment infrastructure to pay for products and services, and an infrastructure that reconciles payments among various institutions. Other infrastructure elements include the PSTN, wireless communication infrastructures, the Internet, and public e-commerce services (portal services, e-commerce hosting, etc.). The following highlights certain key elements of the technology infrastructure.

- *Wireless network infrastructure:* The need for mobility, remote access, and the ubiquity of the Internet have triggered the emergence of the wireless networks. Wireless networks have played a vital role in facilitating the exchange of short messages and wireless telephony for the past few years. Their importance has further risen due to their fast-emerging role in the world of e-commerce. Users can access the Internet and other enterprise computing resources by connecting to various wireless networks.

  The wireless network infrastructure includes various technologies such as GSM, cellular wireless networks, pager networks, and satellite networks. Multiple organizations, standards, and technologies exist to facilitate wireless connectivity for telephony and data connectivity. An enterprise's challenge is therefore to ensure selection of the appropriate wireless service providers and services to offer e-commerce services to its customers and to enable internal users to leverage the enabling capabilities of the wireless infrastructure.

- *Operational infrastructure:* Ongoing operations of e-commerce systems require the existence of an operational infrastructure. An operational infrastructure hosts applications, provides ongoing support for those systems and applications, and ensures the upkeep of all components required for the ongoing functioning of any IT system. The existence of such an infrastructure within the enterprise or external to the enterprise is quite vital for e-commerce systems. As mentioned earlier, numerous external data centers have surfaced in the past few years that provide such support. An organization's challenge is therefore to choose the right mix of internal and external data center hosting services for the support of its e-commerce systems.

- *Internet domain infrastructure:* An Internet-enabled e-commerce system requires the seamless and flawless operation of the Internet's domain infrastructure. Also referred to as the Domain Name System (DNS), this infrastructure allows for a rapid lookup of e-commerce sites on the Internet. Establishing an e-commerce presence on the Internet requires a basic understanding of the issues in registering domain names and the multiple registrars that exist for providing such registration services. Chapter 6 elaborates on the DNS infrastructure and its future directions.

- *Application hosting:* Though not an absolute requirement for the hosting of e-commerce systems, an application hosting infrastructure provides small organizations a quick entry into the e-commerce arena. Application hosting service providers assist in the development of basic services required for engaging in online e-commerce activities. Development and integration activities include Web page design, integration with various payment entities for processing of payments, registration of the site's domain name, and so forth. Chapter 5 provides examples of various service providers.

- *Payment infrastructure:* E-commerce systems would be incomplete without the existence of a payment infrastructure. The payment infrastructure facilitates payments among parties and reconciles payments. The payment infrastructure allows consumers to pay using traditional payment methods such as credit cards, debit cards, and checks. Businesses, on the other hand, use the Automated Clearing House (ACH) infrastructure for transfers between accounts. The Fedwire payment infrastructure facilitates payments of large sums between financial institutions.

  Organizations offering e-commerce systems need to interface their Web sites with these payment networks and infrastructures to allow for automatic processing of payments. Chapter 7 elaborates on various forms of payments and respective payment networks.

- *Internet infrastructure:* Internet access is the cornerstone of e-commerce systems. ISPs and various other NSPs facilitate access to the Internet through various channels. An enterprise's primary challenge is to choose appropriate ISPs and NSPs that will form the primary infrastructure for the future rollout of its e-commerce systems. Chapter 3 highlights appropriate issues in the selection of ISPs and NSPs.

- *Public key infrastructure (PKI):* E-commerce systems by their nature require tighter security controls. Public-key cryptography offers opportunities for secure e-commerce solutions. However, doing so requires a PKI. An organization has multiple options for leveraging on PKI services. As will be explained in Chapter 8, organizations can rely on an external PKI infrastructure, develop their own PKIs, or jointly develop an enterprise-specific PKI in conjunction with external PKI service providers.

- *Legal infrastructure:* Conducting commerce electronically has numerous legal ramifications that enterprises need to consider when designing e-commerce applications. Various government regulations put the onus of addressing those requirements on online sites. Such considerations include incorporating privacy statements on the site, copyrighting certain content before publishing on the Web, and so on. Chapter 9 highlights such issues.

- *Network architecture:* An enterprise's network architecture has to be flexible to accommodate connectivity from the general public, remote employees, and business partners. The network and related applications should build in appropriate security controls and performance to support this diverse set of users. Such issues addressed at an infrastructure level, as opposed to each individual e-commerce system's level, provides better performing and secure applications. For example, an enterprise can lease certain network connections from a network service provider for all of its e-commerce systems needs, as opposed to having each e-commerce project individually negotiate different service level agreements with various NSPs.

1.5   E-Commerce Business Domains and Respective E-Commerce Systems Requirements

E-commerce systems deliver value to organizations on three different fronts. First, organizations can use e-commerce systems to provide consumers with an electronic medium for buying, selling, merchandising, and delivering content over various electronic computing networks. Second, e-commerce systems streamline an organization's internal operations by facilitating remote intelligent collaboration, empowering staff with the right information at the right time to conduct business activities, and enabling an organization to get closer to its customers. Third, e-commerce systems provide an economical and a robust medium for organizations to conduct business with other organizations that are part of the same value system. This enables the enterprise to form virtual value chains, thus optimizing their product and service throughput. This section highlights and analyzes those business requirements that drive the design of e-commerce systems for all three business domains. Table 1.1 compares the business requirements of the three types of systems.

1.5.1   Business-to-Consumer Application Domain

The business-to-consumer domain is concerned with offering online products and services to a wide range of Internet-connected public domain customers. Examples of business-to-consumer Web sites include shopping Web sites, financial institutions' Web sites, and news Web sites. These Web sites offer consumer banking, online sales, shopping, and other services. The nature of this domain thus attracts a large base of users. Besides, design considerations for these online businesses encompass the preferences and profile of the general

Table 1.1

General Business Requirements and Characteristics of Three E-Commerce Business Domains

E-Commerce Entity	Business-to-Consumer Domain	Intrabusiness Domain	Business-to-Business Domain
Users	General population niches (individual consumers)	Corporate employees	Corporate users of organizations (e.g., trading partners)
Number of users	Low → Very High	Low → High	Low → High
Transaction volumes	Low → Very High	Low → High	Low → High
Amounts transacted per transaction	Low → High	N/A	Low → Very High
Payment system(s)	Credit cards, electronic checks, electronic cash, electronic wallets, etc.	N/A	ACH, FedWire, Electronic checks
Nature of content	News, product images, product and service descriptions, tools, calculators, etc.	Business applications, internal department content, policies, procedures, etc.	Business applications
Primary business components of e-commerce systems	Shopping cart and catalog, programs, credit card payment processing, Web-enabled customer service programs, etc.	Document management programs, groupware applications, etc.	Catalog applications, Web-based requisitioning, financial EDI applications, supply-chain applications, etc.
Security risks for business transactions	Low → High	N/A	Low → Very High
Security risks for information assets	Very High	Low → Very High	Very High
Support of various information appliances	High (driven by customer demographics and their empowerment with devices)	Medium	Small (PCs, NetPCs, etc.)
Use of information appliances	Business transactions, communications, inquiries, etc.	Communications, business transactions	Business transactions

user population, making usability, friendliness, and nature of content the primary features of such Web sites.

The nature of content in this domain is geared to the general public. This includes information on products and services that an organization markets to its customers and other relevant information. For example, a shopping Web site publishes information about its products and services. This information may include a description of the products and services along with appropriate static and video images. Similarly, to attract more customers, financial institutions offer all types of customers the ability to open accounts online and provide customers with access to various forms of financial tools and calculators free of charge.

Security in this business domain employs basic techniques as the number of transactions is within acceptable risk parameters for each customer. For example, a business that sells content online may require customers to authenticate to the Web site to access and view content. Security in this context employs basic schemes (e.g., authentication is usually based on a simple static password, where customers enter an ID and a password to get access to an online business's site), or security may require more sophisticated system authentication schemes. Such content includes specialized newsletters, industry reports, entertainment services, games, and others. However, organizations may have to employ stringent security controls within the organization to safeguard customer's private information (e.g., name, addresses, contact information, preferences).

Other Internet security technologies enable customers to engage in financial transactions. These security technologies enable customers to use their traditional modes of payments, such as credit cards, debit cards, and checks, to pay for products and services over the Internet. New forms of payment, including electronic cash and electronic checkbooks, further facilitate secure means of payment. Furthermore, to ease the process of ordering over the Internet, electronic wallets facilitate tracking of expenses and prevent repeated entering of payment information by securing card numbers and other payment information in electronic wallets.

With Internet usage dramatically increasing, users span all demographics, including both genders, high-income and low-income users, children and adults of various educational levels, and so on. Appearance of Internet domain names in normal advertising channels such as television, radio, and newspapers. indicates the popularity of the Internet among all users. In addition, the ubiquity of the Internet does not limit users' access by geographical boundaries. Online businesses are scurrying to attract users to their sites and to turn them into loyal customers by employing intelligent e-marketing initiatives.

Due to the potential number of customers over the Internet, scalability of e-commerce systems for the business-to-consumer domain is a major concern. Popular Web sites can attract millions of visitors and customers. Amazon.com, for example, is a popular Web-based bookseller that attracts millions of customers. Design of such systems therefore requires large servers or clusters of servers to handle the large user load. Similarly, various other shopping Web sites presenting popular brand products attract millions of users. This user load, facilitated by the Internet paradigm, has no parallel in the history of computing.

The nature of business-to-consumer transactions rarely involves large sums of money. The primary reason is that consumers rarely spend large sums of money (relative to business-to-business payments) on the Internet. Furthermore, the usual payment tokens available to consumers (e.g., credit cards, electronic checks) very rarely support large amount transactions (e.g., in millions of dollars).

Customers in the business-to-consumer paradigm can trigger e-commerce transactions through a multitude of information appliances. These include PCs, handheld computers, wireless telephones, pagers, and TV sets. The ability of powerful, high-speed network technologies enables connectivity of these appliances to the Internet thus attracting a larger customer base through diverse channels. Although these appliances range in their ability to offer Internet-enabled services, continual technological advancements are narrowing the gap.

1.5.2 Intraenterprise Application Domain

E-commerce related technologies, when implemented appropriately within the enterprise, form an organization's central nervous system. Such a system enables an organization to effectively communicate with all of its constituents, facilitates coherent thinking among an organization's staff, and helps an organization to leverage upon people's past experiences and knowledge base to propel the organization intelligently toward its objectives. This is facilitated through establishing a robust e-mail and groupware infrastructure, supporting virtual and remote connectivity to enterprise databases, and linking an enterprise's information assets in a manner that facilitates optimal knowledge sharing. An organization's intranets, which facilitate access to a large information base, are examples of such an infrastructure.

The intraenterprise e-commerce domain provides information sharing and collaboration functions to the enterprise's internal users to facilitate internal operations and to nurture the knowledge management culture of the organization. Typical content that an organization provides to its internal users in this domain includes policy and procedures manuals, a telephone directory, company news, and human resource services. Another example in this regard is a document management system within the enterprise intranet that lets staff access an organization's information bases through various information appliances. Other examples of applications within this domain include groupware applications and other intranet-based applications. Applications in this domain do not usually involve financial transactions unless various departments establish payment mechanisms for interdepartment remittances.

Generally, intranet-based content is accessible to all an organization's staff. However, an organization's departments can establish security schemes to regulate access to restricted information. For example, the finance department may publish reports on the intranet to enable its employees access to those reports, but it may restrict this information from other organizations that have access to the intranet. Organizations can use various extranet technologies to establish such regulated access over the intranet.

Users within the enterprise primarily access enterprise systems and databases through PCs. However, remote and virtual computing has fueled the use of various information appliances, including handheld devices, laptops, pagers, and other gadgets. These information appliances facilitate Internet connectivity, remote connectivity to an enterprise's systems, and use of personal productivity applications.

Internal systems catering to internal users should scale to support an enterprise's entire set of users. For a large corporation, this number could be large. This means that Internet-based systems, internal knowledge repositories, e-mail systems, remote connectivity servers, Internet connectivity, and so on should scale to support these users. Unavailability of an e-mail system for example, can seriously disrupt an enterprise's internal operations.

The nature of content in this domain is targeted primarily toward an organization's internal employees. Content includes internal policies and a knowledge repository pertaining to various issues (e.g., product development, educational material, company-wide events). Sensitivity of such content requires protection from external Internet users.

1.5.3 Business-to-Business Application Domain

The business-to-business domain covers systems that interact with peer businesses to carry out critical tasks of a business process. Activities in the business-to-business application domain include logistics management, inventory control, buying and selling of enterprise assets, and more. EDI applications therefore fall into this category. Other business processes in this category include interbank remittances, goods procurement, and supply-chain management applications. Organizations are leveraging the business-to-business e-commerce domain to integrate their value chains with other organizations. This is enabling suppliers to transact with organizations more cost-effectively and efficiently. Customers indirectly benefit form this model as cost savings achieved through business-to-business collaboration are passed on to the consumers of those businesses. The general availability of the Internet and its improved reliability over the years have pushed all businesses to exploit the advantages of the business-to-business domain.

Content in the business-to-business commerce domain is specific to strategic partners and businesses. Examples include product catalogs, viewing of necessary inventory information, industry-specific news, and so forth.

Entities in the business-to-business domain include corporations, banks, and various government agencies. Historically, only large corporations transacted business-to-business payments electronically. The primary reason was the high costs associated with EDI networks that facilitate connectivity and transfer of electronic payments between large organizations and banks. Such transactions were also referred to as financial EDI. However, the availability of the Internet has opened an inexpensive and flexible channel for both small and large organizations. The emergence of new security technologies is further accelerating the trend to move to the Internet.

The number of users within the business-to-business domain is usually much smaller than in the business-to-consumer domain, which involves millions of users. Users in the business-to-business domain for an enterprise primarily consist of trading partners' users and financial institutions. While transacting business services, an enterprise appoints a few individuals and authorizes them to engage in business transactions on the organization's behalf.

Users in this domain include an organization's internal users as well as staff belonging to external organizations who require access to the organization's systems. Internet- and e-commerce–based technologies have enabled enterprises to provide external staff with controlled access to an organization's systems. Such external access streamlines an organization's business processes. For example, corporate customers can directly access their organization's portfolio in a bank's databases as opposed to requesting electronic files from the banks and processing them in-house.

The large amounts involved in the business-to-business domain, however, require more reliable security technologies. For example, secure socket layer (SSL)–based Internet sessions are commonly used to handle consumers' credit card transactions even though they do not fully provide end-to-end security.

However, for business-to-business transactions, organizations rarely rely on regular SSL sessions to transact large sums of money. The primary reason is that the common use of an SSL session does not support client authentication and merely authenticates the server to the client. In processing transactions of large amounts, authentication of all parties is vital. For this purpose, organizations use either SSL, where both clients and servers mutually authenticate each other, or rely on alternate VPN-based solutions or traditional payment networks.

Financial transactions between organizations are either preestablished between entities (e.g., registration for ACH payments) or employ special software usually triggered through PC- or mainframe-based applications. The use of mobile information appliances to trigger such payments is still not popular due to the infancy of the mobile information appliances market and the various issues related to security processes and procedures.

The use of mobile information appliances for other business processes within the business-to-business domain, however, has continuously been on the rise. These appliances allow users both to work remotely and access internal systems and databases, and to access trading partners' systems and databases to trigger various transactions. For example, as will be further explained in subsequent chapters, a traveling salesperson can order merchandise by accessing a trading partner's stock replenishment systems and placing appropriate orders.

1.6   E-Commerce Technology Architecture

As can be observed through previous sections, embarking on e-commerce initiatives involves linking the business processes and activities of an organization's entire value chain. This includes processes for building the appropriate products and services, acquiring products and services from external trading partners, selling them to external customers, and various other administrative processes.

Specifically, a full-fledged e-commerce initiative requires the following high-level activities:

- Enabling an organization's business processes with sophisticated and value-added information appliances. For example, PCs or dumb terminals are no longer the sole means to engage in information processing transactions and information access. Other devices, such as personal digital assistants, wireless telephones, pagers with access to the Internet, and a host of others enable such access (Chapter 2).

- Connecting to high-powered computing networks. Various data networks have surfaced to facilitate the seamless flow of information and provide connectivity to a wide base of users. These include wireless networks supporting data and voice simultaneously, the Internet, and pager networks (Chapter 3).

- Hosting Web sites for selling, billing, and servicing customers requires deployment of various access gateways (Chapter 4).

- Deploying backend applications that interface with e-commerce Web sites (Chapter 4).

- Attracting customers to a business (Chapter 4).

- Communicating with customers to fulfill requests. (Chapter 4)

- Intelligent segmentation of an organization's networks to facilitate building of virtual value chains by enabling internal and external users regulated access to information and other services (Chapter 5).

- Establishing a network and systems management infrastructure to effectively monitor and control the organization's IT infrastructure (Chapter 5).

- Interface to public portals. Portals facilitate easy access to various types of information and services (Chapter 5).

- Intelligent partnering with external organizations, such as Internet service providers (ISPs) and Application service providers (ASPs) to deliver and host systems (Chapter 5).

- Interface to the Internet domain infrastructure (Chapter 6).

- Intelligent acquisition of technology components, including Internet-based middleware and directory services (Chapter 6).

- Deployment of groupware solutions to maximize inter- and intraenterprise collaboration and communication (Chapter 6).

- Interfacing with electronic payment networks to enable payment processing and reconciliation (Chapter 7).

- Understanding of various security technologies to deploy secure e-commerce systems (Chapter 8).

- Interface to a public-key infrastructure for operating e-commerce applications. Engaging in transactions over the Internet requires innovative security technologies to guard organizations' and users' privacy (Chapter 8).

- Legal consideration for conducting e-commerce (Chapter 9).

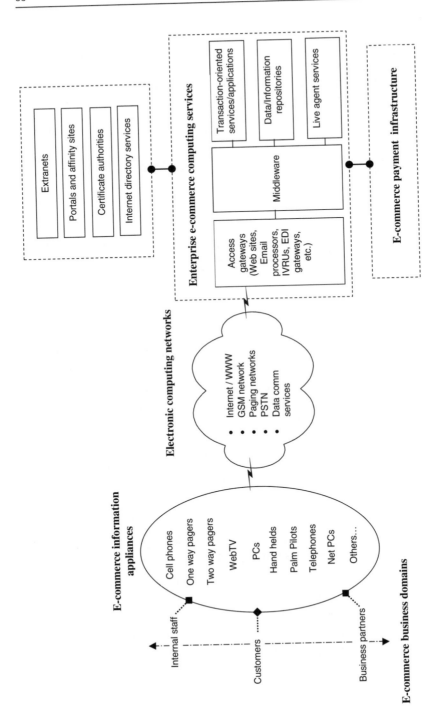

Figure 1.1 E-commerce systems technology architecture.

Figure 1.1 illustrates an e-commerce systems technology framework that comprises these elements. These components form the thrust of the discussion for the remainder of this book. Besides highlighting the primary building blocks of e-commerce systems, the descriptions of the architecture components in later chapters further highlight the inherent synergies required between the various components that result in full-fledged functioning e-commerce systems.

Notes

[1]   Tapscott, Don, *Blueprint to the Digital Economy*, New York: McGraw Hill, 1998.

2

Information Appliances

Information appliances enable users to reach e-commerce services and applications through computing networks such as the Internet, pager networks, and GSM networks, to name a few. Information appliances extend beyond personal computers and terminal-type devices to an array of non-PC and mobile devices. The continued maturation of computing networks has enabled these appliances to support richer information and content processing.

The ability of information appliances to connect with various e-commerce services has driven their penetration in all facets of e-commerce. Covering simple scenarios of end users involved in the exchange of information, ranging to their use in complex business processes, these devices provide all parties with the window as well as the tools to access information and touch various e-commerce functions. For example, customers can self-track shipped packages using various information appliances.

The information-appliance market continues to grow at a very rapid pace and is expected to maintain its momentum. This explosion stems from an information appliance's ability to allow people access to a wide variety of content and services that helps them perform their functions effectively and efficiently on both a personal and a business level. Earlier use of information appliances was limited to enterprise employees and restricted access to host-based content and services. However, in the past few years the use of information appliances has matured to include access to a wide range of content and services including the following:

- Simple content such as text and graphics.
- Rich content such as video and audio entertainment.

- Interactive content that allows users to request additional information on displayed content. For example, interactive TV enables users to click on a moving image and request associated information.
- Simple communications, such as e-mail or voice communication.
- Interactive communication. For example, users can communicate interactively through voice, Internet relay chat, or videoconferencing.
- Personal applications.
- Robust business applications and services.

2.1   Overview and Background

Figure 2.1 illustrates the various maturity stages of information appliances. By definition, dumb terminals connected to a host computer also fall into the category of information appliances. Dumb terminals, through applications that reside on the host computer, allow users to read and update content that resides on the backend host. Dumb terminals therefore represent the first era of information appliances that empowered users to reach content, which in turn enabled users to perform job-related functions. However, the penetration of dumb-terminal–based computing was limited to the confines of the enterprise. The only electronic content that noncorporate users received was through one-way radio and television transmission in the form of audio and TV signals, respectively.

The dawn of PC-based computing in the early 1980s marks the next milestone in the maturity of information appliances. The power of the PC opened the doorway to the information economy by bringing information services within the grasp of corporate and noncorporate users alike. The power of the microprocessor coupled with the small size of the computer and associated information storage qualified the PC to truly fit the description of an information appliance. Users could run intelligent personal, business, and scientific applications on these appliances.

Internetworking of PCs with each other and with other host-based systems pushed PCs into a new paradigm, as remote information access coupled with local information processing provided wondrous opportunities to corporate users. The networking era pushed the PC as an information appliance into the next stage of maturity. Parallel advancements in networking technologies enabled enterprises to further explore the potential of these information appliances by empowering their employees with intelligent information access and enhanced communication abilities. Later, the dawn of the Internet pulled noncorporate users as well by providing them with access to content and other

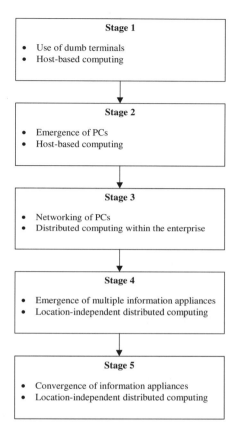

Figure 2.1 Evolutionary stages of information appliances.

computing services. In the Internet landscape, the PC has continued to enjoy its role as the primary information appliance.

The end of this millennium marks the fourth era in the maturity of information appliances. The intertwining of people's lives and enterprises' processes with networked information appliances has necessitated increased capabilities and flexibility in the use of information appliances. The industry has responded with alternate gadgets that fulfill the role of information appliances. Smart phones, handheld computers, intelligent cable set-top boxes, and so on are flooding the market and provide complementary services of information and content access, wireless communication, productivity applications, and many others. In this era, enterprises are reengineering business processes around these appliances, as they bring intelligent information access closer to their users without restricting them to the physical boundaries of the enterprise. Customers, who are equally empowered with these appliances, are propelling enterprises to deliver innovative services that customers can access through these appliances.

Just as the fourth era exploded to empower businesses and consumers with a multitude of information appliances that differ in functionality and other parameters, the fifth era in the maturity of information appliances will be the era of convergence. Information appliances will converge to provide more functionality. Users or enterprises will not depend on the type of information appliance to devise their life styles or business processes. Rather, information appliances will mature to suit the life styles and business requirements of the users. For example, users carry pagers for their short messaging needs, mobile computers for their computing needs, and a mobile wireless telephone for their wireless telephony needs. Information appliances will converge to provide these functions in one appliance. In the present era, mobile users who require access to the Internet's rich content have to wait to get to their PCs. The fifth era will bring additional computing power to these information appliances without limiting them to specific functions.

2.1.1   Services Provided by Information Appliances

Within the enterprise, information appliances stand at various business and technical maturity levels. At a business level, appliances serve three primary needs, and the business should assess an appliance's value before rolling out the appliance to corporate staff. The first need addresses personal productivity issues by empowering users with personal information management (PIM) applications. The traditional PC, with its select applications, debuted a few years ago as this type of a tool before enterprises embraced it for critical business applications. HPCs (handheld PCs) and wireless telephones are contemporary examples of tools that provide their users with various PIM applications, such as address book organizers and time management tools. When used in a standalone mode, the unavailability of these applications or the appliance that provides these applications may have an impact on a person's productivity levels but does not necessarily critically hurt business performance.

Providing access to office applications such as word processors, spreadsheets, and e-mail is the second need that information appliances address for business users. These office applications have become a necessity for the end user. Unavailability of these applications or the appliance that supports them may at times be critical to a business process. Enterprises that depend on e-mail for interoffice communications may find it quite difficult to perform certain business functions. Also, the unavailability of office applications may impede staff from preparing presentations and other such documents. However, unavailability of these applications and appliances are rarely showstoppers as people usually find another way to perform such functions (e.g., use a peer's appliance to send e-mail or draft a document) [1].

The third need that information appliances address is that of providing access to mission-critical applications. At this stage these appliances are critical for carrying out a business process, and any unavailability may result in severe consequences. A bank teller's computer terminal (PC, network computer, etc.) is an example. Unavailability of the terminal will prevent the teller from serving customers appropriately and will accordingly affect business performance and customer satisfaction levels. Another example is that of a ticketing agent using a network computer for travel reservations. Unavailability of the network computer will prevent the ticketing agent from making reservations for travelers.

A consumer, on the other hand, has different needs. Enterprises benefit from a consumer's use of appliances in two different ways. First, enterprises can offer differentiating services to their consumers. Enterprises offer such applications to gain a competitive edge over their counterparts. However, unavailability of such applications is less likely to hurt a business's bottom line with any great magnitude. For example, Internet banking is a service that many would still consider as a differentiating service, as very few banks offer this service. Similarly, organizations that provide Internet-enabled customer service may gain a competitive edge over their competitors and thus attract and retain customers. However, the lack of such a service will not drive the general population away from the enterprise.

The second benefit that enterprises can gain from their consumers is by offering them essential services through various information appliances. These services form the core for an enterprise, and their unavailability can seriously impact customer service and drive customers away. Cable subscribers, for example, demand and expect to receive cable TV service whenever they turn on their TV sets. Similarly, customers expect that an enterprise will offer a customer service department that will have an interactive voice response (IVR)–based system to listen to their requests, complaints, feedback, and so on. Customers also expect that their bank's ATM machine will always be available to provide necessary services. Multiple instances of unavailability of a bank's ATM machines will have the customer shopping for anther financial institution.

To summarize, information appliances provide value to the enterprise from two primary perspectives. First, an enterprise can use information appliances to empower its corporate staff and business partners by enabling access to business-specific computing services, offering innovative communications channels, and providing access to content that empowers them to perform various enterprise services more effectively and efficiently. Second, by recognizing its customers' empowerment with these appliances, an enterprise can offer various backend-computing services and provide value-added content to those customers.

The objective of this chapter is as follows:

- Highlight necessary business-process enabling and reengineering characteristics of information appliances as they relate to various e-commerce systems;

- Introduce the subject of content and various content forms used on information appliances (focus will be on content presentation; later chapters will cover content creation);

- Delineate the popular information appliances used in industry and their business applications;

- Introduce the operating systems and underlying computing technologies that power these information appliances;

- Review the primary business factors that drive the choice of information appliances for various e-commerce business processes.

Subsequent sections of the chapter address each of those objectives.

2.2   Enabling Characteristics of Information Appliances

The emergence of a wide range and many types of information appliances provides users with numerous opportunities to access value-added services in the e-commerce arena. The appliances or devices also provide enterprises with equal opportunities to reengineer their business processes. Organizations can equip their staff with appliances to boost their effectiveness and efficiency by enabling increased exposure to their services and products. Customers equipped with these devices can access an enterprise through alternate channels. Organizations thus can expose their services and products through non-physical channels. A typical example in this case is that of Amazon.com, which has no physical presence. Rather, the virtual channels it offers attract a larger customer base than the traditional brick-and-mortar bookstores do.

Depending on their characteristics, information appliances provide users with various functions. Collectively, information appliances enable all types of users to perform functions that fall under the categories of communication, personal productivity, information access, and business transactions and services. Communication functions enable users to contact other users through various means and forms. E-mail, voice communication through traditional telephone, and Internet access are examples of such communications. Personal productivity applications enable users to stay organized, thus boosting their productivity levels. Calendaring software and personal organizers are examples of these types of applications. Information-access applications enable users to access information through various channels and download them to their information appliances as a means to better inform them and enable them to make more intelligent and better-informed decisions. Finally, specific business applications enable users to execute business transactions and enable other value-added services.

This section explores the various ways in which information appliances empower enterprises to use the four primary functions and to engineer intelligent e-commerce systems. The section demonstrates the various strategic benefits that such e-commerce–enabled systems bring to the enterprise and its customers.

- *Window to the Web:* With the Web becoming the primary platform for the new economy's business, information appliances provide various means to access Web content and its services. These value-added services are available through most of the information appliances that exist in the market today such as PCs, GSM telephones, and palm computing devices such as Palm VII. Information appliances facilitate the means of conducting various business-to-business, business-to-consumer, and intrabusiness transactions through the Web. For example, enterprises can offer services to enable consumers to order books on the Web as they are traveling on a train using their handheld computers. Similarly, a corporate employee can check the inventory in a business partner's databases by signing on to the partner's extranet through a Web interface.

- *Streamlining of business processes:* Information appliances enable enterprises to achieve tighter integration between the various steps in a business process. Information appliances facilitate this by providing timely access to information and the ability to perform appropriate data-related services. For example, field support engineers can access corporate data through their mobile terminals while they are in the field, which enables them to make appropriate decisions. Similarly, a collaboration session between a call center agent and a customer over the Web enables the customer service representative to push valuable mortgage information and material to the customer, thus helping the customer to decide expeditiously to enter into a loan agreement with the financial institution. This tighter integration of processes reduces cycle time for the sales force and customer service representatives, which translates to reaching additional prospects and servicing more customers, respectively.

An enterprise can extend this integration of business services to the end customer as well. For example, an enterprise can offer services that empower the customer to inquire about bank account balances and trigger money transfers through their wireless telephone's messaging capabilities instead of requiring them to visit the local branch to perform such operations.

- *Enhanced customer communication:* Business in the e-commerce paradigm mandates a many-fold improvement in the quality of communication with an enterprise's customers. Exposing an enterprise's service representatives through various information appliances facilitates tightening of that relationship. For example, enterprises can provide the appropriate infrastructure that enables customers to invoke an interactive customer session using a multimedia PCs' Web interface, and while their request is in the queue, customers can continue to work on their PCs. Once connected to the agent, customers can talk to the agent through an Internet telephony session.

- *Location independent information access:* Arming enterprise employees with various information appliances lets employees connect to their corporate databases and make immediate decisions irrespective of their geographical location. For example, a sales employee can remotely access its company's databases using a multitude of appliances to provide pertinent information to its customers and make a sale. This eliminates steps such as "I will be in touch tomorrow with the information," and thus reduces the cycle time to close the sale. More importantly, this enables the salesperson to make the sale when the customer is interested and not defer the sale until a time when the customer's interest in the product or service has diminished.

- *Support of various content formats and types:* Mobile information appliances have matured considerably to provide access to rich content. For example, Windows CE–based appliances enable access to rich content including text and audio and video clips. This enables enterprises to offer enhanced services to their customers, such as shopping through the Internet when customers are away from their PCs. In this context, enterprises can offer rich content to customers to attract them in buying products and services. For example, enterprises can deliver shockwave-enabled multimedia content to demonstrate their services to customers through various appliances.

- *Accessibility to various networking backbones:* All popular information appliances can connect to the Internet through various network backbones. This ubiquitous access through multiple appliances and networks brings e-commerce services closer to the end user. Conventional modes allowed access only through PSTN networks. Now mobile users can connect their information appliances to the Internet or other data networks through wireless, cable, and other networks. Chapter 3 covers these networks in more detail.

- *Evolution to intelligent appliances:* The emergence of powerful networking and content-handling technologies are transforming information appliances into intelligent appliances. Initiatives are underway to enable intelligent agents to reside on mobile appliances that continuously request and present personalized information to the user. For example, e-speak is an HP initiative that enables information appliances to become intelligent by working and requesting services on behalf of the end user [2].

- *Support for personal information management (PIM) applications:* Information appliances empower users with personal productivity applications, such as address book organizers, scheduling applications, and other office applications. When used in a networked environment, these appliances allow group scheduling and conferencing. For example, executives can remotely connect to their administrative assistants' information appliances, thus enabling synchronization of the schedule information. A telephone services representative can schedule services of field technicians by remotely connecting to their appliances and issuing service instructions through the attached notes on a schedule.

  Availability of applications such as word processors, spreadsheets, and other applications on mobile appliances has resulted in a many-fold boost in user productivity. Users, for example, can reduce the cycle times by composing a memo or report as they are traveling and upload it to their primary information appliance when they get to their homes or offices. Similarly, users actively use these appliances to manage their financial accounts. Banks, for example, are exploiting this opportunity by rushing to offer customers the ability to download their bank statements in a format appropriate for particular financial packages.

- *Access to strategic corporate applications:* The fueled growth of information appliances is driving developers to deliver enhanced applications that allow users access to enterprise resource planning (ERP) data. For example, various vendors are marketing products that enable users to access ERP data through PalmPilots and other HPCs. Other products allow users to use terminal emulation programs through Windows CE appliances to access corporate legacy applications.

Initiatives by major database vendors have pushed "light" versions of enterprise databases to information appliances. Oracle8i Lite and Sybase's SQL Anywhere Studio are examples of popular databases that install on handheld PCs, allowing users to locally access corporate applications and update data on the appliances. Later, users can remotely connect to the enterprise databases and upload a copy of the database. This allows the users to use corporate applications without remote connectivity, thus saving on network services costs while allowing them to work in a mobile environment.

- *Facilitating paperless environments:* Sophisticated mobile appliances such as pen tablet computers enable field workers to collect data that automatically uploads to the corporate database and thus helps eliminate paper work and the need for data re-entry. These appliances can provide great value to healthcare and field applications, which require workers to be continuously on the move to collect data.

- *Availability of data services for rugged conditions:* Information appliances have matured to cater to the needs of enterprise workers who work in rough weather and rough physical environments. Typical examples include steel engineers roaming in steel plants that typically have high temperatures and other rough physical conditions. Other examples include construction company engineers, who require access to data in dusty environments. Tablet computers have surfaced that allow enterprises to equip their workers with computers that suit such rugged conditions.

- *Convergence of communication functions:* Voice and data communication functions converge on some information appliances. A user can use the same wireless telephone to make a telephone call or request data services such as Internet or paging services. This feature of information appliances, in addition to providing some of the aforementioned value-added services to users, also eliminates the need to carry multiple appliances. Today, most information appliances provide specialized functionality, but with ongoing technological innovation, information appliances are packing more functions. For example, earlier pagers provided only a one-way communication stream. New pagers allow a page-message recipient to respond to the message by using a small keyboard. Similarly, palmtop computers now can receive pages.

These trends are enabling users to carry fewer devices while providing them access to more functions.

2.3   E-Commerce Content

The power of the information appliance stems from its ability to access and present valuable content to its end user. Without content, there is no need for an information appliance. It is therefore vital to understand the meaning of *content*. This section focuses on content issues as they pertain to the information appliance. Later chapters will address other issues related to e-commerce content.

Content is any information that provides value to the end customer. Any information, such as news, sports clips, health-related topics, and information from corporate databases constitutes content for various types of customers. Users usually invest heavily to access value-added content. This investment is either in the form of building the appropriate IT and business infrastructure or in the form of subscription to content services. For example, customers pay a monthly subscription fee to get sports news or other types of news delivered to their pagers, cellular telephones, and other information appliances.

Enterprise users, on the other hand, access content that resides in their organization's databases and ERP systems to enable them perform their business duties. For example, a bank officer accesses the bank's systems to retrieve content specific to a customer to fulfill a customer's request. With penetration of e-commerce in the business-to-consumer arena, enterprises are opening their systems and resources to enable customers to directly access content that resides in enterprise databases. Customers, for example, can directly access a stock trading company's systems to place stock orders. Similarly, enterprises are creating, publishing, and selling content that includes entertainment services such as live shows, music, and videos through the Internet.

The increase in the creation and delivery of value-added content can be attributed to number of factors. Some of the factors associated with the increased demand of value-added content are as follows:

- *Increase in network capacity:* A continuous increase in enterprises' and public networks' capacities has enabled delivery of rich content. Information in the form of text, audio, video, and multimedia can be delivered cost effectively to customers and various businesses. This is driving enterprises to deliver compelling content to user's appliances for a wide range of business processes including marketing and advertising campaigns.

- *Reach of the Internet:* The reach of the Internet is another reason for the popularity of content. Never before has it been possible to pull information from varied sources, derive intelligence and knowledge from such data, and offer it as value-added content to pertinent customers. Enterprise and end users alike can access the rich resources of the Internet and the Web to access value-added content. Enterprise users can use the Internet to access their corporate databases to access pertinent content.

- *Accessibility to content:* Customers are empowered to receive content through various information appliances. Today, if a busy worker is unable to catch the evening news due to a hectic schedule, various services by network operators promise to deliver this content as a value-added service to their customers on their favorite information appliances. For example, pager and wireless telephony networks collaborate with various content organizations (e.g., MSNBC) to deliver appropriate content to interested customers. This content includes news, health care industry topics, and other popular topics.

- *Maturity of hardware:* The maturity of appliance hardware has further fueled the creation and delivery of rich content. Faster microprocessors and colorful display technologies pushed the creation of appropriate content. Today, PCs and workstations offer rich displays to their users. However, smaller mobile appliances are unable to parallel larger information appliances in their presentation of content. Recent advancements in information appliance hardware and associated operating systems, however, have brought certain features such as colorful displays to users' appliances.

2.3.1　Understanding Content

With all types and forms of content distributed so widely, it is vital to understand the typical life cycle of content. The life cycle typically includes creation of content, delivery of content, and presentation of content to the end user. The following paragraphs describe the process.

2.3.1.1　Content Creation

Creation of content depends on the specific content type. Following are some of the means used for creation of content:

- *Data updates in files and databases through applications and programs:* Someone using an application, such as an order-entry application, updates databases with appropriate order and customer information. This information or content is available for later retrieval and use.

- *System updates:* Various systems can automatically update information and databases. For example, logical controllers connected to steel manufacturing equipment can automatically update databases, which later constitute valuable content for engineers.

- *Recording of audio and video:* Various tools can be used to record audio and video clips that can be stored in files and later be retrieved as value-added content by customers.

- *Live audio and video:* Tools are available that can capture live audio and video streams, encode in appropriate digital formats, and deliver them to the user through specialized media servers.

2.3.1.2   Content Delivery

Content delivery pertains to mechanisms that enterprises use to deliver content to end users. Three methods of content delivery reign in the IT world. They are as follows:

- *Database applications:* This is one of the oldest methods of content delivery, dating back to the use of dumb terminals. In this method, database applications (sometimes coupled with transaction monitors) provide users direct access to content in the databases. The transition to client/server computing retained this model of content delivery except that database applications and associated components were distributed to multiple processors.

- *Web hosting:* Hosting of content on a Web site is the predominant method of delivering content to external customers and partners. However, with the proliferation of portal and security technologies, enterprise users are also moving toward this method of accessing content. In this method a Web-hosting engine sits between the user and database applications to offer users a standardized interface and method of accessing relevant content.

- *Webcasting:* Webcasting enables the playing of live or delayed audio and video content through the Internet. Webcasting allows live video and audio streams to be sent to a user through specialized multimedia technologies. Special encoder servers hook to live audio and video streams and deliver content to a user's information appliance. Users requesting Webcasting sessions require special software players on their Web browsers. Popular session players include RealVideo and Media-

Player. Webcasting plays video and audio content through a Web server that supports multicasting. Multicasting allows one sender (the Web server) to transmit content to a specific group of users.

Subsequent chapters provide further details on the aspects of content creation and delivery. This chapter focuses on content presentation due to its direct applicability to information appliances.

2.3.1.3   Content Presentation

Content presentation is the last leg in presenting content to user's information appliances. Content presentation provides meaning to content. For example, a downloaded video presentation on an information appliance has no value unless appropriate content-handling applications (e.g., RealVideo and Media-Player) are available to play that video clip to the user. Similarly, a content dump from a corporate database to a user's information appliance is useful only if there are appropriate tools on the information appliance (e.g., a client program presents that content to the user in an intelligible format).

From a business perspective, applications for content presentation on a user appliance can be grouped into three general categories:

- *Vertical applications:* These applications focus on specialized content presentation needs. A call-center application providing customer service representatives access to content on a calling customer or a bank teller's application presenting content on a customer's accounts are examples of such applications. Businesses usually develop these applications or acquire them from vertical solution developers to fulfill a specific business need.

- *Horizontal applications:* These applications find generalized use across a wide base of users. Office applications, PIM applications, and so on reside on various information appliances and provide users the ability to access content.

- *Public applications:* Web browsers and their associated software (e.g., plug-ins, applets) are public applications as they are the most widely used types of content-presenting applications. The industry clearly envisions an era where Web browsers will provide access to all forms of content, which today are accessible only through vertical and horizontal applications. For example, users will be able to access corporate database content through Web browsers. Advances in content technologies and content-handling technologies will converge on the use of

Web browsers that are moving to provide standardized content presentation functions for all types of content.

2.3.2 Content Handlers

Content handlers facilitate presentation of content to users' information appliances. Various content handling applications have surfaced in the past that have facilitated access to remote corporate databases, locally stored content, and content available on the Internet. Figure 2.2 illustrates four categories of applications that are described in further detail in the following sections.

2.3.2.1 Terminal Emulators

Terminal emulators are applications that run locally on an information appliance and facilitate communications with a mainframe or a server using specific data streams such as 3270 or 5250. However, the business logic of the application runs on the mainframe computer and not on the information appliance. Terminal emulation software is still actively used in enterprises to provide access to corporate applications and business content stored in databases.

2.3.2.2 Appliance Applications

These applications run locally on the information appliance by using local processing power of the appliance. Applications in this category provide access to a vast array of local and remote content. Numerous applications fall into this category. These applications use database connectivity software to pull content from remote databases or display locally stored information. Applications developed for the Windows environment are examples of such types of applications.

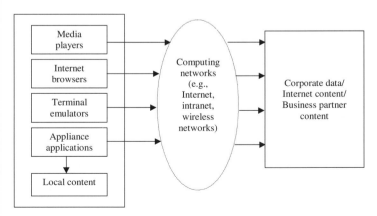

Figure 2.2 Content-handling applications.

2.3.2.3   Internet Browsers

Internet browsers are application programs that execute on the information appliance and provide access to Internet- and intranet-based content. Internet browsers incorporate technologies that enable them to access various types of content (e.g., HTML, DHTML, XML). Browsers designed with appropriate Internet technologies can present both graphical and multimedia content.

Mosaic was the first popular browser for the Internet. Today, Netscape Navigator and Internet Explorer are the two most popular browsers for the PC market. Internet and intranet browsers also exist for other platforms such as UNIX, OS/2, Macintosh systems, and others.

With the increased popularity of mobile appliances, browsers have emerged to support those appliances as well. These browsers are also referred to as microbrowsers. Microbrowsers are especially designed to accommodate the weight, size, and memory requirements of such appliances and have a very small footprint to enable them fit the information appliance. Up.Browser, from Phone.com was designed for wireless telephones. Device Mosaic from Spyglass is another example of a browser that runs on set-top boxes and other mobile and wireless devices. Sony uses Device Mosaic in its set-top boxes. Pocket Explorer for Windows CE systems is another such example. Other appliances such as Palm VII provide browsers with "Web clipping" features that facilitate basic information access from the Internet without any graphical formats and features.

2.3.2.4   Media Players

Browsers are not designed to handle all types of content. For specialized content, content developers furnish appropriate media players. Media players are windowed graphical applications that run locally on the information appliance and present various types of specialized content. Some media players have become de facto standards in the industry, and users can download them from multiple sources on the Web.

Some of the popular media players to handle various types of Internet content include:

- *Windows MediaPlayer:* This multimedia player from Microsoft supports various forms of content, including MP3, ASF, WAV, MPEG, and other popular formats.
- *Shockwave player:* Shockwave enables users to access rich multimedia content that includes games, CD-quality audio and music, and presentations.

- *QuickTime player:* This player handles Apple's QuickTime multimedia content.

- *Cosmo player:* Platinum's Cosmo player enables users to view VRML content. Users can download these players and run them inside their browsers.

2.3.3 Content Technologies

Content has many forms and types. The first era of information appliances characterized by dumb terminals provided enterprise users with access to business-specific content that resided in legacy databases. With the move to PC-based computing, powerful applications emerged on the PC that enabled users to access more compelling content. Office documents created from word processors, slide, and spreadsheet programs gave birth to various forms of content. In general, content exists in the following formats:

- Text in plain text files and databases;
- Stored images in databases;
- Audio files;
- Streaming audio files;
- Video files;
- Streaming video files;
- Interactive content.

This section will review various technologies related to content types and handling of such content. Numerous technologies exist related to content presentation, but this section focuses on Internet-related technologies.

2.3.3.1 Standard Generalized Markup Language (SGML)

SGML is a standard developed by the International Organization for Standardization (ISO) in 1986. SGML defines the rules to define and develop a markup language such as HTML or XML. *Markup* refers to special characters (also called tags) inserted in a document that define the output of the document (how the document is printed, displayed, and so on). For example, a defined character (or tag) within a document may refer to displaying the line in the center of the page. HTML and XML are popular markup languages that employ the rules set forth in the SGML standard.

2.3.3.2   Hypertext Markup Language (HTML)

HTML is a document authoring language based on the rules defined by SGML. HTML uses tags to describe content and display them on applications called browsers. Using various tags, HTML describes the data that is displayed by the browser along with the styling information of the content to be displayed (e.g., centering a line in the document). Browsers incorporate the rules to define HTML tags and interpret them to display HTML content to the user's information appliance displays. However, HTML is not "aware" of what the data is. This makes it difficult to extract information based on the data elements that are embedded within HTML.

2.3.3.3   Dynamic Hypertext Markup Language (DHTML)

DHTML is an extension of HTML that enables development of richer HTML pages by enhanced presentation, flexible content positioning, and a richer set of fonts. This is facilitated by the use of technologies such as style sheets. Style sheets define the layout and style for displaying content of a Web page. This includes line spacing, margin widths, and so on. For example, users can create a style sheet for displaying a newspaper type of document. Cascading Style Sheets (CSS) enable the use of multiple style sheets for one Web document. For example, part of a document may use one style to display catalog information, whereas another part uses a different style to display press releases.

Interactive TV shows are also adopting the use of HTML. Interactive TV shows broadcast users appropriate HTML files (sent either along with the video signals or through an Internet connection) to enable users to engage in an interactive video session and request additional information related to the program or invoke other e-commerce services such as ordering goods through the "online-TV" experience.

2.3.3.4   Extensible Markup Language (XML)

XML is the next step in displaying Web content to users. However, unlike HTML, XML supports intelligent user interaction by defining an appropriate structure to the Web page. Due to its ability to understand the structure of content, users can perform intelligent searches and effectively navigate the Web for information. XML separates content from style, unlike HTML, which mixes style in with the definition of a document. XML has great potential to be used in EDI transactions. Applications such as browsers have to be intelligent to interpret XML data format. All browsers include appropriate support to display XML data to users.

2.3.3.5   Extensible Style Language (XSL)

XSL, which is similar to CSS, is an enhanced styling specification for Web pages that applies to XML pages as well. However, XSL uses the XML notation for specifying style, unlike CSS that uses its own methods. Besides, XSL provides transformation capabilities that allow presentation of XML data in various views. XSL includes these features due to the various pattern-matching constructs built into the language. Users, for example, can display press releases by date or by the type of events.

2.3.3.6   Synchronized Multimedia Integration Language (SMIL)

With the dawn of multimedia content, the problem of integrating various forms of content surfaced. Although Web browsers in conjunction with associated plug-ins provide the ability to present various forms of multimedia content to a user's information appliance, coordination of all elements such as text, graphics, video, and so on requires extensive programming. SMIL is an initiative that promises to provide a structure for multimedia content thus facilitating appropriate handling of such content. SMIL also facilitates the means by which a content producer can associate multiple versions of content with a Web page depending on the user's access conditions. These conditions (e.g., access through small mobile appliances or a network's limited bandwidth) may inhibit the user from accessing rich content. By allowing multiple versions of content, the user receives content appropriate for their conditions.

2.3.3.7   Wireless Markup Language (WML)

WML is the new name for Handheld Devices Markup Language (HDML) and fully conforms to XML specifications. WML is a derivative of the SGML language and is very similar to HTML. WML surfaced to provide Internet browsing capabilities to wireless devices. WML displays the text portions of a Web page on a user's microbrowser. WML caters to the limitations of mobile information appliances. Some of the limitations are reduced display sizes (4-inch diagonal screens with a 4:12 aspect ratio), slower bandwidth, and limited user navigation capabilities. A standard similar to WML is compact HTML. Compact HTML is a subset of the HTML specification, which is used to support Web browsing on small information appliances that have display, memory, and other restrictions explained earlier. Compact HTML does not support the following common HTML features: JPEG image, table, image map, multiple character fonts and styles, background color and image, and frame style sheet [3].

2.3.3.8   Wireless Application Protocol (WAP)

WAP enables the use of Internet applications over wireless handsets. WAP surfaced to offer a solution to the slow bandwidth associated with wireless networks and smaller display sizes of mobile information appliances. WAP's specification allows appliances to run on any wireless data link that include CDMA, TDMA, GSM, and CDPD protocols (covered in Chapter 3). WAP resolves these issues by letting network operators (e.g., cellular network operators) install servers at their sites that translate content from the Web to a format that is appropriate for the mobile devices. WAP has enabled the development of microbrowsers specific to each mobile device that display Web content on the device's display in an appropriate format. The WAP architecture is robust enough to handle multiple types of mobile information appliances equipped with touch-screens, full keyboards, and software keyboards. Numerous mobile telephone vendors are providing WAP-enabled functionality in their handsets.

2.3.3.9   Graphics Interchange Format (GIF)

GIF is a file format for storing graphics files. GIF is a very popular method used for sending images on the Internet. GIF uses a lossless compression scheme, which means that the compressed image does not lose any of the original image's characteristics. The decompressed image therefore is identical to the original. The JPEG standard, by contrast, is a lossy standard. GIF employs Unisys's Lempel-Ziv-Welch (LZW) compression technology. Animated GIF is an extension of GIF that contains a series of images that when displayed on the user's screen appear as a moving image.

2.3.3.10   Joint Photographic Experts Group (JPEG)

JPEG is a standard for the production of still image files of varying quality. By allowing the use of various compression techniques and image encoding and decoding speeds, JPEG addresses issues of delivering digital content to users in a networked world that have bandwidth restrictions. JPEG's compression techniques result in a loss of image quality when compared to the original image. However, JPEG compresses images by losing image quality that cannot be detected by the human eye. JPEG supports multiple algorithms for image compression, and each algorithm varies in its compression techniques by producing varying levels of image quality.

2.3.3.11   Virtual Reality Modeling Language (VRML)

VRML is a standard that displays 3-D content to the user and allows the user to interact with such content. Just as HTML enables access to 2-D content for users, VRML provides users with a three-dimensional interactive experience on the Web.

VRML has numerous applications in the industry. Real estate agents, for example, can guide potential home buyers through an Internet session that allows users to enter rooms and activate various controls (e.g., turn door knobs).

Due to certain limitations of presenting real 3-D images on the Web, a consortium of companies that includes Microsoft, Sony, Oracle, and Apple have started to work on another standard called Extensible-3D (X3D). This standard will bring 3-D animation to e-commerce–based services such as marketing and advertising. The proposed standard caters to the limitations of bandwidth on the Internet and promises to provide 3-D content in such environments.

2.3.3.12 Musical Instrument Digital Interface (MIDI)

MIDI is a protocol that includes command sets that enable the playing and recording of musical content on computers. Unlike simple audio files, MIDI caters to the complexities of musical sound formats and facilitates information exchange between various musical appliances and computer sound cards. MIDI achieves reduced file sizes as it stores information on how music or sound is produced as opposed to representing musical data directly in a file. MIDI files are stored in .MID, .MFF, or .SMF file formats. All devices that support the MIDI standards, including computer sound cards, can produce and play MIDI sounds. Microsoft and IBM's WAV standard is another standard for storing sound files on a computer.

2.3.3.13 Waveform Audio File Format (WAVE)

Waveform Audio File Format is one of the first standards proposed by Microsoft and IBM that facilitated sound encoding and decoding for personal computers. It is still the most common format used for delivering audio content over the Internet. Files of this format have .WAV extension.

2.3.3.14 Moving Pictures Expert Group (MPEG)

This group controls the standards for digital video compression. The group has developed various standards that include MPEG-1 and MPEG-2. MPEG-2 offers much better quality video than the MPEG-1 standard with a resolution of 720×480 and 1,280×720 at 60 frames per second (fps). The quality is similar to video images seen on a VCR and TV and competes with standards such as NTSC and HDTV. Digital Versatile Disks (DVD) use the MPEG-2 standard.

MPEG-1 supports a lower resolution, but due to its history and reduced file sizes it is used extensively especially when offering video content to users over the Internet. MPEG-1 has a resolution of 352×240 at 30 fps, which is adequate to sense a video image by the human eye.

MPEG competes with other digital video standards such as QuickTime and Video for Windows. QuickTime is an Apple standard with a file extension of .MOV, whereas Video for Windows is a Microsoft standard with file extensions of .AVI. However, as MPEG is a standards organization, it is considering incorporating QuickTime technologies and codecs. Additional standards are surfacing within MPEG that promise to provide better quality and video with reduced file sizes. Advanced Audio Codec (AAC) is one such standard that promises to deliver traditional MPEG audio quality at half the size.

2.3.3.15   QuickTime

Apple devised the QuickTime system, which enables playing of digital video on user's information appliances and competes with Microsoft's Video for Windows and DirectShow [4] standards. QuickTime systems employ .MOV and .QT extensions. QuickTime also supports video compression standards of MPEG. QuickTime has achieved a very high penetration rate among Web hosting services and Web users. The primary reason for its popularity is due to its cross-platform compatibility across platforms, including Microsoft Windows systems. Another reason is that QuickTime supports multiple file formats that include .AVI, .WAV, and other popular file formats.

2.3.3.16   Video for Windows/ActiveMovie

Microsoft's Video for Windows and ActiveMovie standards are similar to Apple's QuickTime for playing video content. Video for Windows and Active-Movie use the Audio Video Interleave (AVI) file extensions. AVI enables the production of multimedia animation files and requires ActiveMovie software player for playing AVI files. The primary advantage of AVI files is its software model in which enterprises can deliver video content without requiring users to have special hardware devices or cards on their information appliances. AVI files have 320×240-resolution and deliver video at the rate of 30 fps, which is adequate for viewing video clips but does not approach the quality of full motion video. These standards compete with Apple's QuickTime standard.

2.3.3.17   Advanced Streaming Format (ASF)

ASF is a steaming media technology that allows playing of media streams as the user starts downloading them on the information appliance, instead of waiting until the entire file is downloaded. ASF is a Microsoft standard and Microsoft media players like MediaPlayer support the ASF standard along with others, such as MPEG, WAV, and AVI.

2.3.3.18 Channel Definition Format (CDF)

This technology allows users to select channels on the Web to view information while allowing content publishers to push information periodically to users who are linked to that channel. Various media companies use this technology to push information such as sports or health information to users who subscribe to those media channels. Users' browsers must be able to support this technology to be able to receive such content. CDF technology is based on the XML standard.

2.3.3.19 Open Software Description (OSD)

OSD is a specification jointly developed by Marimba and Microsoft that builds on the XML technology to describe various components of software. It enables intelligent delivery and installation of software components through the Web as opposed to downloading the entire software to users' information appliances.

2.3.3.20 Resource Description Framework (RDF)

RDF is another XML application that provides a seamless, standard, and transparent infrastructure for accessing data across varied sources including the Web, a user's hard disk, and other sources. RDF accomplishes this transparency by cross-referencing and cross-linking content and its components.

2.3.3.21 Advanced Television Enhancement Forum (ATVEF)

ATVEF is a new format to provide interactive TV content to viewers through their TV sets. This specification defines blending of video and data (HTML-based and other) into one format that can be transmitted through the cable network. The ATVEF membership includes representatives from computer, cable, and consumer electronics organizations. This specification will allow standardization of creation and delivery of content for interactive TV throughout the industry.

2.3.3.22 WWW:MMM

As interest in accessing the Internet and other data services continues to grow, content developers are focusing on delivering appropriate content as well. Nokia, Motorola, and Ericsson have launched a new initiative called WWW:MMM that identifies content for mobile appliances. WWW:MMM is a logo that content providers can license to display on their content.

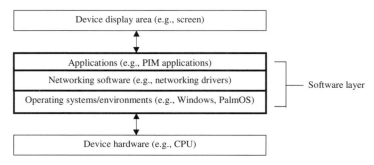

Figure 2.3   Software architecture of information appliances.

2.4   Choice of Information Appliances

There are numerous devices on the market that easily fit the description of information appliances. Recent advancements have introduced information appliance–type devices that are included in noncomputer equipment, such as washing machines. These appliances will send alerts to appropriate service organizations that will dispatch technicians to rectify defects in those appliances. This section, however, focuses only on popular information appliances that are proliferating to provide data services within the business-to-business, business-to-consumer, and intrabusiness e-commerce paradigms.

Figure 2.3 illustrates an information appliance's typical architecture. On top of the appliance hardware, an operating system and an operating environment manage all the programs on the information appliance and associated functions. Numerous networking technologies, either loaded as part of the operating system or separately, provide various network connectivity functions that include connection to printers, other appliances, and data networks. This section will not focus on data networks, as they are the topic of the next chapter. Innovation in technologies also enables appliances to connect at the service layer, thus enabling appliances to request network services directly from other appliances or sources. Residing at the top layer are the content handler applications that retrieve and display content to users on the appliance's display, depending on the type of content.

The following text describes several popular information appliances, and Table 2.1 summarizes various information appliances and typical applications.

2.4.1   Wireless Telephones

The primary function of wireless telephones is to enable users to establish voice communication with other users on the PSTN or a wireless network. The advancement of wireless telephony has delivered better voice quality and enhanced security features, thus offering users a viable alternative to traditional telephones. To use telephony or other services, subscribers have to choose the appropriate wireless telephone that can interoperate with the services of the wireless network service provider (NSP). For example, a subscriber usually requires a GSM type of telephone when traveling from the United States to Europe because most European networks are based on GSM. Various manu-

Table 2.1
Typical Information Appliances and Their Applications

Information Appliance	Popular Operating Systems	Primary Applications
PCs and Notebooks	Windows 9x Windows NT Mac OS	Business transactions Internet access using browsers Access to corporate data Full-strength personal and business applications
Pagers	FLEX	Sending and receiving of short messages Simple PIM applications
GSM/Cellular phones	EPOC JavaOS for consumers GEOS	Wireless voice telephony Sending and receiving short messages Sending and receiving e-mail Storage of addresses Internet access using microbrowsers Access to corporate data PIM applications
HPCs	PalmOS Windows CE Java OS	Business transactions Internet access using browsers Access to corporate data PIM applications Communication Access to corporate data
WebTVs	Java OS/PersonalJava	Sending and receiving e-mail Interactive access to content through TV
Car devices (Microsoft's AutoPC)	Windows CE	Sending and receiving e-mail PIM applications

facturers provide wireless telephones that have the ability to switch between frequency bands and analog or digital modes.

The use of wireless telephones in e-commerce–based computing has been on the rise since wireless NSPs and wireless handset manufacturers started to offer data services over wireless networks. Similar to telephony services, the use of data services on a wireless handset requires the proper appliance that can interoperate with the wireless NSP's services. The user also requires a subscription to the appropriate data services through the NSP. Access to certain data services usually requires the activation of Subscription Identity Module (SIM) cards on the wireless handset for GSM telephones.

In general, wireless handsets support many data services. Most wireless handsets run on the EPOC operating system. Some from Nokia use the GEOS operating system. Screen display for some wireless telephones, such as those from Nokia, is an impressive 640×200, thus allowing the usage of robust applications and data services similar to HPCs. These services include Internet access, Web browsing, e-mail access, fax, and short message services. Various solution providers have developed microbrowsers that enable Web browsing through the wireless handsets. Device Mosaic developed by Spyglass and Up.Browser developed by Phone.com are common examples of browsers used on the wireless handsets. As mentioned earlier, users are required to subscribe to all networked data services available from appropriate wireless NSPs. Chapter 3 provides a detailed explanation of cellular and wireless services.

Newer wireless telephones have advanced to include a wide range of PIM applications. These include applications for calendar/event reminders, phone books, calculators, and so forth. Users can employ these applications locally and synchronize them with applications on other appliances such as PCs or HPCs. Furthermore, the increasing memory size of these handsets has fueled the development of customized applications by various developers for vertical industries. These include applications that enable medical staff to review procedures and field technicians to access corporate databases and review maintenance history of equipment, as well as wireless trading applications that empower users to place stock orders from their wireless handsets.

2.4.2   Pagers

Pagers enable users to send and receive short text messages and alerts. The first-generation pagers were one-way, allowing users to receive text messages. These pagers are still quite popular and offer users a very inexpensive way to stay accessible. Current trends in paging networks also enable users to periodically receive specialized content (news, stock quotes, financial information, and so on) on their pagers. Users can also elect to receive e-mail messages in full or in part (e.g., sender and subject heading) on their pagers. All pagers primarily support text messages. However, new offerings by certain paging organizations have enabled users to receive voice messages on their pages and support downloading of graphical images.

Next-generation pagers enable users to send messages using a small keyboard that is attached to the pager. Similar to wireless telephones and associated data services, the use of appropriate data services through pagers requires a subscription to appropriate data services through the pager NSP. Chapter 3 provides a detailed explanation of the data services from the pager NSPs.

Two-way pagers range in their functionality as well. Less sophisticated devices enable users to send preprogrammed messages from the devices. Better devices enable users to send e-mail to other users. If the e-mail recipients have similar paging devices and paging services, they can also receive those e-mail messages on their pagers. Using appropriate pagers and paging services, users can also manage Internet e-mail. For example, Motorola provides sophisticated pagers with an enhanced e-mail client that allows the pager to interface to a special messaging server at the pager NSP that interfaces to the subscriber's e-mail server and allows the subscriber to manage e-mail on the e-mail server through an IMAP interface.

Motorola has been at the forefront of the pager industry providing hardware and software for the paging network infrastructure and the paging devices. With the introduction of Motorola's FLEX operating system, application developers can develop applications for pagers as well. These applications include PIM applications (address book, scheduling applications, etc.), which, similar to wireless telephones, enable synchronization of information to other information appliances such as a PC or an HPC. The availability of software development kits from various vendors allow the development of business applications for banking, shopping, and other uses to enable users to exploit e-commerce services from the palm of their hands.

2.4.3 PCs/Workstations

PCs and workstations provide access to feature-rich applications and content. Of all the information appliances, these appliances offer the most functionality. PCs were among the first few information appliances to surface in the IT industry. These are immobile appliances that usually rest on a desktop [5]. Primary reasons for this are the size of the PCs and their requirement for a full-fledged and continuous power supply. PCs are the most common form of information appliance and run all types of simple and sophisticated applications.

PCs require a regular-sized keyboard and support large display sizes, memory, and disk space. Notebook and laptop computers are smaller version of PCs that provide better mobility features but have a smaller display size. Most PCs operate on Intel or other x86-compatible CPUs. Apple systems use Motorola's family of CPUs (G3 and G4 processors). However, Apple has far fewer users than the Intel-based PCs. PCs support a wide range of network connectivity options including dial-up, LAN, and WAN connections. Enhanced networking software also allows these workstations to be used in wireless network configurations.

PCs are ideal for use by knowledge workers who require access to powerful personal and business applications. PCs can handle rich content, including video and audio streams, rich graphics files, and full TV programming. Some PCs also have special graphics subsystems that further enhance their power to display even richer content.

2.4.4 Handheld PCs (HPCs)

HPCs are mobile devices that let users perform numerous functions that were previously available only on full-fledged immobile PC systems. HPCs provide the most mobility features and, unlike notebook computers, HPCs are small enough to be held in the palm of a hand and have longer battery lives. HPCs have small display areas and vary in their ability to display various forms of content. Screen resolution varies from 160×160 for a PalmPilot to 480×240 for a Window CE–type device. This poses challenges in displaying rich content such as video streams and rich graphic files. Also, certain HPCs lack support for color displays. Some common uses of these devices are for maintaining personal schedules and contact information, performing calculations, taking notes, and so on. However, recent trends indicate a continuous rise of these appliances for mission-critical and office types of applications.

HPCs fall into two primary categories. The first category includes computers that run on Microsoft's Windows CE operating system. This category is commonly referred to as HPCs. The other category runs on the PalmOS or the EPOC operating system, from 3COM and Symbian, respectively. These devices are called handhelds [6]. Windows CE systems have regular keyboards, whereas the other category leverages stylus pen input for receiving user input.

HPCs provide various network connectivity options. Most HPCs have the ability to print information on regular printers directly. Others require synchronization with desktop software from where information can be printed.

HPCs also allow connectivity to various wireless networks for accessing data services that include Internet browsing and e-mail functionality. However, most network functions require users to load special software and attach hardware accessories to the HPC to perform those functions.

Windows CE HPCs allow users to access their favorite office applications such as word processors and spreadsheet programs. However, as these applications occupy a small footprint on these devices, their functionality is limited. PalmOS systems, on the other hand, are not compatible with Windows CE devices and require their own suite of applications. However, due to the popularity of the PalmPilot, numerous applications have emerged, including word processors and financial software packages that run on the PalmOS platform. PalmPilot continues to enjoy its popularity even though users are more accustomed to using Windows applications on their regular desktop and notebook computers.

HPCs provide enterprises with more powerful solutions for running vertical applications as these appliances are dedicated to computing tasks, unlike wireless telephones whose primary function is to provide users with telephony functions. A bigger display and more memory enable enterprises to deploy richer and more robust applications. Similar to wireless telephones, these devices ideally suit an enterprise's field force. In a business-to-business paradigm these appliances empower workers by running local applications on their appliances. For example, the field sales force can update data in the local databases and then connect through the Internet and upload the data in corporate databases.

2.4.5   Set-Top Boxes/Internet Receivers

Set-top boxes let users access the Internet and Web from their TV sets and also allow users to interact with that content. For example, while viewing a TV sports program users can click on a player and request additional information on that player's statistics, or while watching a program, users can request other Web links about that TV program. This interaction is facilitated through the WebTV Networks, Inc., which is a Microsoft-sponsored network that provides users with interactive TV features. Microsoft also introduced the software for the WebTV unit, and the software is available for other set-top boxes as well.

The WebTV-based Internet receiver has an optional wireless keyboard that enables users to surf the Web. In some cases, users can also use the remote control device of their TV sets. Connection to the Internet is made by using conventional telephone cables through a user's ISP network services. However, initiatives are underway that will bring Internet connectivity to the Internet receiver through the coaxial cable (available through the user's cable company) thus providing users with high-speed Internet access.

Set-top boxes (also referred to as Internet receivers) translate digital video signals into analog signals thus enabling users to receive digital TV broadcast on conventional TV sets. The set-top box includes software that lets users access Internet-enabled interactive content. Set-top boxes that are appropriately programmed for interactive TV enable users to request various services and provide them with enhanced communication and collaboration functions. Depending on the service leased through service providers such as WebTV Networks, Inc., users can get either a simple method to browse the Internet through their TV (WebTV Classic Service) or interact with TV content through the Internet (WebTV Service Plus). Microsoft charges users a monthly subscription fee for these services [7]. In the first scenario, users receive regular Internet content on their TV sets; in the second scenario, users get special content (interactive content) with links to the Internet and the Web.

There are several ways in which users can combine the TV viewing experience with the Internet. First, they can use MS WebTV for Windows and attach an appropriate TV tuner to their computers and watch TV on their computers. The TV tuner provides TV access, while the WebTV software and appropriate network services provide an interactive experience. @Home provides another service to Windows and Mac users in which users receive high-speed access to rich content over the Internet through cable TV lines. This content includes sports, entertainment, news, music, games, and rich collaboration features such as Internet chat and others. @Home supplies its users with customized versions of Netscape Communicator and Microsoft's Internet Explorer Web browsers to facilitate access to its content [8].

In the other method, users who watch television through regular TV and do not have access to the Internet can use a WebTV-based set-top box in conjunction with the WebTV network to get e-mail, Internet access, and interactive TV links features.

Interactive TV content includes features such as interactive chat, collaboration, electronic greeting cards, electronic TV guides, interactive sports, and other rich content. Broadcasters have to broadcast appropriate content to enable users interact with the content. Various broadcasting organizations such as NBC have jumped on the bandwagon to provide such content. Currently, most interactive TV offerings take users from one screen to another. However, technologies are evolving that will enable users to access other features, such as e-mail or interactive content information, on the same physical screen by splitting the TV screen into smaller subsections. Internet access will mature fully to parallel a PC connection as users will be able to attach printers to their TVs to print appropriate information that is displayed on their TV sets.

Using set-top box browsers such as Device Mosaic, which is available for Sony set-top boxes, users can use WebTV and interactive TV functionality for various e-commerce functions. These include browsing their bank statements, shopping, entertainment services, and accessing special electronic programming guides.

2.4.6   Network Computers

Network computers are PCs with no hard drives that are connected to a network. These appliances provide a truly thin client solution to enterprises connecting users to backend legacy and server-based applications. This provides users with connectivity to transaction-based corporate applications and to the Internet. The new breed of network computers support Java-enabled application access by including a Java Virtual Machine (JVM) in their core.

Network computers have several advantages as a corporate information appliance compared to PCs and other full-fledged workstations. Some of the advantages include:

- Reduced total cost of ownership, as there are fewer requirements to install software on the computers;
- Upgrades to software affect only the backend servers;
- No security concerns as no data resides on the network computer;
- Access to legacy databases using conventional terminal emulation modes such as 3270 and 5250;
- Internet access;
- JVM-enabled network stations support Java applets;
- Ability to emulate Windows applications for users who sparingly require access to such applications.

Network computers (or Windows terminals) come in different flavors. Some connect to the server, download the core of the operating system from the server in their memory, and then act as terminals for that server only (e.g., AS/400, AIX, Windows). Others provide a built-in core operating system in their memories. For example, certain network computers that support Windows NT server applications provide Windows CE in their core. This minimizes boot time for these computers as there is no need to download operating system software.

Certain implementations of these appliances include IBM's Network Station and NCD's ThinSTAR. These information appliances suit transaction-intensive applications such as those used by bank tellers who require continuous backend access and have little requirements for the use of a PC. Deployment of these appliances also suits environments where the general user population requires access to the Internet and backend data access but does not need to store data files on the terminals (e.g., public libraries).

2.4.7   AutoPC

AutoPC is a Windows CE–enabled information appliance that lets users access various value-added services from the comfort of their cars. These services include retrieving driving directions, accessing a car's stereo system, providing wireless capabilities for retrieving e-mail, accessing PIM functions, and so on. These functions are available to the user through voice activation, which, for example, triggers an AutoPC to retrieve an address book entry and reads e-mail or the address book entry to the driver.

AutoPC is based on the Windows CE operating system, and various manufacturers provide devices that support AutoPC functionality. Clarion AutoPC is an example of such an in-dash personal computer. When installed, the device provides users with native AutoPC capabilities. The device incorporates a set of fixed commands that triggers the device to furnish user's requests. In cooperation with various content developers, AutoPC provides various value-added services. For example, Clarion interfaces with a NAVTECH database from Navigation Technologies Corporation to provide users with detailed driving directions in select cities.

2.5   Information Appliance System Software and Technologies

The granularity, depth, and sophistication of content necessitates innovative system software and technologies to support the various functions of information appliances. This section reviews popular operating systems (regular and real-time) and technologies that power the new generation of information appliances to support presenting this content to the user.

2.5.1   Operating Systems and Environments

Operating systems are system-level programs that run on the information appliance and control all aspects of the execution of programs on user appliances. These include controlling user interfaces, multitasking activities, memory management, interface to other appliances, and so on. Numerous operating systems have surfaced in the IT industry to power information appliances and server systems. The past couple of years, however, have seen the emergence of new operating systems that power mobile appliances that have size and weight restrictions. In general, operating systems for mobile appliances have the following additional characteristics:

- Support for small display sizes;
- Support for variable input modes such as touch-screen, pen, TV/VCR remote controls, and so on;
- Small footprint in appliance's memory;
- Handle real-time tasks to cater to various appliance functions;
- Economical in power utilization;
- Support low interrupt latency;
- Provide robust thread synchronization mechanism.

The following sections briefly describe the various operating systems and environments that run on the information appliances described in the earlier sections.

2.5.1.1   MULTOS

MULTOS is an operating system for smart cards. One popular implementation of MULTOS is on Mondex smart cards used for electronic cash. MULTOS is designed for the use of multiple applications, thus enabling smart card software developers to load multiple applications (e.g., business-specific applications, electronic cash, customer-related applications). MULTOS employs the necessary security features that enable safe use of these smart cards.

2.5.1.2   PalmOS

PalmOS is 3COM's operating system for the Palm computing platform. PalmPilot is one of the most popular hand-held devices in the HPC category. Many applications have surfaced that run on the PalmPilot. Accordingly, many software developers and software development environments have also emerged that enable development of appropriate applications for this platform.

PalmOS's kernel is based on a real-time operating system (RTOS) called AMX, which is developed and licensed by Kadak Products Ltd., a Canadian vendor. 3COM licenses the RTOS to run over its proprietary hardware, and it is not portable to other mobile devices such as wireless telephones and other HPCs. This is unlike other mobile operating systems such as EPOC and Windows CE, which are positioning themselves to be the leaders in the mobile information appliance arena.

PalmOS is a multitasking operating system. However, due to licensing agreements and other restrictions, PalmOS does not allow third-party applications to use its multitasking capabilities. In addition, PalmOS is not optimized to handle rich content such as MPEG or other rich graphic files, unlike Windows CE, which has better graphics handling capabilities.

PalmOS's biggest strength is its large number of supported applications and wide support for the development of such applications. Numerous applications have surfaced in the market that include sophisticated networking applications, enabling the PalmPilot to communicate with other information appliances, and PIM applications with enhanced data synchronization capabilities, enabling the PalmPilot to easily exchange data with other HPCs and applications that support data exchange with high-end desktop applications, such as Microsoft Money.

2.5.1.3   PersonalJava

The PersonalJava initiative is an effort by Sun to bring the Java environment to portable devices. PersonalJava is one of the four members of the Java family (the others being Java for high-end desktop and enterprise systems; Embedded Java for printers, pagers, and wireless telephones; and JavaCard for Smart Cards). PersonalJava is an operating *environment* (as opposed to an operating system) for information appliances such as set-top boxes and mobile devices like smart telephones and HPCs. PersonalJava includes Java Virtual Machines (JVMs) and Java APIs with customizations made for consumer devices that have limitations of display sizes, memory restrictions, and in general a smaller hardware.

PersonalJava runs on various RTOSs such as EPOC. However, PersonalJava is optimized to run on Sun's own RTOS, called JavaOS. The PersonalJava environment is a subset of the native Java environment, and it does not include functions that are not required for the mobile appliances. The environment provides specialized functionality to cater to the inherent characteristics of the smaller devices that were mentioned earlier. A key feature of PersonalJava is its support for portability. This allows the same application to run on all devices that support the PersonalJava run-time environment. For example, an application developed for a set-top box that supports the PersonalJava environment could also interoperate on another manufacturer's set-top box that supports the PersonalJava environment. Moreover, since PersonalJava is a derivative of the Java environment, applets developed for PersonalJava can run on high-end appliances such as PCs. This facilitates enterprises to develop content that can run independent of the appliance.

2.5.1.4   JavaOS

JavaOS comes in two different flavors. JavaOS for consumers is comprised of the ChorusOS kernel and PersonalJava technologies from Sun Microsystems. This operating system enables development and delivery of Java-enabled information appliances to consumers. Certain vendors such as Ericsson have already developed mobile handsets that incorporate JavaOS technology and have delivered microbrowsers that run on this platform. However, other initiatives by the industry, such as the ones described earlier, are also partnering with Sun to deliver Java-enabled platforms based on other operating systems.

JavaOS's other variation is JavaOS for Network Computers. It is optimized for network computers and handles network computer-related issues such as remote client system administration.

2.5.1.5   Windows CE

Microsoft launched Windows CE as the operating system for mobile and embedded devices. However, CE's primary penetration is in the handheld computer arena. Windows CE inherits similar features of the other Windows platforms and Microsoft has built CE from the ground up to be a full-fledged operating system for mobile information appliances. Windows CE's modular architecture allows developers of various mobile appliances to derive a customized version of CE that is optimized for their mobile appliances. This feature provides developers with the opportunity to build optimized operating systems to eliminate functions not required for their appliances, thus installing a minimal footprint on the mobile appliances.

Windows CE is a 32-bit operating system that fits in approximately 200K of ROM of the mobile appliance. The design of Windows CE makes it interoperable with various hardware architectures. Multiple versions of Windows CE are available for different mobile devices to cater to the CPU differences between the devices (e.g., Intel/AMD's x86, Hitachi's SH-3 and SH-4, NEC's MIPS). Windows CE achieves hardware transparency by including a thin layer of code called the hardware abstraction layer (HAL) between the

Windows CE kernel and the hardware layer. The primary purpose of the HAL is to handle hardware-specific functions.

Windows CE builds on Microsoft's Win32 programming environment but contains only a subset of the Win32 API due to Windows CE's support of smaller-sized appliances. Windows CE is a multitasking operating system and, like other mobile operating systems, provides multithreading capabilities. This allows Windows CE devices to react to multiple events and application input. Windows CE's memory architecture differs from other Windows platforms, as in the case of mobile appliances where there is no disk storage, memory is partitioned into storage memory and program memory. Storage memory stores data and user applications, whereas program memory stores system-related applications such as stack and memory heap–related functions.

Windows CE provides extensive communication support enabling Windows CE devices to interact with printers, modems, and so on, as well as with LANs using the TCP/IP protocol suite.

2.5.1.6   EPOC

EPOC32 is the latest 32-bit operating system primarily targeted for wireless telephones and handheld devices. Psion, an HPC vendor, initially designed the core of EPOC. With the increased popularity of the operating system, Nokia, Motorola, Ericsson, and Psion formed a joint venture called Symbian, which is now responsible for all aspects of the operating system.

EPOC's architecture is hardware and platform independent. EPOC's operating system mimics the desktop operating systems in various ways. In its kernel, EPOC includes a HAL, which allows EPOC to be portable among various platforms. EPOC is a multithreaded operating system that allocates separate address space for each executable process. EPOC's reentrancy, low interrupt, and thread latency features enable it to run real-time communication software using the same processor (as opposed to having another one), a key feature in mobile operating systems due to the limited size of information appliances. EPOC incorporates most networking technologies such as IrDA and TCP/IP that enable supported appliances to communicate with other devices, the Internet, and corporate intranets. Also, like desktop operating systems that incorporate stability features, EPOC incorporates exception handling features that include memory, resource, and stack clean-up activities.

EPOC32's architecture is modular and includes a hardware-independent graphical interface called EIKON. EIKON incorporates special functions to handle special input from the appliance's keyboard and pen input. EPOC32's applications have two primary components. The GUI portion specifically handles the graphical display features of the application, whereas the application's functionality is handled by the "engine support" component of the operating system. This modular architecture of the EPOC32 system makes the graphical portion of EPOC32 optional because it could be replaced with an alternate graphical environment.

EPOC supports applications that include rich content, such as sound, color, and pictures. Although EPOC's vision is to become the dominant player in the entire mobile market, early trends seem to indicate that EPOC will primarily hold a larger niche in the wireless telephone industry. EPOC has also implemented a Java environment for EPOC that allows the execution of Java applets and applications on Java-enabled and EPOC-based information appliances. Due to the portability of Java, applications developed on other platforms can easily be ported to an EPOC-based information appliance. Such porting, however, does not accommodate certain device-specific limitations such as small display size, reduced support for colors, or absence of a pointing device.

2.5.1.7 Windows 9x Operating Systems

Windows 9x systems and its predecessors (3.x systems) have been the dominant operating systems for the past few years for PC and notebook computers. Windows 9x systems focus on the consumer industry and include features, tools, and technologies to support that market segment. Windows 9x systems have the following features that make PCs powerful information appliances:

- Windows 98 provides a unified view of local machine content and Internet content by providing a standardized user interface called the ActiveDesktop. Users get the same feel when browsing through the contents of both environments.

- The operating system provides enhanced and easier Internet connectivity options, thus providing users with minimal set up procedures to access Internet content and services.

- Windows 98 packs more multimedia tools and technologies as the operating system is geared more for the consumer market. These include WebTV for interactive content, Windows Media Player, and DVD.

- The robust architecture of the operating system, which was designed to run on powerful microprocessors, allows users to run full-strength applications.

The Win32 programming model is at the core of all Windows applications that run on information appliances ranging from desktop PCs to mobile devices such as HPCs. These Windows applications run on various Windows platforms, including Windows 95, 98, NT, and CE platforms.

2.5.1.8   Windows NT

Windows NT is primarily the OS for the enterprise that has a need to run corporate applications at the user's desktop. Primary reasons behind the popularity of Windows NT as an information appliance for corporate use are the following:

- NT's architecture allows execution of full-strength applications that allow users to access all forms of content on the appropriate information appliances. Users can fully exploit applications with multimedia and content-streaming features.

- Built-in security in NT caters to the complex security issues inherent in a business environment.

- A robust architecture provides better performance and reliability, which is required for business applications, than Windows 9x systems.

- Windows NT's architecture provides the ability to link to other technologies such as enterprise Computer Telephony Integration (CTI) applications, which enables seamless integration of Windows NT appliances with various business processes.

2.5.2   Information Appliance Connectivity Technologies

Device connectivity primarily refers to functions that involve data transfers from one device to anther and printing. A user who connects to a printer using his or her Palm Pilot is an example of such connectivity. Service connectivity, on the other hand, implies that devices can interact with each other and computing servers to deliver value-added services to respective users. For example, a user connected to the Internet to browse data through his or her cellular phone is an example of service connectivity. This section covers device-to-device connectivity technologies. Networking technologies are the topic of the next chapter.

2.5.2.1   Jini

Sun Microsystem's Jini is a networking technology that allows mobile information appliances to communicate with each other and request and furnish network and application services. Jini extends distributed computing to a new level in two primary ways. First, Jini devices can communicate without the presence of a network server and deliver value-added services to their users. Second, Jini-enabled appliances can communicate without the need of special appliance-specific device-driver software. Jini, however, requires the presence of a JVM on the device. Jini-enabled appliances can locate each other on the network and request appropriate network services. This technology thus can enable the development of enhanced value-added services for consumers as well as enterprises.

Jini is a layered software architecture that includes four components. The boot, join, and discover component lets an appliance poll the network and registers its services to the network when it is turned on. The directory server is the second layer that registers the presence of the information appliance on the network. JavaSpace lets appliances expose their services and make them available to other appliances on the network. Remote Method Invocation (RMI) enables appliances to actually pass data from one appliance to another.

Jini is a relatively new technology, and vendors and various information appliance developers have embarked on aggressive initiatives to roll out Jini-powered appliances. Initiatives are underway, for example, by wireless handset developers to incorporate Jini technology into their handsets that will enable the handsets to communicate with other devices. For example, a handset will communicate with a printer enabling the user to request printing of a document from the handset. Handsets will also be able to communicate with information appliances at home. For example, as users are approaching their homes, they can turn on the Jini-powered microwave oven in their homes using their wireless handset.

2.5.2.2 Universal Plug and Play

Universal Plug and Play is a new Microsoft initiative and an extension to the Plug and Play standard that lets devices interface with each other seamlessly. Plug and Play allowed computers to self-configure once devices were connected to them. The Universal Plug and Play standard extends to a network level and is similar to Jini in that it provides service connectivity for various mobile information appliances. Information appliances, for example, can discover each other and request appropriate networking services without regard to the transport medium (e.g., Internet or wireless). This standard is language and operating system independent (i.e., does not require a Windows environment) and provides connectivity between various information appliances. Apple has always supported similar features in its Macintosh line of operating systems. MacOS, for example, supports certain core technologies of Universal Plug and Play. One example is Automatic Private Internet Protocol Addressing (APIPA), which defines the mechanism by which network-enabled devices announce themselves over the network without the presence of a central server—for example, a Dynamic Host Configuration Protocol (DHCP) server.

2.5.2.3 HP's JetSend

JetSend technology by HP allows for information exchange between various information appliances through any transport medium such as the Internet, phone lines, and infrared. For example, users can print information from their mobile appliances by sending data to a printer, or users can take notes through one device and upload that information to another appliance, such as a PC or a notebook. Other examples include pointing a JetSend-enabled device, such as an HPC, to another JetSend-enabled device, such as an electronic whiteboard, and transfering information through a wireless connection to the HPC. Developers of information appliances have to incorporate this technology into their devices to JetSend-enable them.

2.5.2.4 Universal Serial Bus (USB)

This technology allows devices to plug and play with a user's computer without the need to add device specific adapters to the computer. USB is a network transport that allows for information exchange. If an information appliance such as a PC supports USB, plugging in new devices requires minimal configuration as the operating system handles those operations automatically. FireWire is a similar initiative that was originally developed by Apple and is defined by the IEEE 1394 standard. Both protocols provide the same functionality but work with different devices. FireWire has certain advantages over USB. For example, it is faster than USB.

2.5.2.5 Infrared Data Association (IrDA)

IrDA is a wireless information exchange initiative that enables information appliances located within a certain range to communicate with each other. Both information appliances have to be equipped with special transceivers. Most mobile appliances incorporate this technology. For example, users can point their PalmPilots at each other and transfer information such as an address book entry from one PalmPilot to the other. IrDA has certain limitations, primary of which is its short range. Appliances have to be within one meter for IrDA appliances to work. A new initiative called Bluetooth expects to rectify those issues.

2.5.2.6 Bluetooth

Bluetooth is a new wireless information exchange initiative that enables information appliances located within a certain range to communicate with each other using radio links. Ericsson, Intel, IBM, Nokia, and Toshiba are the primary sponsors of the Bluetooth initiative. Bluetooth is similar to IrDA but extends the range of information appliances from 1 meter to about 100 meters. Bluetooth, therefore, may replace IrDA as a standard for sharing information.

Bluetooth enables information appliances located within a certain distance to exchange data. Typical examples include printing documents and PIM data synchronization between HPCs, pagers, and cell telephones. This eliminates the need to connect information appliances with cables to perform such functions. Similar to Jini, Bluetooth appliances detect each other without regard to the other appliance's underlying operating system.

2.5.2.7 e-speak

e-speak is an initiative by Hewlett Packard (HP) that enables information appliances to become intelligent by doing work and requesting services on behalf of the end user. Like intelligent agents, e-speak enables information appliances to perform intelligent tasks by performing a series of brokering, mediation, discovery, and composition steps. For example, an e-speak appliance could search for products based on certain specifications, order products, and have them shipped to the customer. Unlike Jini, which provides only connectivity to e-commerce services, e-speak builds another layer of intelligence on the information appliances.

2.6 Matching Information Appliance Features With Business Process Requirements

One of the first steps required to enable a firm's business processes with various types of information appliances is to assess the business value that they will provide to the underlying business process. When considered collectively, information appliances support a wide range of functions that suit varying personal and business conditions. However, applicability of each appliance depends on the user, functional usage (business or personal), physical conditions (e.g., location, environment), and business process requirements or the end user's personal preferences. Enabling a firm's business processes with these appliances thus requires a thorough analysis of business and technological factors. This entails analyzing the applicability of the information appliance to the business process and providing the essential technological and business infrastructure to provide content that will suit the information appliance. Building this infrastructure is also necessary to enable enterprise customers with the information appliances that they own or use.

Introducing information appliances into a business process or building the appropriate infrastructure can range from being quite simple to very complex.

For example, equipping staff with HPCs to help them boost their productivity is not a very complex process since the activity requires distribution of the appliances to appropriate staff members. No major infrastructure modifications are necessary. However, significant modifications to the technology and business infrastructure may be necessary if an organization intends to equip its staff with HPCs that they will use to connect to corporate databases through a wireless service and use proprietary vertical applications to access and update content on those databases. Similar infrastructural changes may also become necessary if the enterprise embarks on an initiative to provide its customers direct access to its in-house content to enable them conduct certain business activities. For example, empowering customers to update their contact information in a financial institution's database through their wireless telephone's Internet browser requires a significant modification to the appropriate technology infrastructure.

This section presents the various steps that a firm should undergo in enabling particular business processes with a particular information appliance. These steps will ensure that the firm considers all factors before making an appropriate decision. The primary steps include a thorough requirements analysis, considering the necessary modifications to the business processes and appropriate technological infrastructure modifications.

2.6.1 Requirements Analysis

Analyzing requirements involves understanding the feasibility of using the information appliance for the given business conditions, the functional needs (usually for content access) of the target community (staff or customers), the costs of enabling the business process with the appropriate information appliance, and applications required for the information appliance to enable users the access to content and appropriate services.

2.6.1.1 Assessing Information Appliance's Market Penetration

Not all information appliances have equal market penetration in all communities. An enterprise may find it challenging to justify investment in building IT services that it plans to offer through information appliances that do not have an adequate market penetration. Considering the scenario for a business-to-consumer paradigm, GSM telephones have a far better market penetration in the European markets than the United States. Providing GSM services, such as stock quotes and bank account balances, in the European markets thus will yield greater acceptance than if the service was provided in the United States. Similarly, offering WebTV-based services such as interactive content will benefit only the small segment of the United States and Japanese market that has WebTV coverage.

Considering the intrabusiness paradigm, if an enterprise has an adequate penetration of older PCs, depending upon requirements, it may be more cost-effective to provide certain users with a Windows-terminal–based architecture to give them access to powerful Windows applications through their older PCs rather than to upgrade to new PCs.

2.6.1.2   Analyzing Customer Demographics

Studying customer demographics is vital before an organization decides to support any information appliance for any of its business processes. For example, if an organization discovers that most of its customers own multimedia-enabled PCs, the organization may offer them Internet telephony-based communications through the Internet channel. Offering such a service to a customer segment or niche that does not own these devices will not result in customer contact through that channel. Similarly, an organization's decisions to invest in infrastructure that will send its customers SMS updates to their wireless telephones may not be appropriate if the target customer segment does not own wireless telephones.

Analyzing user demographics is also appropriate for intrabusiness and business-to-business paradigms. A few years ago, with the rocketing popularity of PCs as a primary information appliance, many enterprises embarked on an aggressive initiative to equip all employees' desktops with hefty PCs. This effort did not yield appropriate ROI because for most users who use transaction-intensive applications, such as in banking and airline reservation systems, the microprocessor power and the large disk space remained largely unused. A better solution, as enterprises discovered later, would have been to equip them with network terminals with access to backend host applications.

2.6.1.3   Industry Applications Supported on Information Appliances

Depending on the information appliance supported and the services that an organization envisions to provide to its customer base, knowledge of applications (content handlers) on the target information appliance is of paramount importance. For example, a bank may decide to offer its customers the ability to download bank statements through the Internet and the ability to load the data in their favorite financial package. Knowing that its customers primarily use Quicken for their financial management needs, the bank may offer its customers downloading of this information in Quicken format, as opposed to other formats, such as MS Money. Doing so may also require that the bank negotiate technical and business arrangements with the financial package vendors to authorize the bank to offer financial statements in appropriate file formats.

2.6.1.4 Cost of Ownership

Total cost of ownership (TCO) is the total cost associated with acquiring and maintaining all aspects of the information appliance. TCO is a critical factor that should drive the decision for the insertion of an information appliance in a business process. TCO should be measured against the value that such appliances provide to the organization. Where ownership of such appliances becomes a necessity, especially in the case of office essentials and mission-critical applications, enterprises take numerous business and technical steps to drive such costs lower. TCO includes some of the following parameters:

- *Acquisition of information appliances:* This refers to the actual cost of the information appliances. This can range from a few hundred dollars for an HPC to thousands of dollars for a PC or a specialized workstation. When calculated for a business's entire set of users, the cost can be enormous.

- *Acquisition/development of software for information appliances:* Most appliances come with bundled software packages. Depending on the business need, acquisition of additional software may become necessary. For example, equipping the enterprise field staff with HPC to access corporate databases for specialized content may necessitate the development of customized applications, potentially resulting in enormous development costs. Conversely, acquiring an appliance for a firm's users, such as a PalmPilot for PIM applications, will entail negligible software costs as the appliance comes prepackaged with a suite of basic tools and accessories. Similarly, to provide paging capability to their users on the PalmPilot, the enterprise will have to acquire the additional software, hardware, and network accessories required to equip a PalmPilot with such capabilities and thus raise acquisition costs.

- *Backup units:* Introducing an information appliance for a mission-critical application necessitates stocking spare appliances to cover for any contingencies. For example, certain package shipping organizations equip their field staff with portable appliances that workers use to record receipt confirmations, delivery dates, and other pertinent information that is critical to ensure delivery. Malfunction of such a device will require the appropriate support office to furnish a substitute appliance to the field worker.

- *Accessories:* Introduction of appliances in an enterprise for critical applications also triggers the need for other accessories and services. For example, an organization that decides to equip its sales staff with Internet-enabled HPCs will require other appliance accessories to enable Internet connectivity.

- *Increase in support staff:* An enterprise needs to increase its staff to support its business employees with various appliances. This support is especially critical for field staff that use new-generation information appliances such as tablet notebooks.

- *Software support and upgrades:* An enterprise may have to invest in appropriate technical infrastructure to periodically upgrade software on the information appliances. This infrastructure also provides the ability to remotely manage those information appliances. These issues fall in the realm of systems management, and Chapter 6 covers them in more detail.

Cost is also an important consideration for opening services to external customers who own an information appliance such as a PC. Revenue generated from opening up an access channel through an information appliance should outweigh the costs incurred of supporting that access channel. However, due to the early experimentation of the e-commerce paradigm, not many organizations are profiting from opening of virtual access channels. In these early periods of e-commerce, organizations' primary focus is to build a larger customer base, and organizations are willing to take revenue hits. Numerous organizations that offer Internet-based services bear proof of this fact. They have yet to post profits, but by attracting and retaining loyal customers, these organizations hope to profit in the future.

2.6.1.5 Maturity of the Appliance

Technology innovations are continuously bringing sophisticated appliances that show a lot of promise. However, their mainstream deployment requires a thorough analysis of their ability to withstand the business environment's inherent complexities and issues. Some of the issues in this regard are the following:

- *Lack of appropriate software for the appliance:* A new information appliance may not support a wide range of software applications. Enterprises, for example, are struggling to use HPCs for mission-critical applications. The lack of applications in the market is preventing many enterprises from introducing these appliances to their corporate staff for mission-critical applications.

- *Reliability and stability issues:* An appliance may not be mature enough to fulfill a business's requirements. For example, a business may invest in electronic writing pads for its administrative assistants but may observe that the appliance's ability to transfer text to another appliance such as a PC or HPC is not adequate for the business's needs. Apple Newton is an example of a technology that failed to mature as an appliance and was pulled out of the market, thus hurting organizations that invested in this appliance.

- *Technology immaturity:* Technologies are continuously maturing to resolve technical issues associated with information appliances. For example, the use of infrared technologies for printing is still in its infancy stages and suffers from various incompatibility issues.

2.6.1.6   Matching Enterprise Services With Device Applications

Due to the nascent mobile information appliance industry, an enterprise should perform a thorough analysis of the available applications (content handlers) and other related features before deploying the appliance for mainstream use. Factors that an organization should consider include:

- Business applications that an appliance can support;
- Bandwidth of the connecting network (e.g., wireless) for remote communications;
- Display sizes to ensure that the content presentation is adequate for the appliance;
- Battery life of the appliance to suit business process requirements.

2.6.2   Business Process Considerations

Introducing new information appliances within the enterprise and offering services to support a multitude of information appliances within and outside the enterprise necessitate consideration of various business factors. These include change in business processes and work flows, devising contingency business processes, training issues, and the need to establish strategic linkages with external organizations for leasing services and supporting the necessary infrastructure.

2.6.2.1 Reengineering and Change in Work Flows

Introducing new information appliances requires staff to adjust to the new ways of conducting business. The business should thus assess the change in business processes and assess necessary implications. Consider the use of EDI for business-to-business transactions. Most organizations employ EDI-based networks to transact with other business entities. With earlier EDI implementations, users had to access a partner's applications and systems from appliances located within the confines of the enterprise. However, with reengineered processes, users can use various mobile appliances and trigger business transactions while they are on the road. For example, in the merchandising industry, field staff took inventories in retail stores, reported back to the field office for updating inventory databases, and accessed partner's systems through EDI networks to order merchandise. The flexibility of remote and mobile access has redefined this business process. For example, certain retail stores equip their workers with mobile devices that they use when stacking goods on the shelves. If the worker detects the need to order goods from the manufacturer, the worker enters a request on the device, which automatically triggers a backend EDI request to the manufacturer. Workers, therefore, do not have to go to the field office to access the EDI networks and can support additional stores.

2.6.2.2 Devising Alternate Work Flows and Processes

Unavailability of information appliances for critical business processes could prove to be a showstopper unless the organization devises alternate mechanisms to ensure continuity of business. For example, if a field worker collects meter readings in a remote area through an information appliance that automatically uploads the readings to the field office, the worker should be trained to follow contingency procedures if the device malfunctions. Similarly, a medical nurse who records a patient's information with a pen-based tablet computer should have access to other online terminals to enter the pertinent information should the appliance become defective or the service becomes unavailable. An enterprise, therefore, should define these alternate business processes before making these appliances part of workers' daily processes.

2.6.2.3 Establishing Help Desks

Enabling corporate staff with new information appliances necessitates establishing an appropriate technical help desk infrastructure to aid staff through various technical issues. For example, if an organization introduces Internet browsing through wireless telephones to a certain segment of its sales staff, the help desk should be qualified to guide users through appropriate issues. Issues include type of telephones supported for Internet browsing, setup procedures, and guidance through appropriate links and channels to get appropriate answers. As Internet browsing through a wireless telephone requires network services from a wireless access provider, the help desk staff should also have appropriate service links to the network service provider to resolve users' issues.

For enterprises that offer their external consumers access to services through such appliances, the help desk should guide users through software setup procedures, (e.g., browser configuration) and help customers understand the technical limitations associated with using an enterprise's services through certain appliances (e.g., wireless telephones).

2.6.2.4   Training

Delivering information appliances to an organization's employees could result in extensive training requirements, especially if the services fall into the mission-critical category. Training applies both to the use and operation of the appliance and training in software. An enterprise should not expect its staff to be able to operate these appliances and use loaded software without any issues on the day of delivery

For example, the use of tablet notebook computers for mission-critical applications requires users to become accustomed to a new way of performing their duties. Use of pens to write on a screen and use of appropriate functions requires training for corporate staff. Using PCs equipped to handle Internet telephony and Internet relay chat (IRC) calls requires appropriate training to handle customer's requests. Enterprises should expect to invest in training sessions to help their users become familiar with such appliances.

2.6.2.5   Establishing External Linkages

Introduction of certain appliances within a business process may require establishing of external linkages with other organizations for various services. This requires that organizations negotiate costs and service level agreements to ensure availability of services. Services include some of the following:

- Wireless network services;
- Content hosting organizations;
- Call center hosting organizations;
- Technical support.

2.6.3 Technology Infrastructure Considerations

Depending on the complexity of the business processes, enabling an enterprise with appropriate information appliances may have a considerable impact on the organization's IT architecture. These infrastructure changes include but are not limited to application development, establishing technical and business support centers, security, and other core infrastructural modifications. The following sections discuss such issues in more detail:

2.6.3.1 Application Development

It is quite common for enterprises to develop applications for information appliances such as PCs. Most enterprises and the industry in general have matured to support application development for PCs. However, enterprises are not equally adept to handle application development for mobile appliances due to their relatively recent introduction to the market.

For business-to-consumer processes, application development has additional challenges and issues. For example, developing an appliance application (to be loaded on the appliance) that will provide consumers with access to an enterprise's services may not prove to be good business sense since it requires distributing software to the target community that carries those devices. This may inhibit customers from accessing enterprise services because it involves getting and loading special software package on their PCs or HPCs. Early comers such as certain banks developed special software and provided it to customers for accessing their online software, and it did not prove to be very successful. Recent trends either provide access to such services through public software (e.g., browsers) or provide applets or plug-ins that users can download through their browsers to conduct appropriate business services.

2.6.3.2 Establishing Innovative Call Center Solutions

Setting up call centers to support customers through various information appliances such as pagers, PCs, telephones, and so on, requires a tremendous investment in the technology infrastructure. Costs are especially high if customer calls are accepted through information appliances such as PCs connected to the Internet and the organization supports customer calls through Internet e-mail, chat, and telephony. Setting up such an infrastructure requires a large investment in appropriate hardware and software (e.g., Internet telephony gateways that route calls to intelligent PBXs). Chapter 4 discusses issues related to the backend infrastructure development for such channels.

2.6.3.3 Content Handling

Offering content to users on their information appliances through a Web site, for example, requires substantial investment in creation of content. For organizations providing audio and visual entertainment to their customers, content creation adds another dimension to these challenges. Creating such content requires recording it through conventional means, encoding it in appropriate formats (e.g., AVI, MPEG), and then hosting it on an appropriate Web site. Organizations providing live and streaming media services to their customers (e.g., CNN, ABC) have to provide an infrastructure that receives the live content through satellites or other means, feeds it to a media server that provides appropriate encoding facilities, and transmits it to the end user's Web player installed on the appropriate appliance.

Providing content to customers also requires keeping customers attracted to a Web site by continually maintaining the content. An organization has to set up appropriate processes that span activities from collecting content to creating content and updating it continuously. This may also require that an organization set up appropriate links to relevant sites that continuously update information on an organization's Web site. An example in this regard is an organization that receives news feeds and stock quotes from other sites and posts them on its Web site.

2.6.3.4 Establishing the Technical Infrastructure for Maintenance

Establishing the appropriate technical infrastructure to support access through an appropriate information appliance requires investment in the technical infrastructure of the firm as well. Depending on the nature of the information appliance that an organization supports, an organization may have to invest in appropriate systems management tools to monitor Web site performance, software distribution to the appropriate appliance, directory services, and other such costs. Later chapters cover appropriate technical details related to these issues.

2.6.3.5 Security

Supporting various types of information appliances requires the formulation of a renewed enterprise computing virtual network. This is because most appliances such as pagers, HPCs, and wireless telephones require public network services for their operation. Public networks are the topic of the next chapter and include wireless cellular networks, the Internet, and paging networks. Deploying mission-critical business e-commerce systems on those networks requires a thorough scrutiny of the security of those networks, network operators, and transactions. These security issues are critical enough to impact an organization's bottom line. Certain appliances are not suitable for sending sensitive information due to the inherent vulnerabilities in the appropriate networks. For example, the use of SMS messaging in GSM telephones may not be very appropriate for transmitting financial information, such as account balances, as the information is temporarily stored in the GSM service providers' databases. Similarly, enabling customers to place stock transactions through certain types of information appliances such as an Interactive Voice Response (IVR) system may have some inherent risks such as repudiation of transactions. This is due to the security risks associated with the PSTN network. Chapter 8 discusses security issues pertaining to all aspects of e-commerce systems in further detail.

Notes and Web Sites

[1]   If the entire infrastructure that supports these appliances and applications is affected, the business impact may be severe. However, this discussion focuses solely on information appliances.

[2]   www.hp.com/e-services/technologies/e-speak

[3]   www.w3.org/TR/1998/NOTE-compactHTML-19980209.html

[4]   ActiveMovie was the earlier name of DirectShow.

[5]   Notebook computers are an exception as they provide limited mobility.

[6]   Other devices are also referred to as personal digital assistants (PDAs) and Palm-sized devices.

[7]   www.webtv.com

[8]   www.home.com

Additional Web Sites

www.apple.com
www.hp.com
www.microsoft.com
www.palmpilot.com
www.sun.com
www.symbian.com

3

E-Commerce Systems Computing Networks

The e-commerce computing paradigm facilitates access to widely distributed information, furnishes the medium for carrying out remote business and personal transactions through various information appliances, and in general provides a flexible means of communication among various parties. This wide array of functions is facilitated by various network backbones that essentially include the Internet/Web, wireless cellular networks, paging networks, and the public switched telephone network (PSTN). These e-commerce computing networks serve as the pipelines for distributing e-commerce services.

Data convergence, or the ability to transmit various types of data over a particular network, is facilitating a push toward a seamless integration of these e-commerce backbones, thus offering users ubiquitous access to e-commerce services. Before this trend, each network specialized in transporting specific data streams, thus limiting users' access to various functions. However, various network backbones are catering to the transportation of various data types. These network backbones have evolved from transporting basic voice services (e.g., the PSTN network or the cellular network) and basic data services (on the primary Internet backbone) to the transmission of compelling content over both wireless and wireline network links. For example, cellular telephone networks are now capable of transporting text data as well as supporting native telephony streams.

These public networks have also positioned themselves to offer an enhanced array of services that individual consumers and businesses can either lease or buy to build a web of e-commerce services for their customers. An enterprise can lease various network services from the network service providers (NSPs) to build the virtual network necessary to deliver e-commerce services to its customers and users.

Advances in the reach and capacity of e-commerce computing networks has pushed their potential to deliver numerous e-commerce value-added services for consumers as well as businesses. In the business-to-consumer model, for example, online malls enable consumers to shop for various goods and services. The high bandwidth of the Internet enables them to engage in real-life experiences such as speaking interactively to a shopping agent. These networks also facilitate the delivery of rich content such as music, video, and other multimedia services. Physical location of computers or other information appliances no longer restricts the consumer to a fixed location. Wireless networks—which traditionally supported only mobile telephony services—have advanced to deliver mobile data services, enabling access to the Internet and other value-added services in various geographical locations. The flexibility of the networks in general have empowered users to communicate interactively, trigger value-added transactions, and search for information unavailable in the best encyclopedias of earlier periods.

The power of the Internet augmented by mobile wireless and paging networks has proven equally beneficial for enterprises both large and small. The Internet is replacing proprietary networks to become the backbone of an enterprise's business-to-business EDI transactions. Business processes reengineered around public computing networks have enabled enterprises to optimize their external linkages with suppliers, distributors, and other partners. Enterprises are actively leveraging the enhanced channels of communication offered by these networks to empower their staff with rich information and content, thus enabling them to work more effectively and efficiently.

The previous chapter highlighted various features of information appliances that suit different needs and e-commerce scenarios. It is vital to reiterate that the maturity of those information appliances has been largely dependent upon the maturity of the computing networks to carry appropriate content and applications to those devices. Were it not for the networks' ability to transport various forms of content at required latency and bandwidth, those appliances would have been limited to their local use. For example, the absence of wireless data services would have limited the use of the PalmPilot to local applications and services.

The complementary nature of the public computing networks that enables them to collectively offer remote connectivity, support for mobility, and high bandwidth network connections is the primary reason behind the continued success of e-commerce initiatives. The ultimate maturity of these networks will arrive when network providers are able to deliver users cost efficient and standardized connectivity to their networks without limiting user's access to specific e-commerce services. For example, mobile users roaming from one region to the other will be able to access rich content (e.g., watching a live news feed from a news content provider) using the Internet access appliance that they are carrying. With current technology this is not currently possible in a cost effective manner due to the bandwidth limitations of the wireless network.

It is also vital to recognize that each of the three public networks have unique characteristics. The Internet, for example, is the primary source of e-commerce content. Wireless networks support user mobility but do not host content. Wireless and paging networks thus serve only as a delivery vehicle for various e-commerce services. Understanding the characteristics of each of these networks is therefore vital in formulating an enterprise's virtual network architecture. The Internet, on the other hand, provides appropriate network services for transporting content and hosting content.

Looking from a consumer and business perspective, life would have been much simpler if there was one network that would enable all parties to conduct e-commerce activities irrespective of their inherent complexities. However, the current state of technology, politics, and other business issues renders this solution impractical.

To implement e-commerce systems accessible through various information appliances, an enterprise needs to link to various e-commerce network backbones and respective network services. This is possible by renewing an enterprise's internal network infrastructure, leasing services through various NSPs, and interconnecting various network services through intelligent gateways. This chapter reviews the various public computing networks that are transforming the e-commerce terrain. The chapter covers the network topology and services of the Internet, as well as GSM/cellular, and paging networks. Although PSTN is the largest global public network, the chapter discusses PSTN only when applicable to the delivery of data services. The final section of the chapter provides valuable insights in formulating an enterprise's network architecture.

3.1   Network Computing Technologies

Advances in networking technologies are the prime catalysts fueling e-commerce initiatives. The emergence of various networking technologies and protocols has driven the establishment of diverse network backbones. These backbones facilitate the inexpensive, global distribution of e-commerce services to appropriate parties. Choice of networking technologies is especially relevant to organizations building global public network infrastructures, and to enterprises that build private network infrastructures in order to engage in e-commerce services. These choices also apply to customers who seek innovative means to connect to networking backbones. This section introduces diverse technologies and trends that underlie the building of computing networks and discusses their significance with regard to the quality and functionality of services to the end user.

Table 3.1
Networking Technologies

Networking Technology	Throughput	Typical Uses
ISDN	64 Kbps per B channel (data) Multiple B channels yield additional capacity (e.g., 2B = 128 Kbps)	High-speed data access for multimedia applications; supports data convergence; SOHO workers
SONET	Available in multiples of OC-1 (51.84 Kbps) OC-2 = 103.68 Kbps or 1.03 Gbps OC-3 = 1.5552 Gbps	For applications requiring high data integrity, superior transport capability and flexibility, high-resolution imaging, catalog applications, video on-demand solutions
T1 (DS1)	24 64 Kbps circuit = 1.544 Mbps	Enterprise WAN connectivity
T3 (DS3)	44 Mbps	Enterprise WAN connectivity
Ethernet	10 Mbps	Used for LAN connections
Fast Ethernet	100 Mbps	Used for LAN connections supporting large number of users that require more bandwidth to access LAN applications
Gigabit Ethernet	1 Gbps	Faster extension of Fast Ethernet technology
FDDI	100 Mbps	Used for faster LAN implementations
PSTN	64 Kbps	Telephone network used for voice and data calls (e.g., connection to the Internet)
Cable connection	Up to 30 Mbps; users usually get 2 Mbps	Used for cable TV viewing and connections to the Internet
Frame Relay	56 Kbps–T1 speeds (1.544 Mbps)	Used for connecting an organization's WAN with a network service provider such as an ISP
ADSL	512 Kbps–8 Mbps (Upstream rate = 16-640 Kbps; downstream rate = 1.5-9 Mbps)	High-speed Internet connections to enable access to rich content.
ATM	25 to 622 Mbps	Used primarily for Internet backbone.

The various network technologies have surfaced for different reasons. The move toward e-commerce is fueling the emergence as well as the deployment of these technologies in all facets of e-commerce–enabled businesses. Some of the primary reasons are as follows:

- Increase the overall capacity and bandwidth issues of the Internet, thus enabling high-speed data transmission rates;
- Enable end users restricted by local loop issues to connect to high-speed networks like the Internet at high throughput;
- Enable the integration of video, voice, and data;
- Enable deployment of video, voice, and data applications on various networks;
- Provide cost-efficient methods to handle "bursty" data;
- Increase networking traffic;
- Support mission-critical networked applications;
- Reduce costs associated with present day networks—for example, for organizations that maintain different networks for voice and data;
- Maximize the offering of network-based services to the entire user population;
- Provide service guarantees;
- Provide security for e-commerce systems;
- Support increased user mobility;
- Satisfy increased demand for low-latency applications such as VoIP and video applications.

Table 3.1 summarizes several common networking technologies and their characteristics.

3.1.1   Wireline Technologies

This section provides an overview of the various wireline networking technologies that relate to the building of public and private networks.

3.1.1.1   Integrated Services Digital Network (ISDN)

ISDN is a network of protocols and services designed to support multiple applications and services (integrated services) over a single digital network. ISDN initially surfaced to enable users restrained by the POTS analog lines and local loop constraints to take advantage of high-speed applications and services over a digital network. ISDN thus provides an end-to-end high-speed digital network solution over the PSTN copper lines. ISDN multiplexes data and control channels (B and D channels, respectively) on one physical cable. The B and D channels provide transmission speeds of 64 Kbps each.

ISDN service is available to users from service providers as basic rate interface (BRI) or primary rate interface (PRI). ISDN BRI includes 2 B+D channels, whereas the PRI interface provides 23 B+D channels. Residential or small office home/office (SOHO) users deploy the BRI service; large businesses at the enterprise level deploy the PRI service.

ISDN is quite suitable for residential access but requires installation of special adapters at the user's premises. Its use in the United States is limited for the remote and mobile worker, however, as an ISDN connection requires special infrastructure, and not all locations (e.g., hotels) have ISDN deployed at their premises.

3.1.1.2   Frame Relay

Frame relay is a high-speed connection standard that operates in the range of 56 Kbps to 1.544 Mbps. Frame Relay is similar to the traditional X.25 packet switching technology, but without the tremendous overhead of error correction protocols. Frame Relay service provides a logical connection for the duration of the connection and thus provides a cost effective way to use bandwidth.

Frame relay is ideal for small to medium enterprises whose data volume is not high but who require Internet access at higher speeds than dial-up connections can provide. Various NSPs offer Frame Relay services and the use of this networking technology for building enterprise WANs is continuously on the rise.

3.1.1.3   Synchronous Optical Network (SONET)

SONET is a set of interface standards at the physical layer that facilitates high-speed data transmission over fiber optic cables. Synchronous digital hierarchy (SDH) is the European version of the SONET standard. SONET's speed is measured by the optical carrier (OC) protocol, which facilitates data transmissions at various rates. The base speed of SONET is OC-1, which provides speeds of 51.84 Mbps. Higher speeds are multiples of OC-1. For example, OC-2 operates at twice the speed of OC-1, OC-3 at triple the speed of OC-1, OC-48 at 48 times the speed of OC-1, and so on.

SONET's primary application is in building of high-speed data networks over long distances. Various inter-exchange carriers (IXCs) are building faster data networks based on SONET/SDH technologies. The Japan–U.S. Cable Network initiative, for example, is an undersea fiber optic network spanning the Pacific Ocean that uses SDH technology and is geared to provide throughput in the range of 80 GPs, expandable to 640 GPs [1]. Numerous telecommunication companies such as Sprint, AT&T, PSI Net and others are part of this project to enable delivery of global high-speed e-commerce services. SONET technologies constitute the next-generation Internet and other high-speed networks. Businesses, too, can request SONET access from various NSPs to obtain higher Internet access speeds.

3.1.1.4   Dense Wave Division Multiplexing (DWDM)

DWDM is another networking technology that boosts the capacity of fiber optic networks by multiplexing multiple data channels for transport in a fiber cable. DWDM's applications are also used by long distance carriers and organizations that provide disaster recovery services. DWDM rings can operate over SONET rings.

3.1.1.5   Fiber Distributed Data Interface (FDDI)

FDDI is a set of protocols that define the transmission of data over a fiber optic cable. With two rings at its core, FDDI can achieve bandwidth of 200 Mbps. However, the use of one ring (when the other is used as a backup) cuts the bandwidth to 100 Mbps. FDDI's primary use is in the building of high-speed LANs that can extend to 100 miles.

3.1.1.6   Asynchronous Transfer Mode (ATM)

ATM uses cell-switching technology to interconnect networks. ATM switches traffic between various networks by breaking data into small fixed chunks of packets called cells. This allows ATM to provide very fast data transmission between networks. ATM is a connection-oriented service, which means that the path between two transmission points is determined before transmission of data. This differs from other data transmission means in which data packets travel to their destination without regard to the network path that they take to get to their destination. The ATM switching function is handled by the hardware and thus provides rapid transmission rates. ATM operates on all physical media, from twisted pair to fiber cables. ATM also operates on SONET rings to provide broadband networking functions.

ATM is finding its primary use in the building of next-generation backbone networks. Various NSPs are actively using ATM switching technologies to build their networks. These backbones will enable the integration of video, data, voice, and other rich content that require high bandwidth and are time sensitive. ATM's primary disadvantages are its high cost and deployment complexity.

3.1.1.7 Switched Multimegabit Data Service (SMDS)

SMDS is a public, packet-switched, connectionless data service that is available at speeds ranging from 1.177 Mbps to 34 Mbps. Various RBOCs offer the SMDS service. SMDS offers alternative means for high-speed Internet access or the establishment of high-speed data links between multiple organizations. The primary advantage of SMDS is its ability to accommodate an enterprise's needs for bursty traffic, thus reducing costs. Unlike leased lines, SMDS eliminates the distance-based charges of telecommunications organizations.

3.1.1.8 Digital Subscriber Line (DSL)

DSL primarily resolves bandwidth constraints experienced by home users and small businesses. DSL competes with ISDN and cable modem technologies to provide broadband access to such end users by resolving bandwidth issues associated with "local loop" constraints. DSL offers high-speed data connections over local copper wires to enable users access to high-speed networks such as the Internet. For this reason, DSL is also referred to as the "last mile" technology, as it speeds up the connection between the user's home or business to the Local Exchange Carrier's (LEC's) central office. LECs are working aggressively to provide DSL connectivity to cover a wide geography by making appropriate installations in their central offices.

xDSL refers to the multiple implementations of DSL. ADSL (Asymmetric DSL), for example, uses different speeds for downloading and uploading signals in which download data streams are faster than the upload data streams. Similarly, other implementations of DSL differ in various ways such as the manner in which a phone company implements DSL at the customer site, achieving higher bandwidths over smaller distances, and so on. Some of DSL's variations are ADSL, CDSL, UDSL, VDSL, HDSL, HDSL2, and RDSL.

DSL's primary advantage is that it enables users to telecommute by allowing access to corporate applications at acceptable speeds, providing users access to richer content, and providing a phone connection on the same copper link without the need to install two telephone lines. DSL's disadvantages are rooted in the difficulties installing it, as DSL requires additional hardware at the customer's premises. DSL is relatively more expensive than the dial-up mechanisms, and not all phone companies offer these services.

3.1.1.9   Ethernet Set of Standards

Ethernet technologies are primarily used to deploy small LANs or LAN backbones. Traditional Ethernet technologies are referred to as 10BaseT systems and operate at 10 Mbps. Ethernet technology is ideal for smaller LANs and small user populations.

Fast Ethernet is the next generation of Ethernet, providing speeds in the 100 Mbps range. It is suitable for providing the backbone of larger LANs. Gigabit is another recent evolution of the Ethernet technologies and provides speeds in the range of 1 Gpbs.

3.1.1.10   T Technologies

T1 lines are one of the most popular digital networking technologies. A T1 carrier multiplexes 24 channels together with each line carrying 64 Kbps, thus forming a total bandwidth of 1.544 Mbps. NSPs also lease T1 lines in increments of 64 Kbps, called fractional T1.

3.1.1.11   Signaling System 7 (SS7)

The SS7 protocol, defined by the International Telecommunications Union (ITU), defines distribution of the PSTN traffic through the wireline and wireless networks. ITU defined this protocol in the early 1980s. All functions in the traditional telephone network related to call initiation, physical location lookup, forwarding through switches, and so on, are handled by the SS7 protocol. Cellular wireless networks use the SS7 protocol to interface with the PSTN.

3.1.2   Wireless Technologies

3.1.2.1   Frequency Division Multiple Access (FDMA)

FDMA is the air interface technology (between a user's handset and the wireless station) employed in wireless cellular networks. FDMA works by dividing the frequency spectrum (hence the term *frequency division*) into 30-khz-wide channels. Each user's voice session uses one dedicated channel. FDMA provides the foundation of the AMPS network, which is the oldest and largest wireless cellular network in the United States. The division of the available frequency spectrum into fixed size frequency channels limits the number of users that the wireless network can accommodate. Digital-AMPS/Time Division Multiple Access (D-AMPS/TDMA) technology surfaced to resolve this issue and others.

3.1.2.2   Digital-AMPS/Time Division Multiple Access

D-AMPS is the digital version of the AMPS system and uses Time Division Multiple Access (TDMA) technology. The two terms are synonymous. D-AMPS/TDMA introduces a time-sharing mechanism in the FDMA spectrum by enabling three users to share a given frequency. This expands the capacity of the wireless network to handle additional user calls. AMPS network has used the TDMA technology to increase its capacity. TDMA has the additional advantage that it converts voice signals to digital format before transmitting them on the analog frequency spectrum. This digitization enables the offering of data services over the AMPS network.

3.1.2.3   Code-Division Multiple Access (CDMA)

CDMA is an alternate air interface technology for wireless networks. CDMA is a spread spectrum technology, which means that each user has access to the entire frequency spectrum and is not limited to a specific frequency. The wireless system differentiates between users by assigning a unique code to each user, hence the term *code division*. Reference to CDMA usually implies the IS-95 standard. CDMA digitizes a user's voice signals before the handset transmits them over the analog signal. Qualcomm, Inc. pioneered CDMA and numerous wireless service providers now operate a CDMA-based network. Both TDMA and CDMA use the same frequencies as the traditional AMPS network.

Wideband CDMA (W-CDMA) is a new technology that promises to expand the bandwidth and capacity of wireless networks. This technology enables both circuit switched technologies (e.g., PSTN-type voice calls) and packed switched technologies (e.g., Internet data). W-CDMA technology targets the requirements of the 3G (third generation) wireless initiative. The 3G initiative is geared to offer enhanced and high-bandwidth services (e.g., high-speed Internet access, delivery of video content, and other such services) through the wireless channel.

3.1.2.4   Cellular Digital Packet Data (CDPD)

CDPD offers digital packet transmission over analog wireless networks such as AMPS, the original analog wireless network in the United States. Since the U.S. market has a large installed base of the analog wireless telephone network, CDPD allows subscribers to those networks to use data services. CDPD also works on TDMA. However, this technology suffers from limited deployment.

CDPD is a specification that brings data services to the AMPS network. The specification recommends the use of unused frequencies on the AMPS network to deliver data-based services to its users. CDPD employs familiar open protocols such as TCP/IP, X.400, and others, thus enabling this technology to interface seamlessly with networks like the Internet and other packet switched data networks. CDPD provides data services in the range of 19.2 Kbps, which is adequate for applications such as e-mail and Internet browsing of less graphics-intensive sites. CDPD is the glue that enables mobile users equipped with mobile information appliances to connect to the Internet or corporate networks.

3.1.2.5   Post Office Code Standardization Advisory Group (POCSAG)

POCSAG is the traditional protocol used in the paging industry. POCSAG operates at low speeds of 2400 bps. However, certain limitations with this protocol are pushing paging companies to adopt Motorola's FLEX standard for paging.

3.1.2.6   FLEX

FLEX is the next-generation paging protocol developed by Motorola. Compared to its predecessor, POCSAG, this high-speed messaging protocol provides better error recovery capabilities than the other protocols thus enabling reliable delivery of messages. The FLEX family of protocols includes other protocols that enable paging service providers to offer enhanced wireless paging services to their customers. The ReFLEX protocol allows for two-way messaging, whereas InFLEXion allows for voice messaging.

3.2   Internet Backbone

This section will elaborate upon the Internet's network architecture and delineate the various network services that the Internet offers for commercial, educational, and other research activities. An organization's network architecture formulated to meet a firm's e-commerce requirements is bound to include various Internet-based services.

The Internet is the primary network behind a wide range of global e-commerce initiatives. Similar to other networks, the Internet's original function was the delivery of data from one point to another. However, the past few years have seen the Internet transform into a full-fledged computing platform. Three primary developments have led to this phenomenon. First, a steady increase in the Internet backbone capacity is enabling the delivery of richer content from one point to another. This delivery mechanism has empowered individuals, businesses, governments, and other institutions to leverage the Internet's power to reshape their business processes. Second, the Internet enables standardized hosting and access to content, thus delivering innovative services to the end user. This has enabled businesses and other parties to provide and sell various forms of content. Finally, the reach of the Internet is creating fast and inexpensive means of reaching customers and business partners in global markets. These factors have positioned the Internet at the heart of all e-commerce initiatives.

The high capacity and reach of the Internet has enabled enterprises to provide and sell various forms of content to their customers. Content is information as well as other value-added services, such as shopping, access to entertainment, and so on. The quality of content continues to improve as the Internet's core acquires more capacity for the speedy delivery of content. This delivery of high-quality content is enabling enterprises to offer better consumer services. Services include marketing by empowering the customer with more information on a company's products and services, building electronic channels for the purchase of goods and services, and providing an innovative means of post-sale support for those products and services. On the business-to-business front, enterprises are shifting away from leasing dedicated networking lines to transact with various businesses. Instead, they are exploring ways to leverage the Internet's networking capacity in order to migrate processing of those functions to the Internet backbone networks. Likewise, on the intraenterprise front, organizations are migrating groupware functions to the Internet, thus facilitating powerful communication and means of collaboration, which in turn is enabling employees to produce appropriate goods and services effectively and efficiently.

Having reviewed Internet value-added services in general, the following summarizes the various characteristics of the Internet that apply to the offering e-commerce services to various users. This will enable readers to make appropriate decisions when formulating their enterprise's network architecture.

- *Reach:* The Internet's reach in general is second only to the PSTN service. For data services, the Internet enjoys the largest reach. Customers can access the Internet through various channels.

- *Messaging applications:* The Internet provides the ability to transmit messages between various points of the globe.

- *Quality of content:* As a standard, the Internet does not support the delivery of rich content. For example, a regular Internet subscriber using a dial-up modem cannot expect access to video-conferencing capabilities. Although the Internet backbone offers enormous capacity, the absence of high-speed links in various legs of the Internet (e.g., local loop) requires special technologies and access means to facilitate access to such content.

- *Reliability:* The Internet by default is not a reliable medium for the delivery of messages and other content. For example, a user sending a regular SMTP e-mail message to the other user does not have any guarantees about its delivery. However, numerous technologies have surfaced to fill these gaps. Enterprises have to choose specific technologies for various situations to ensure reliability of service. For example, organizations can lease services from certain ISPs that deliver reliability of services.

- *Security:* Similar to the other issues, the Internet by default does not provide any security. However, various technologies and services allow enterprises to employ security solutions to resolve these issues.

The Internet is a network of numerous interconnected networks that enable the sharing of information. It has gone from being a mere network to a full-fledged computing platform, offering its users a wide variety of network, application, and data services. Just as the PSTN is ubiquitous, and the primary channel for voice communications, current trends clearly put the Internet at the forefront of data services that will eventually include voice communications.

The fabric of all wireline public networks is based upon connections between various telecommunication entities—regional bell operating companies (RBOCs), Internet service providers (ISPs), local exchange carriers (LECs), competitive LECs (CLECs), and interexchange carriers (IXCs). Before explaining these terms, it is vital to understand the concept of a local area transport area (LATA). LATAs are geographical regions within the United States that distinguish local calling from long distance services. All calls within the LATA are termed *local calls,* whereas connections between two LATAs are *long distance calls.*

An LEC controls the network and its interconnections within a LATA. A LATA can have multiple LECs serving the area. The deregulation of the telecommunications industry (Telecommunications Act of 1996) triggered the emergence of other telecommunication companies that operate at various local, regional, and national levels. Most LECs emerged after the break-up of AT&T in 1983. RBOC is another term for LECs that resulted from this break up. However, there were other LECs that existed independently. GTE is an example. LECs have traditionally controlled the local loops.

The deregulation act requires LECs to relinquish control of the local loops and sell local services at wholesale prices to a new entity called CLECs.

CLECs can thus offer the same services as an LEC in respective local areas. Users are therefore no longer required to request local loop services from their traditional regional telephone companies.

IXCs provide long distance services between various LATAs. AT&T, Sprint, and MCI are examples of popular IXCs. However, the deregulation act fueled the emergence of other IXCs, including Qwest, Level 3 Communications, and others. The newer IXCs are laying the foundation for next-generation backbones that will enable users to integrate video, audio, and data traffic on the Internet. These IXCs claim to offer superior telecommunications services compared to the older IXCs, as they are building new networks based upon the latest technologies. Because recent networking technologies (e.g., DWDM) depend upon superior cables to provide superior throughput, it is difficult for the traditional IXCs to uproot all their cables and replace them with better quality cables overnight.

ISPs provide users with access to the Internet and its services. ISPs connect to the Internet through telecommunication lines leased from LECs, CLECs, and IXCs. Some ISPs have also installed their own networks and thus do not depend upon telecommunication providers such as AT&T, Sprint, and others for the use of high-speed lines.

The Internet has its roots in the early 1970s, when the U.S. government established a network infrastructure to share information between various governmental and educational organizations. Referred to as the Advanced Research Projects Agency network (ARPANET) project, it was funded entirely by the U.S. government, and its use was restricted to the connected entities. Thus, no commercial activities were carried out on the Internet.

By the mid-1980s, this network of computers had expanded to include numerous organizations and computing facilities. The National Science Foundation (NSF) established the NSFNET backbone that later was transformed into the primary backbone of the Internet. However, the use of NSFNET in its early days was still limited to noncommercial use.

In 1991, NSFNET's architecture was transformed to include activities of the commercial sector. Commercial Internet Exchange (CIX), an association including General Atomics, UUNET, and Performance Systems International (PSI), Inc., facilitated the introduction of commercial entities called ISPs into the NSFNET backbone. In 1995, the NSFNET backbone infrastructure transformed into a NAP-based architecture that is the basis of today's Internet backbone architecture. The following sections elaborate upon the Internet's topology in greater detail.

3.2.1 Internet Network Topology

The Internet is a large network that connects various networks. Its architecture can best be described by the entities (networks) that connect through their major interconnecting points. Figure 3.1 illustrates the major components of the Internet's topology. As shown in the figure, ISPs and Network Access Points (NAPs) are the two primary building blocks of the Internet. ISPs are large networks that connect various users on their backbone. Various ISPs connect at major interconnection points called NAPs that exchange traffic from various ISPs. Such an interconnection of various networks and traffic exchange forms a large web-like network that enables users connected to one network (e.g., ISP) in order to to communicate with remote users connected to the Internet through other ISPs or similar entities.

The following section will use this overview as a general framework to expand upon the Internet's major components in further detail.

3.2.1.1 Network Access Points (NAPs) and Metropolitan Area Exchanges (MAE)

These are points on the Internet that aggregate regional network's connections for exchanging traffic and form the Internet's primary backbone. Exchange of Internet traffic atNAPs is also called *peering*. As explained earlier, the NAP-based

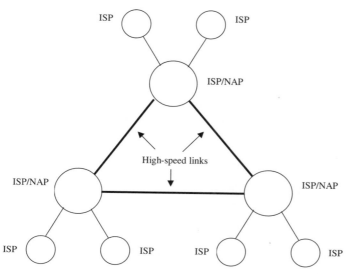

Figure 3.1 Internet network architecture.

architecture resulted from the NSFNET architecture. NSF originally sponsored four primary NAPs within the United States. They are located in Chicago (operated by Ameritech), San Francisco (operated by PacBell), New York (operated by Sprint), and Washington D.C. (operated by WorldCom). The Washington, D.C. NAP is actually a collection of seven interconnection points, each referred to as a Metropolitan Area Exchange (MAE). Two major MAEs are located in Washington D.C. (called MAE East), and California (called MAE West). With the increase in the reach of the Internet and the number of users that connect to its backbone, other organizations are continuously surfacing to provide such interconnections of networks.

ISPs and other organizations connect to NAPs at high speeds. Earlier NAPs connected at mere 56 Kbps speeds. Later, NSFNET requested the upgrade to T1, and then subsequently to T3 network links. With the commercialization of the Internet, various organizations serving the NAP functions have continuously upgraded to higher-speed links. These links range from T3 to OC-12 speeds. All NAPs interconnect at even higher speeds with links in the range of OC-48 and higher. This increase in the Internet's capacity is the primary force behind the development of e-commerce.

NAPs enable ISPs to exchange traffic at very high speeds. NAPs are layer 2 switches, which means they do not route IP traffic directly. Rather, they simply forward ATM cells to other ISPs. Due to the high-speed switching that they use, an NAP can support the interconnectivity of many ISPs. NAPs also provide connectivity to other newer-generation networks. For example, the Chicago NAP—one of the top NAPs in the country—is also the conduit to the Abiline and vBNS projects, which are initiatives for testing the next generation of the Internet backbone. Organizations that connect to NAPs have to obtain their Internet IP addresses from Network Solutions (formerly called InterNIC), or other registrars that have surfaced recently. Chapter 6 covers the topic of Internet registrars in more detail.

3.2.1.2 Internet Service Providers (ISPs)

As explained earlier, the ARPANET and NSFNET networks formed the foundation of the Internet and limited its use to research and other government activities. Later, as the government reduced its control over these networks and encouraged the entry of the commercial sector, numerous networks emerged to offer Internet-related services. These networks, called ISPs, form an integral component of the Internet network architecture. These organizations are controlled by the bylaws of CIX and provide various Internet services to its users. An ISP is thus a large network of computers interconnected by its local high-speed links, which in turn interconnect to one or more NAPs. ISPs allow free peering of traffic to enable a seamless interconnection of all Internet nodes, thus making the Internet look like one big network.

ISPs offer various Internet network services and charge their customers for most of those services. Some ISPs provide free Internet connectivity to their users. Their primary interest is to register a high volume of customers, earn their loyalty, and use this advantage to attract ancillary business revenues (e.g., advertising through banner ads). This concept is very similar to portal services that attract a large customer base and generate revenues by advertisements and sponsorships.

3.2.1.3   User Connectivity Through ISP Points of Presence (POP)

ISPs provide POPs in various locations to enable regional users to access the ISP's network and subsequently provide connections to the Internet. Figure 3.2 illustrates a user's connection to the Internet through a POP. Users gain access to the Internet either through dial-up telephone lines or through a dedicated connection provided by the ISP. Dial-up telephone lines are available through the users' regional LEC or CLECs, which control the first leg of Internet connectivity from the user's premises to the network provider's central office. Users can use analog modems and other technologies to access the ISP through the local loop provider's telephone lines. Users can also connect to ISPs through dedicated leased lines provided by major ISPs. Once connected to the Internet, users can either access content on the Internet, host content for other users, or leverage other Internet services.

3.2.1.4   Advancements of the Internet Backbone

The present-day Internet infrastructure has acquired enormous capacity within the past few years. However, several issues related to network congestion and the unreliability of various value-added services continues to discourage enterprises from deploying sensitive and mission-critical applications. To resolve such issues, the Internet2 project surfaced to provide a test bed for universities

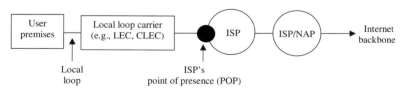

Figure 3.2   User's Internet connectivity through a POP.

and various research institutions. A consortium of universities, research organizations, and networking organizations called the University Corporation for Advanced Internet Development (UCAID) is spearheading the Internet2 initiative. The objective of Internet2 is to test new networking technologies and their ability to support bandwidth-intensive applications that will provide value-added services such as distance-learning, digital libraries, and 3-D imaging. Various networking organizations such as Qwest, Nortel, and Cisco are active members of the project and provide the networking infrastructure and technologies to support its development. The network that provides the networking backbone for the Internet2 initiative is called the Abiline network [2].

The Next Generation Initiative (NGI) is a parallel federal program whose charter is to build the next generation of high-performance networks. The next generation's Internet architecture consists of the Department of Defense's Research and Education Network (DREN), NREN (NASA's Research and Education Network), and vBNS (NSF's Very High Performance Backbone Network Services). The Internet2 program works closely with NGI to ensure interoperability of standards between the two programs.

3.2.2   Internet Network Services

ISPs are the primary source of leasing Internet services and form the core of the e-commerce–enabled business paradigm. These services range from simple e-mail services to a wide assortment of value-added services. This section provides a high-level view of the various services that ISPs offer to their users.

3.2.2.1   Internet Connectivity Services

Internet connectivity is the basic form of service that ISPs offer to their residential and business customers. ISPs maintain various POPs to enable their users to connect to the Internet. For large business owners, ISPs and other network providers offer dedicated Internet connectivity services to bring high-speed Internet access to enterprises in order to accommodate e-commerce services. The following text describes the primary connectivity means to the Internet [3].

- *Analog modem:* The most common means of connecting to the Internet is to use an LEC dial-up connection over regular PSTN lines. Users plug their information appliances to an analog modem that connects to the ISP through the LEC telephone lines. The maximum bandwidth possible through this type of connection is 56 Kbps. This type of connection is usually adequate for users who require access to basic types of information and e-commerce services but have no need to download very large files or rich content such as interactive video sessions.

- *xDSL:* xDSL technologies enable users to access Internet services over regular copper telephone lines at speeds considerably higher than regular dial-up connections. Since most local loops are comprised of copper lines, this form of connection offers residential and SOHO users inexpensive and speedy access to the Internet. xDSL service requires support of both the LEC and ISP.

- *ISDN connections:* ISDN connections are an alternate mechanism for high-speed connections to the Internet. The technical architecture of ISDN allows for multiple channels, allowing the user to connect to the Internet on one channel and receive calls through the other. Recent trends in the United States seem to indicate an increasing preference for xDSL technologies over ISDN.

- *Dedicated connectivity:* Large enterprises that must support large numbers of users and want access to enhanced services require dedicated connections to the Internet. Enterprises can request these high-speed connections to the Internet by leasing links through various ISPs. These links range from T1 links to OC-3 (155 Mpbs) speeds, depending upon the enterprise's requirements.

- *Cable modem access:* Cable networks provide an alternate high-speed Internet connectivity mechanism that enables users to access rich content. Users have access to this high-speed channel primarily due to the ability of the coaxial cable to handle high-speed data transmissions. Another reason is that residential customers already enjoy cable connectivity for receiving TV content, unlike the limitations of the PSTN lines, which limit access to high-speed connections due to the slow copper lines installed at user's premises.

The primary function of cable networks was the transmission of television content through the coaxial cables that connect residential customers in most regions. However, the fast speeds of the cable network (faster than ISDN; in the range of 2 Mbps) triggered a whole new industry of Internet cable service providers, who provide residential customers with Internet connectivity similar to PSTN service.

The traditional cable network architecture consists of a coaxial cable that originates at the cable company (also called as the cable headend). The cable headend is the central location that receives all TV content from satellites and other sources. The cable backbone routes to various neighborhoods through various feeder lines and terminates at the subscriber's premises, at an interface point referred to as the network interface unit (NIU). The signals traveling over the coaxial cable are radio frequency (RF, or analog) signals that operate over a 350 MHz spectrum. This network implementation supports transmission of one-way analog signals from the cable headend to the subscriber's NIU that in turn connect to a subscriber's TV set. No data services are possible with the traditional implementation of the cable network.

To accommodate data services, the cable industry defined the Data Over Cable Service Interface Specification (DOCSIS). Cable providers use these specifications to provide two-way data services (like the Internet) to its subscribers. Cable operators have to undergo numerous changes to support data access services through the coaxial cable network. First, cable operators have to install special signal amplifiers at the cable headend, which enable cables to transmit data from a subscriber's home to the cable companies. Second, the upgrade involves the replacement of the coaxial backbone with a Hybrid Fiber Coax (HFC) network, which essentially replaces the coaxial cables with fiber optic cables. Finally, the cable companies have to install special fiber nodes that distribute signals from the cable headend through the HFC network to various subscriber homes. Fiber nodes distribute regular coaxial cables to subscribers' homes and no replacement is necessary at the subscribers' premises.

Connectivity to the cable network is similar in architecture to PSTN connectivity, as the computer connects to a cable modem that acts as a modulator and demodulator of signals. The demodulator component in the cable modem converts the RF signal into a digital equivalent and routes it to the NIC card installed on the user's computer. This provides inbound data transmission to the user's desktop. Similarly, the modulator translates digital data into appropriate analog signals and tunes the signal to the appropriate frequencies before sending them to the cable headend. Users that have TV cable service can use the Internet connection through the same physical cable connection by the use of a "splitter." The splitter keeps the TV signal and Internet connectivity signals separate as both connections use different frequencies for transmission.

@Home service is the largest Internet service provider in the cable industry and provides its services to various regional cable companies that in turn offer those services to their users. All cable providers by the token of their reach cannot provide Internet access to residential customers. Instead, by employing technology and services from @Home, cable companies upgrade their infrastructure to provide Internet services to their users.

3.2.2.2   Groupware Services

Groupware refers to various technologies that enable users to communicate and collaborate. These communication mechanisms include interactive as well as other means of communication. E-mail, voice over IP, Internet relay chat, and others are some of the mechanisms that fall under groupware. ISPs provide basic groupware functionality for their subscribers. Some ISPs provide more enhanced functions to support users of a large business. Medium-to-large businesses deploy an internal groupware infrastructure to cater to intraenterprise communication needs. Chapter 6 discusses the issues related to the deployment of groupware. The following text presents the basic groupware functions that ISPs provide to residential and business users.

- *E-mail:* E-mail is the basic form of communication that all ISPs provide to their customers. E-mail services are usually limited to POP3 services in which users send and receive e-mail messages using their information appliances. Some ISPs differentiate their e-mail offerings by providing IMAP-enabled mail service that allows users to manage their mail on the ISP's mail server as opposed to downloading it to their POP3 mail client. ISPs extend e-mail services to their business customers by providing large space on their mail servers and transmitting mail messages to the customer's mail server. This enables the business to leverage the infrastructure of the ISP without investing in a full-fledged e-mail infrastructure with the added benefit of centrally controlling all user accounts. ISPs also provide e-mail services customized to a domain name of the organization.

  Certain ISPs also offer sophisticated e-mail processing functions such as "intelligent e-mail processing." This enables the customer organization to sort mail received from customers. For example, all product feedback mail gets routed to one department, while customer complaints are forwarded to another department. However, since establishing an intelligent e-mail infrastructure depends upon a business's internal processes, the effort involves a considerable investment in the technology infrastructure at both the ISP and the business. Chapter 6 will provide details on the implementation of such services within the organization.

- *Extension of IP services to mobile information appliances:* Certain ISPs offer extension of their IP services to various information appliances. For example, IBM Global Services has extended its IP Remote Access Service to support PalmPilot devices from 3Com and the Windows CE operating system from Microsoft.

- *Internet fax:* This service offers customers access to fax capabilities over the Internet. Customers can send e-mail messages and related attachments to a fax machine, or they can broadcast fax messages to various fax machines and e-mail addresses. With all traffic routed over the Internet, this offers customers a cost-saving initiative along with data convergence on a network. ISPs usually deliver these services by installing sophisticated fax servers that provide organizations with various means of sending and receiving faxes through the ISP's IP backbone. Individuals and organizations can fax directly to a particular number or through installed PBXs. Certain ISPs deliver enterprise-level faxing solutions, which allow organizations to link to the ISP's fax server and empowers them to manage and control all fax activities.

- *Streaming protocol services:* This refers to multimedia services running over IP networks and includes applications such as Internet telephony and video-conferencing applications. For Internet telephony, also referred to as voice over network (VON), ISPs offer services to allow their customers communicate to others through the Internet as opposed to the PSTN, thus offering cost savings initiatives. ISPs offer this service to both residential and corporate users. For enterprise-wide deployments, organizations can use VON services from a major ISP and deploy telephony servers at both ends of the call. This enables the enterprise to use IP telephony for their voice communications and helps the business reduce the cost of their voice calls, which are normally routed over the PSTN circuit-switched network. Internal deployment of VON services requires businesses to install special equipment at the business premises that have PBXs installed. This architecture enables the routing of voice calls through the Internet. Chapter 4 will elaborate the primary implementation issues.

3.2.2.3   Enhanced E-Commerce Services

- *Collocation services:* This service offers enterprises the choice of outsourcing Internet and other e-commerce services off-site. This provides enterprises with the ability to leverage the mature network infrastructure of the ISP.

- *VPNs:* VPN services allow business users to establish secure channels. ISPs offer these options for businesses that require secure channels of communication over the Internet.

- *E-commerce services:* ISPs provide infrastructure services that enable users to host personal Web sites. Users can offer various products and services through these Web sites. These solutions are usually viable for small businesses that do not have a large customer base and have minimal customer support requirements.

3.3 Wireless Network Backbones

This section elaborates upon the cellular mobile services provided by various wireless NSPs. Mobile wireless networks emerged primarily to provide mobile telephony services through the wireless ("air interface") medium. Unlike making voice calls using a regular telephone over the traditional telephone network, the user's telephone does not connect to any wireline equipment. For example, a user's PSTN telephone connects to the wall jack that in turn connects to the telephone network. In wireless networks, users can roam in a large geographic area (within a city, nationally, or even internationally) and use their mobile handsets to make and receive calls. The appropriate geographic area, however, must have appropriate mobile service coverage provided by the user's wireless NSP.

Developments in the past few years have enabled the wireless NSPs to provide data services as well. These data services include sending short messages to other users on the wireless network, sending Internet e-mails, and so on. Consumers can thus access numerous value-added e-commerce services by exploiting these services.

Numerous wireless technologies have facilitated the advancement of wireless networking. However, unlike the standardization achieved in the case of the PSTN, the wireless market suffers from multiple standards, which segment the wireless NSPs into various camps with each NSP standardizing on technologies of their choice. The evolution of these technologies has also resulted in different generations of networks. The first generation of networks which still prevail in a majority of countries employ analog technologies. The second generation followed the introduction of digital standards and technologies. The interoperability problem worsened in this era as different organizations in different countries pushed different digital standards. The following summarizes the various standards that emerged in various countries:

- Emergence of an analog standard in the United States (AMPS);
- Analog standards that emerged in European countries (NMT, ETACS);
- Surfacing of multiple digital standards in the United States (TDMA, CDMA);

- Standardization of a digital standard in Europe (GSM standard);
- Adoption of a different version of the European digital standard (GSM) in the United States;
- Adoption of a set of digital standards in the United States (PCS).

The term *wireless* refers to the transport medium, which is the air interface. The present generation of wireless networks supports various data services and enables users to connect to the Internet, which is the primary hub of e-commerce services. Wireless therefore enables the end user to transact various e-commerce services such as receiving short messages pertaining to their bank account balances, triggering money transfers pertaining to bill payments, browsing the Internet, sending and receiving Internet e-mail, and so on. The challenge for enterprises, therefore, is to build and provide appropriate gateways and interfaces that enable users to access various e-commerce services. This empowers users to access such services when they are mobile and do not have easy access to any wireline networks. Enterprises can also benefit by directly using the wireless services. Wireless data services can provide numerous benefits such as a reduction in operating costs and improved customer service response.

From the perspective of offering appropriate network services (the focus being on delivery) to its users, wireless networks have the following characteristics:

- *Reach:* Mobile wireless NSPs provide varying but limited geographic coverage to its subscribers. Unlike ISPs, wireless NSPs differ in their data service offerings depending upon the specific technology that they employ. Although the use of wireless services is continuously on the rise, wireless subscribers accessing the Internet lag considerably behind users accessing Internet services through wireline networks.

- *Messaging applications:* Wireless NSP's primary offering is voice communication. Most digital networks also enable sending and receiving of e-mail messages.

- *Quality of content:* Limited bandwidth of the wireless data network (14.4–19.2 Kbps) allows users to send and receive modest size files. However, with current technology implementations wireless networks are not suitable for the delivery of rich content (e.g., video clips).

- *Reliability:* The transition from analog networks to digital networks improved the quality of signal transmissions considerably. However, wireless technologies used in the mobile wireless networks do not provide any reliability guarantees. It is common, for example, for sub-scribers to lose data connections in the middle of data transfers.

* *Security:* GSM and CDMA networks use robust authentication and encryption algorithms over the air interface. However, there are no special mechanisms to secure the data on the wireline networks of the wireless networks.

The wireless telephony market in the United States has been gaining momentum since the early 1980s, when the ubiquitous Advanced Mobile Phone System (AMPS) network emerged as the standard means of wireless communication. AMPS therefore falls in the first generation of cellular communication systems. As explained earlier, AMPS builds upon FDMA technology to provide wireless telephony services to its users. Counterparts of AMPS in Europe are the ETACS and NMT standards.

D-AMPS emerged to offer a digital wireless standard that builds upon the AMPS standard. The main reason for the introduction of D-AMPS was to counter certain limitations of the AMPS system, primarily its inability to provide data services. D-AMPS uses the TDMA technology and therefore accommodates both FDMA (for compatibility) and TDMA technologies. D-AMPS—now referred to as TDMA—is a primarily U.S.-based standard used by carriers within the United States. Handsets that operate on the D-AMPS network are also interoperable with the AMPS network.

In parallel, the GSM initiative surfaced to standardize the wireless telephony market in Europe. The original system was the GSM 900 standard, which resulted in the two standards of DCS 1800 (later renamed GSM 1800) and PCS 1900 [4]. However, the GSM 900 standard has a greater installed base due to its history. There are minor differences between the three versions. PCS 1900 is the North American implementation and is used by wireless carriers in the US. However, it should be noted that PCS 1900 in the United States at times also refers to a combination of digital standards that include GSM 1900, TDMA, and IS-95 (the IS-95 standard allows the mobile terminal to work on both the AMPS and the CDMA system). All these standards are air interfaces and have different technology implementation issues such as frame structure, data field size, and transfer rate. All three standards are incompatible.

3.3.1 Wireless Topology

Although the standards are all incompatible, the overall topology of all wireless networks is very similar. A user handset communicates with a wireless base station in the user's geographical area (also referred to as a cell). The base station

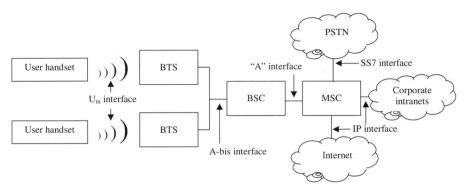

Figure 3.3 GSM wireless network architecture.

then communicates to the main wireless subsystem—which also provides interfaces to the PSTN for routing PSTN calls—and interfaces with the Internet for connectivity to Internet services. Since GSM is an international standard and the most widely deployed network in the world, this section will focus on the GSM network's topology. The reader should note that networks differ in their implementations and this is by no means an in-depth explanation of the intricacies of wireless networks. Readers should refer to specialized textbooks for such descriptions. *GSM Networks: Protocols, Terminology, and Implementation* [5] provides a good description of the GSM network. Figure 3.3 illustrates the topology of a GSM network. The following describes each of the components in further detail.

3.3.1.1   Wireless Cells and the Use of Frequencies

In a typical wireless system (AMPS, GSM, PCS, etc.), a geographic area is divided into multiple radio cells. The wireless subsystem allocates frequencies for each user conversation. As there are a fixed number of frequencies, the cellular system reuses frequencies. However, only nonadjacent cells share frequencies to prevent two calls getting on the same frequency. The cellular structure of the wireless systems enables mobile handsets or terminals to move from one cell to another and the inherent structure of the network allows for the handoff between one cell to another.

3.3.1.2   Base Transceiver Station (BTS)

A BTS covers wireless users within its geographical boundaries. A BTS is similar to a POP on the Internet. Similar to a POP, a BTS is the wireless user's primary and frontend interface point to the wireless network through the mobile handset. Unlike a POP, the user does not have to be aware of the BTS's presence. The mobile terminal automatically connects to the nearest BTS that is within the range on assigned frequencies. BTSs have fixed broadcast ranges that range from quarter of a mile to approximately 25 miles.

As mentioned, a BTS covers all wireless users within its range. Each BTS consists of various transmitters and receivers that operate on their allocated frequencies. BTSs employ multiplexing techniques to allow various users within their range to simultaneously make or receive calls. The air interface between the mobile terminal and the BTS is termed U_m. This interface uses a combination of FDMA and TDMA wireless technologies to communicate with the BTS. As users move from one BTS to the other, the call automatically transfers to another BTS. This essentially means that the BTS covering the other cell takes control of the user's call. This handing-over operation is transparent to the user but is a technical and complex process. In earlier wireless networks, one transmitter covered a large area but the call was lost whenever the caller left that service area.

3.3.1.3   Base Station Controller (BSC)

A Base Station Controller (BSC) controls a group of connected BTSs that are in turn connected together using a wireline network. A BSC concentrates all calls from the BTSs that it controls. On the other hand, a BSC also routes all calls to the appropriate BTSs as they come from the PSTN. BSC is also responsible for handing over the calls from one BTS to another, as a user moves from the range of one BTS to another. BSC interfaces with the BTSs through the *A-bis interface* and interfaces with the MSC through the *A-interface*.

3.3.1.4   Mobile Switching Center (MSC)

The MSC is the primary hub of the GSM network. MSC performs various functions primary of which is providing a bridge between the GSM network and a PSTN network for telephony and the Internet or other intranets for data traffic. The MSC interfaces with the PSTN via the SS7 telephony standard. The MSC also stores various subscriber-related information in its databases and controls various functions involved in subscriber handoffs, in addition to handling visitor subscribers (subscribers of other wireless networks roaming into another network). MSC stores three databases that it uses for various functions. These databases are the HLR, VLR, and AUC.

The HLR (Home Location Register) contains subscriber-related information pertaining to the subscriber's validity, the list of subscribed services, current location, and mobile terminal status. The GSM network interrogates the HLR every time the subscriber receives or makes a call. The HLR distinguishes between the mobile terminal (wireless handset) and the subscriber. The mobile terminal contains a Subscriber Identification Module (SIM) that stores subscriber membership information including the International Mobile Subscriber Identity (IMSI). The HLR houses the authentication key related to the subscriber that only the Authentication Center (AUC) can decipher. IMSI is used to retrieve information from the HLR.

The Visitor Location Register (VLR) is the database associated with the MSC that holds information about subscribers moving about a network. The VLR stores information about the visitor such as the mobile handset's Member Identification Number (MIN). The information stored in the VLR pertains to the visiting user and the MSC interrogates the visitor's HLR to retrieve the visiting subscriber's information. VLR works in conjunction with the HLR to establish a user's identity and to let the visiting mobile user use wireless services in that area.

3.3.1.5   Roaming

Roaming refers to the ability of users to use their wireless handsets in another coverage area. Various wireless networking technologies have resulted in the emergence of multiple wireless network operators that support different wireless standards. Although vendors have started to offer telephones that allow handsets to support multiple technologies and modes (analog and digital), no handset yet exists that can operate from every geographical region.

Network operators that support similar wireless standards also need to have appropriate roaming agreements that will allow subscribers of one network operator to roam to the coverage area of the other operators. In this context, the North America Cellular Network (NACN) and the continued proliferation of the GSM standard provide the largest coverage areas for mobile users.

AT&T wireless pioneered the NACN in 1990. NACN facilitates roaming in networks of other carriers and thus forms the largest cellular coverage area in the United States. The NACN technology also provides international coverage to users covered by the NACN network through the Universal Wireless Communications Network (UWCN), which provides an interface to the GSM network. The NACN also extends the roaming reach of the AMPS network users to GSM coverage areas This allows subscribers covered by the NACN to receive and make calls in GSM networks and to make calls in GSM areas. NACN provides connectivity between various carriers through the SS7 protocol. For a user to roam from one network to another, the mobile terminal's Member Identification Number (MIN) should be active in the NACN-member carrier's switch [6].

Network operators that support the GSM 900 standard facilitate another large coverage area. With the exception of the United States, GSM 900 is an accepted standard in most countries. Roaming agreements between various network carriers in various countries enable users to roam between countries. However, users should check roaming agreements before roaming into other areas.

3.3.2 Wireless Network Services

Wireless networks were designed to primarily deliver voice, unlike the Internet which was primarily designed to deliver data. However, wireless networks have incorporated technologies into their existing infrastructures that enable them to offer data-oriented services to their customers. The availability of data services and applications over wireless networks is enabling the push toward offering of e-commerce services. The Wireless Data Forum (WDF), for example, a trade group of various wireless NSPs and vendors, has committed to an initiative called the Wireless E-Commerce Initiative that will enable the selling of products and services through the Internet. The following sections will review the various data services that wireless NSPs provide to their customers.

3.3.2.1 Connectivity Services

Wireless networks enable mobile data services to users through two modes. First, subscribers to a wireless network carrier can request data services through their handsets. The subscriber must subscribe to the services and carry a compatible handset and appropriate accessories (if required due to the nature of the wireless network or the user's handset). The second mode allows mobile information appliances (i.e., PalmPilot, PDAs, laptops) to connect to a wireless network through a modem or the handset. The following covers the various means by which a subscriber can request data services through the wireless networks using their mobile appliances. Due to the technological incompatibilities in wireless networks, all connection modes may not be applicable on one wireless network.

- *Using a cellular handset as a modem on a wireless network:* This allows the user to connect an information appliance such as a notebook computer to a cellular handset. The handset in turn establishes a data connection between the user's PC and the Internet or other data networks. In this mode, users can transmit e-mail and data through their PC software over the Internet. In GSM networks, the operation works as follows: the user connects their notebook PC to the GSM handset using a special adapter. The GSM handset transmits the digital signal over the GSM network. At the network, the interworking unit (IWU) fulfills the function of an analog modem and converts the signals into a format suitable for telecommunication carriers.

- *Cellular modem:* In this mode, the user connects a handset to a wireless modem. The cellular modem provides connectivity to the wireless network. However, the cellular modem must be compatible with the network operator's wireless frequencies.

- *CDPD connectivity:* Connection through a CDPD network (e.g., AT&T wireless) requires that the information appliance (e.g., notebook PC) has an appropriate CDPD certified TCP/IP stack installed and is hooked up to a CDPD modem. Users can then access various data services (e.g., Internet browsing) through their mobile information appliances.

3.3.2.2   Data Services

Some of the common data services available to users through the mobile wireless network are as follows:

- *Wireless messaging:* This service delivers short messages to the GSM mobile handset. There are three types of Short Message Service (SMS) messages: Mobile Originated (MO), Mobile Terminated (MT), and Cell Broadcast (CB). The limit of SMS messages is 160 characters. MO messages refer to the user formulating and sending messages, whereas MT messages are received by the mobile user on their handset. CB messaging allows the sending of messages to all handsets within a specific cell. CB can be used to give location-relevant information to the user [5]. If the wireless service provider cannot deliver the message due to the unavailability of the handset, the messages are stored at the service center until the mobile station becomes available. The MSC deletes the messages at the expiration of a time-out period.

- *Internet e-mail:* The e-mail service delivers longer messages than those supported by SMS. E-mail capabilities come in different flavors. One method allows users to send Internet e-mail to other users using their handsets. Subscribers can use the e-mail program on their handsets to send e-mail to another user. In another method, users can connect to corporate servers through their mobile appliances. For example, PCs can send e-mail messages to peers who are located on the proprietary networks. Another method enables users to send e-mail to other user's mobile handsets. For example, users can receive e-mail on their PCS handsets in the format user_phone_number@wireless-network.net.

- *Internet browsing:* As mentioned earlier, users can connect a mobile information appliance such as a PDA equipped with an appropriate modem (e.g., PCMCIA Type II modem) and connect to the handset. This allows users connectivity to data networks such as corporate databases or the Internet. Internet access and Web browsing through the wireless medium does not provide enough bandwidth to support rich content applications. The handset's small screen size and slow bandwidth (9.6–19.2 Kbps), as provided by wireless NSPs (due to wireless technology limitations), are the primary reasons for this limited functionality. However, innovation in this arena is continuously improving users' experiences with access to the Internet and related services. Besides, this bandwidth is sufficient to support basic data services such as receiving or sending short e-mail messages.

- *Integration with corporate networks:* The SIM on the GSM handset allows various enterprises to offer vertical applications to their GSM users. For example, enterprises can develop applications and install them on the user's SIMs. These SIMs communicate with an enterprise's corporate databases in client/server mode to fulfill various e-commerce services. Users can then request bank balances, initiate bank transfers, order stocks, and make other such transactions using the data services of their handsets, without interacting with a live agent.

3.4 Paging Network Backbone

The paging network backbone is a satellite controlled wireless communication network that comprises the PSTN, computers digitizing messages, and various transmitters that enable the sending of pager messages to subscribers. Paging backbones also connect to the Internet for the transmission of e-mail and pager messages. This section illustrates the interconnection intricacies a paging network employs in its interface with the PSTN and the Internet backbones to provide basic and advanced paging services to its users. These services include two-way messaging services, guaranteed delivery of messages, push services for specialized content (news and other bulletins), paging of voice messages, e-mail message forwarding, and so on.

Paging is primarily used to send and receive short messages to other users. In the enterprise arena, this capability has opened additional channels for staff communications, thus enabling workers to perform their duties more effectively. Very recently, the ability of the paging companies to interconnect to the Internet for sending and receiving e-mail messages to and from user pagers and other information appliances has significantly benefited the e-commerce–enabled enterprise. Users can send and receive Internet e-mail messages that subscribers can retrieve on the paging device that could be a dedicated pager or an information appliance such as a PalmPilot with plug-and-play capabilities to offer paging functions.

More recently, the need for more information has enabled paging companies to collaborate with various content providers to push value-added content to various users. Health care professionals, for example, can remain informed about health industry developments. Stock traders can receive continuous financial industry updates, thus enabling them to make timely decisions.

As paging organizations continue to develop and leverage advanced protocols for paging, additional paging services will surface in the market. For example, paging companies have started to support two-way messaging as well as voice communications using recent innovations by Motorola.

The general characteristics of the paging network are:

- *Messaging applications:* Paging networks provide simple data communications features such as the sending of text pages and e-mails and generally receive one-way messages. Next generation paging services allow users to send messages through the paging network

- *Reach:* Paging companies enjoy a larger customer base than that of the wireless networks but smaller than that of Internet users. Paging provides limited geographic coverage to its users. Most network operators provide national coverage within the United States. However, their abilities are limited in providing global coverage to their users.

- *Internet connectivity:* Users do not get any direct Internet connectivity. However, the push model of pagers allows paging companies to send specialized content to users' pagers.

- *Quality of content:* Paging companies do not support rich content due to the technology limitations of the paging protocols. However, newer paging technologies allow the sending of voice messages.

- *Reliability:* Paging companies generally do not provide any delivery guarantees. However, in some cases the use of newer paging protocols provides guarantees of delivery.

- *Security:* Paging companies do not guarantee security of messages.

Paging networks deliver short messages over a wireless network to users equipped with pagers and other information appliances. Paging networks operate at designated frequencies they lease from the FCC. All paging networks operate independent of each other. This is because the FCC allocates separate frequencies to the paging companies within a fixed frequency spectrum range, and those companies can operate only at those frequencies. For the end user, this means that a user's pager address or number will not work on another paging NSP's network unless the paging companies provide such abilities to their subscribers.

Paging networks use traditional one-way frequency spectrums to broadcast one-way messages. To enable their networks to handle two-way messaging in order to offer advanced services to their users, paging networks operate on the NPCS (Narrowband PCS) frequencies that they lease from the FCC. NPCS provides next generation wireless messaging services through paging networks.

Subscribers to paging networks do not enjoy the flexibility of global roaming that wireless telephony users have. In some cases, local paging companies partner with paging companies in other countries to provide paging coverage in those countries. Metrocall, PageMart, SkyTel (recently acquired by MCI), MobileComm, and AirTouch are some of the major players in the paging industry in the United States.

3.4.1 Paging Network Topology

Figure 3.4 illustrates the basic structure of a typical paging network that includes the following components. Subsequent discussion elaborates on the details of the paging network architecture.

- Paging terminals;
- Paging transmitters;
- Radio interface of the paging coverage area;
- Gateways to various networks;
- Pagers.

3.4.1.1 Paging Terminal

A paging terminal is the front-end of the paging network. It interfaces with the sender of the page to collect information such as the recipient's address (pager

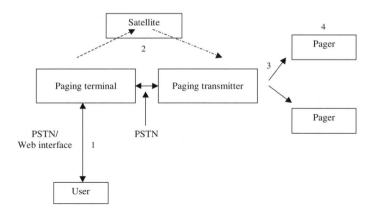

1. The user transmits a message through the telephone or the Web
    interface to the paging network. (The entry point to the paging
    network is the paging terminal.)
2. The paging terminal transmits the pager message to paging
    transmitters through the PSTN or satellite connections.
3. The transmitter(s) transmit the pager message over the radio signal.
4. The pager with the correct code picks the pager message.

Figure 3.4   Paging network architecture and message flow.

number) and the message that the sender wishes to send to the recipient. The
paging terminal validates the subscriber's address and subscription information
in its database before preparing the message for delivery. Once the terminal ver-
ifies the user's identity and service privileges, it translates the message into an
appropriate paging protocol and transmits it to the paging transmitter. For
one-way messaging, pager companies use the POCSAG or high-speed FLEX
protocol from Motorola. For enhanced paging services, paging terminals
employ other protocols. The ReFLEX protocol, for example, is used to support
two-way messaging services, and InFLEXion protocols are use to support voice
communication.

3.4.1.2   Paging Transmitters

Paging transmitters are located in all coverage areas and are responsible for
transmitting the pager messages in their vicinity. Transmitters typically cover
areas from 2 to 20 miles. The paging terminal can interface with the paging
transmitter through many mechanisms. The terminal can either transmit the
message via regular PSTN lines, in which the message propagates to all transmitters via repeaters, or it can transmit the message to a satellite, which then
propagates the message to main transmitters in the subscriber's area that in turn

send pager messages to other local transmitters. The Telocator Network Paging Protocol (TNPP) is commonly used to link various components of the paging network to provide wide area coverage.

3.4.1.3   Radio Interface

Transmitters transmit a user's page over the assigned radio frequency of the paging network through radio signals. All messages have a unique code called the cap code that is associated with each pager. All pagers in the vicinity of a transmitter pick the signal. However, only the pager with the appropriate cap code picks the message.

3.4.1.4   Gateways to Various Networks

The first generation of paging networks supported one-way short message communications between two people. Recently, the paging companies have provided interfaces with the Internet for sending and receiving e-mail and accessing specialized content. Paging companies provide Internet e-mail gateways that forward and receive e-mail messages. Paging companies also receive specialized content from various sources and transmit (push) this content to its subscribers. The paging company collects this information through the Internet or through dedicated network connections between the paging organization and the content providers.

3.4.1.5   Pager Device

Pagers receive messages through the paging networks. Pagers have installed receivers that receive the paging signal and decode it into information to display on the user's pager screen.

3.4.2   Paging Network Services

Paging NSPs provide the following services to their subscribers.

3.4.2.1   Connectivity Services

For connectivity to paging networks, paging companies work on specific allocated frequencies and their subscribers can use network facilities only in areas covered by the paging network. There are no roaming agreements as in cellular wireless networks. There are instances, however, where paging companies collaborate with other paging networks to offer paging coverage in the other network's coverage area thus expanding the network operator's coverage area. Certain paging companies in the United States, for example, collaborate with other companies in the North American hemisphere to expand their subscriber's coverage in those areas as well. However, if subscribers are not in their coverage areas, special services from the paging network operator enable their users to retrieve messages from their mailboxes. PageMart is an example of a company that has built partnerships with international paging companies to enable their users to receive pages when traveling to "external" countries.

Users can send paging messages to others by accessing the paging network through the following means:

- *Direct PSTN connection:* Using this method, the sender calls a specific number through a telephone connected to the PSTN that delivers the call to the paging network (paging terminal).

- *Operator dispatch:* Alternately, users can call a live agent or operator on the paging network that notes a sender's message and sends it to the recipient through the paging network.

- *Internet/Web connection:* Various paging service providers provide Web pages through the Internet to enable users to send pager messages to subscribers of the paging network. The Web page resides on a Web server that in turn interfaces with the paging network.

- *Modem connection:* Users can access the paging network by dialing into a modem pool at the paging network. A user interface enables them to send messages to a subscriber or broadcast it to multiple users. The users in this case do not need an Internet connection to send the page(s).

3.4.2.2   Data Services

Following are the various types of services that paging companies provide to their users:

- *Message delivery:* In addition to sending one-way messages, various major networks also support two-way message delivery to users. This means that users equipped with two-way pagers can receive as well as compose and send messages. As mentioned earlier, advanced protocols enable users to send voice mail messages to other users.

- *Message delivery guarantees:* Certain paging networks also provide various message-delivery guarantees to their subscribers, a service that traditional paging services do not provide. Enterprises can use this service to enable customers to initiate e-commerce transactions (e.g., financial money transfers) on backend systems.

- *Internet e-mail:* Users who have access to the Internet can send text messages to users by composing and sending an e-mail message. Typically, this is accomplished by sending an e-mail to an address with the format users-pager-number@pager-network.com.

- *Paging notification:* In this service, a paging company pages a subscriber's pager when someone leaves a message on the person's voicemail.

- *Support for other content:* Certain paging networks offer their subscribers the ability to receive specialized content as well. Paging operators receive content feeds from various content providers through the Internet or dedicated network connections and broadcast that content to subscribers of the appropriate service. Users can also subscribe to specialized content through other content organizations, which deliver content to subscriber's pagers. For example, datalink.net offers services for users to receive customized information on various financial and sports content.

- *Support for multiple paging services:* A few paging organizations also provide wireless transmission of messages to alternate mobile devices. This service eliminates the need for users to carry two devices. Certain service providers, such as PageMart, provide users with the option to plug a paging card into their PalmPilots that enables them to receive pages through the PalmPilot. Special interfaces trigger the organizer upon receipt of the message, in which case the PalmPilot sounds a beeping signal to alert the user.

3.5 Integrated Network Services

As highlighted in previous sections, several new network services are shaping the backbone of the enterprise's virtual network. For wireline networks, these include delivery of voice over data networks, IP-based facsimile, advanced e-mail options, and so on. For wireless networks, these services include accessibility to the Internet, paging services, and other services. For an enterprise this necessitates formulation of its network architecture by piecing together required network services from appropriate NSPs. This translates into multiple issues such as managing relationships with various NSPs, formulating multiple service-level agreements, integration, and interoperability issues. To counter such issues, various initiatives have started to surface that would provide enterprise with a one-stop shopping solution for its network needs. These solutions would provide integrated solutions and free customer enterprises from the worries of maintenance, scalability of networks, and many other such issues.

This section discusses the two most popular initiatives for wireline and wireless networks that promise to provide such solutions. For wireline networks, Sprint has introduced the Integrated On-Demand Network (ION). Wireless Knowledge is a joint initiative of Microsoft and Qualcomm to address the integration of various wireless network services and provide it to customers. The following paragraphs describe them in further detail.

3.5.1   Sprint's Integrated On-Demand Network (ION)

Enterprises continuously struggle to upgrade their networks to keep up with the new business requirements of capacity, wide reach, and cost reduction. Upgrading an enterprise's network infrastructure is both complex and costly. Building an efficient network management solution is almost a necessity for countering post-installation issues. To solve this problem, Sprint has embarked on an aggressive initiative called the Integrated On-Demand network (ION). This network packs all the features required for businesses and customers to build the new e-commerce services without the headaches associated with building piecemeal solutions.

ION is a digital network based on ATM networking technology. A majority of the network is ATM and based over SONET rings. Sprint plans to offer the service to both residential and business users. ION's primary feature is its high bandwidth, which ranges from 1.5 to 45 Mbps. Such a high-bandwidth digital network that is accessible directly from user premises will enable users to access content-rich applications, including digital entertainment, distance learning, and other bandwidth-intensive applications. Business users will benefit threefold from this service. First, ION provides enterprises with the opportunity to integrate voice, data, and video services onto one network. This can provide substantial cost-savings to enterprises that operate and maintain multiple network segments for various uses. Second, general availability to the end user will enable enterprises to offer high-speed content to their users. ION is also poised to offer QoS guarantees for its traffic, thus enabling enterprises to offer applications that require reliability and predictability. Network managers will be able to request appropriate QoS services for various traffic flows. Finally, organizations will be able to outsource their network operations to Sprint, which will provide network management and other systems management functions, alleviating enterprises from the headaches and costs of network management.

Sprint also plans to offer various value-added services on ION. Besides Internet e-mail, fax, and other Internet-related services, Sprint plans to offer customers unified messaging services that will provide access to various types of messages including voice messages, pager messages, Internet messages, and fax messages.

The availability of this network can provide substantial benefits to enterprises, especially in terms of cost savings and maintenance. Since Sprint owns the ION network, an enterprise will be less concerned about the woes associated with network upgrades and maintenance activities that relate to systems and network management. Sprint has started to offer some of these services to select customers in 1999. However, as Sprint and MCI announced a merger deal in late 1999, the merger may affect some of the planned ION services.

3.5.2 Wireless Knowledge Services

Wireless Knowledge [7] is an initiative spearheaded by Microsoft and Qualcomm to bring end-to-end wireless data services through the Internet and corporate intranets. Today wireless devices support limited groupware functionality across the Internet and corporate intranets. This inhibits business professionals from performing various services. For example, enterprises have to devise complex solutions to synchronize a user's mobile handset PIM data with backend databases or other systems. The Wireless Knowledge service provides mobile professionals with easy access to information such as e-mail, calendar, contacts, scheduling information, task lists, and unique corporate information on corporate servers. Wireless Knowledge enables such groupware functionality by facilitating access to groupware servers such as Microsoft Exchange. This frees users from the need to access enhanced groupware functionality through their information appliances connected through traditional wireline connections. By partnering with Wireless Knowledge, enterprises will no longer have to worry about leasing of appropriate wireless and Internet services, or building data solutions and then integrating them to empower their employees with appropriate business content and services. Wireless Knowledge is the first major initiative that promises to pull the wireless world into mainstream e-commerce–enabled business computing by enabling users to connect to multiple data sources and services through the wireless medium.

The Wireless Knowledge initiative supports numerous information appliances including Windows CE–enabled devices, wireless telephones equipped with microbrowsers using HDML, two-way pagers, Windows platforms, and others. Wireless Knowledge offers the Revolv service to enterprises to resolve some of the aforementioned mobility needs. Revolv interoperates with numerous wireless transports, including CDMA, FDMA, GSM, and CDPD. Wireless Knowledge offers these services to enterprises as a technology solution, to be deployed internally. Alternatively, Wireless Knowledge also offers organizations the option to outsource such hosting to Wireless Knowledge.

3.6   Formulating the Enterprise Virtual Network Architecture

Formulating an enterprise's network architecture is quite challenging. The challenge stems from the fact that multiple options exist to build various legs of the enterprise network. This challenge—coupled with varying application requirements—makes it quite complex to choose an appropriate networking strategy. An enterprise has to lease network services from NSPs to build the fabric of its virtual private network. This web of network services extends the reach of the organization to the e-commerce paradigm, thus enabling users to carry out e-commerce transactions. Figure 3.5 illustrates how a firm leases network services from various NSPs to formulate a firm's virtual network.

As illustrated, an enterprise's network has become virtual in nature because—unlike earlier times—enterprises lease network services from external entities. As physical aspects of the network have started to blur in representing an enterprise's network, it becomes increasingly vital to understand appropriate ramifications of these trends.

In formulating an organization's network architecture to handle e-commerce services, the following objectives are of paramount importance:

- Reduced TCO of the network architecture;
- Enables support of next generation e-commerce applications;
- Flexible enough to accommodate growth (scalability);
- Incorporates redundancy;
- Adequately controls network traffic, connections, and problems;
- Easier to maintain.

Formulating an organization's virtual network architecture first requires a clear definition of business objectives and the business processes required to meet those objectives. Chapter 1 stressed the criticality of this prerequisite step in the formulation of an organization's generalized e-commerce systems architecture, of which computing networks are an integral part. Business objectives thus lead to the conception of the required e-commerce system that will support the appropriate business processes.

The next step is to understand the networking characteristics of the appropriate e-commerce systems. These characteristics address issues such as customer volume, traffic patterns, and customer demographics. Business characteristics are thus the prerequisites that help in the derivation of various networking requirements such as a network's reach, capacity, and stability. For example, if the appropriate e-commerce system involves deployment of an

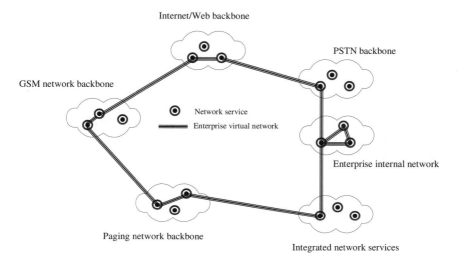

Figure 3.5 Enterprise's virtual network.

e-commerce site to sell products to customers in global markets, then the Web site's network has to handle high volumes of traffic. Similarly, if the appropriate e-commerce system involves the transfer of huge volumes of data, the network should have the appropriate capacity to handle these data transfers.

Understanding networking requirements eventually leads to the formulation of the enterprise's networking architecture, which essentially involves choice of the appropriate networking technology and networking services that the organization either builds internally or leases through external NSPs or ISPs. This section discusses these issues in further detail.

3.6.1 E-Commerce Systems' Network Requirements Assessment

Understanding appropriate networking requirements consists of the following four steps:

- Assessment of network reach;
- Assessment of network capacity;
- Assessment of network stability;
- Assessment of network security.

These steps are explained in detail in the following sections.

3.6.1.1   Assess Network Reach

A network connects users to content. In laying the foundation of an enterprise network, the first obvious step is to identify all potential users that will connect to an enterprise's network for accessing content. A network reach essentially determines whether the organization's network will be localized in a single office or will span an international geography, connecting a wide range of customers, partners, and staff members. To illustrate with a simple example—if the organization intends to allow its corporate users access to the Internet through all its branch offices, it will have to provide appropriate networking connections to all those sites. Similarly, if the organization wishes to provide a specific user group with network access to corporate content through wireless channels, it will have to arrange for appropriate wireless services in those areas.

The physical location (local or remote) of an organization's users (both internal and external) and content hosting sites determine the reach of the enterprise network. The following paragraphs further elaborate upon these two factors. Analysis of these factors leads to the definition of the appropriate physical network topology within the enterprise with external connections to users and service providers. This set of connections collectively forms the virtual network of the enterprise.

Determine End User Presence

Determining end user presence and mode of connectivity drives the formulation of the enterprise's virtual network to enable all users access to an organization's e-commerce systems. As explained in Chapter 1, an enterprise's users fall into the three primary categories—internal users, business partners, and external customers. The various modes of connection for these users fall into the following five categories:

1. The enterprise's internal users connecting to enterprise content through enterprise's internal network;
2. The enterprise's internal users connecting to enterprise content through remote channels and networks;
3. The enterprise's internal users connecting to external content through enterprise's internal and external networks;
4. Business partners connecting to enterprise content;
5. External customers connecting to enterprise content.

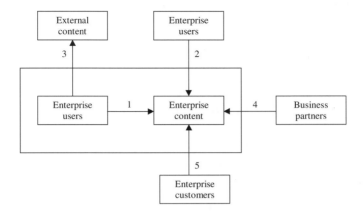

1. The enterprise's internal users connecting to enterprise content through the enterprise's internal network.
2. The enterprise's internal users connecting to enterprise content through remote channels and networks.
3. The enterprise's internal users connecting to external content through the enterprise's internal networks.
4. Business partners connecting to enterprise content.
5. External customers connecting to enterprise content

Figure 3.6 Network connectivity scenarios within an enterprise.

Figure 3.6 illustrates these modes of connection for the various users. The following paragraphs explain common scenarios that relate to each of the preceding categories.

The enterprise's internal users connecting to enterprise content through the enterprise's internal network The first scenario involves the establishment of LAN and WAN connections in order to weave a network within enterprise boundaries, thus enabling users to communicate and access enterprise content. An enterprise usually weaves its internal LANs without the need of external network services. However, to establish a WAN, an enterprise leases various network services,. For example, it leases framework relay connections from various NSPs to extend the reach of the enterprise's network to a wider geography.

The enterprise's internal users connecting to enterprise content through remote channels and networks Remote computing technologies fueled trends of the emerging virtual office. It is quite common for enterprise users to log on to the internal enterprise network in order to access locally-hosted content and services through various remote channels. Several technological options exist to provide remote connectivity to an enterprise staff. Users can either dial-in directly to the enterprise through modem pools or users can use an NSP's remote connection services to remotely connect to an enterprise's services.

The enterprise's internal users connecting to external content through the enterprise's internal and external networks   There are various reasons for an enterprise's internal users to connect to external content. This includes connections through the Internet to access Internet content, accessing partner organization's systems through EDI networks and leasing of various network services for other intrabusiness requirements.

- *Providing Internet access:* An organization has various means of providing Internet access for employees. Access through dial-up involves minimal networking changes at the user's premises. DSL installation requires the installation of a DSL adapter that provides a connection to the appropriate ISP via the LEC's premises. Faster and dedicated connections require special networking gears that include routers, CSU/DSU, and so on. Requesting such connections therefore requires the leasing of services through various NSPs. For establishing business-to-business commerce links (e.g., EDI) through the Internet, organizations can lease secure Internet services (e.g., VPNs) through ISPs.

- *Leasing of wireless network services:* An enterprise has to lease network services from various NSPs to fulfill wireless networking needs (e.g., paging or wireless services). However, not all paging companies will be able to fulfill an organization's requirements for appropriate geographical coverage. Depending upon required users' locations, an enterprise will have to identify a suitable paging organization. For example, very few paging companies provide international paging services. Companies that do provide such services collaborate with paging companies in other countries in order to do so. Besides, providing mobile connectivity to the Internet through various wireless service providers requires the leasing of appropriate networking services from various wireless NSPs. As was discussed earlier, restrictions associated with wireless networks may restrict an enterprise from providing appropriate services. These restrictions apply to users' mobility, the nature of the services, and the performance of the services. For example, based on the current technologies and available wireless services, an organization may not be able to offer its users ubiquitous access to the Internet through wireless handsets.

- *Establishing EDI network connectivity:* Traditional EDI networking connections involve the setting up of dedicated networking connections from the enterprise to various value-added networks (VANs). Other methods involve establishing dedicated connectivity such as leasing T1 lines from the enterprise to the business partner organization. Most organizations that do not intend to move to the Internet for business-to-business commerce continue to use these connectivity configurations.

Business partners connecting to enterprise content   The increased flexibility and security of network technologies is pushing enterprises to provide controlled access to an enterprise's internal content. For example, organizations can deploy controlled extranets to facilitate such connectivity. This involves the establishment of appropriate networking gear such as firewalls and routers to segment networks and provide appropriate security to control business partner access. Chapter 5 provides more detail on extranets and their deployment.

External customers connecting to enterprise content   For business-to-consumer e-commerce transactions, enterprises require the establishment of appropriate network links and services in order to give end users access to enterprise hosted content. The most popular means of doing this is to have users connect to enterprise content through the Internet. However, other scenarios facilitate such access as well. The following describe other such scenarios:

- *Extending networking services to end customers from specialized locations:* Enterprises usually provide content on the public Internet, allowing its end customers access to that content through their local Internet connections. However, at times, an organization enables its customers to access its corporate databases or other services from other locations. For example, a bank may establish video kiosks in various locations (airports, supermarkets, etc.) to enable its customers to contact customer service agents. In such cases, an organization may request dedicated network links from those locations to the enterprise premises. The organization will need to check with local service providers about their ability to provide such connections from those locations. For example, an organization requiring access to dedicated ISDN circuits in a physical location to provide the above services may find that no NSP in that location provides those services. To cite another example, a retail store may install special video kiosks in various regional locations, enabling customers to shop locally without the store having to erect a brick-and-mortar outlet. This application provides great value to compensate for the limited Internet access and bandwidth available through customers' homes and providing such connections to users who do not have residential Internet access. An organization may lease dedicated ISDN connections from those locations in order to content hosting sites to provide such services.

- *Assessing end customers' access to networks:* Although not directly related to the formulation of an enterprise's network architecture, an organization needs to assess whether customers will have the required network capabilities to access an organization's content. For example, an organization intending to provide Internet-based video-streaming services should ensure that customers have no geographical restrictions to access to broadband networks such as cable, DSL, or ISDN.

- *Wireless support for accessing enterprise data:* Providing such access requires the establishment of physical links from the wireless NSPs to the enterprise's content hosting location(s). This scenario is possible in situations where organizations offer delivery of special information or content to its customers through wireless networks. In such cases, an enterprise collaborates with a wireless service provider and establishes dedicated and secure connections from its WAN to the wireless network operators' network operations center. A typical example is that of a bank that offers customers the ability to request bank balances through their SMS-enabled GSM telephones. Another case involves granting special corporate users such as sales agents the ability to access corporate databases through their Internet-enabled cellular telephones, thus reducing the cycle time to close a sale.

Determine Content Hosting Sites

A physical location at which an enterprise hosts its content is another factor that determines an organization's network reach. An organization usually hosts its content in specialized data centers or network operations centers (NOCs), where organizational processes and procedures control the content for external access. Traditionally, organizations centralized its content within its premises in these controlled data centers. However, evolutionary changes in the definition of content and advances in the distributed computing paradigm enabled organizations to host content throughout the Internet.

For content hosted in external data centers, organizations require network links to connect its users to those sites. This connectivity is required for activities such as updating content that customers and partners may access and also for providing controlled content access to its own internal employees.

3.6.1.2 Assess Network Capacity

Having determined the span of an organization's network, the next step is to assess the network capacity required to support an organization's e-commerce systems. User volume, traffic content, and network traffic utilization drive the requirements of an enterprise's network capacity. The following paragraphs elaborate upon these factors.

User Volume

User volume determines the load of the network, which is crucial in the design of a network. Inability to access an e-commerce system's services due to network congestion will only drive end customers away, or in the case of an intra-enterprise system, frustrate internal customers and partners. For example, assessing internal users' volume to access Internet services will determine the choice of the network connection from the enterprise's premises to the appropriate ISP. ISPs offer varying degrees of connections that can range up to OC-48 speeds. Similarly, in the design of LANs and WANs, an organization's choice of a specific network type is also dependent upon the volume of the users that will access various value-added services over the enterprise network.

Traffic Content

Traffic content refers to the nature of the content that a network carries from one point to another. This includes traffic such as voice, video, or a high volume of data transfers. In determining traffic content, an organization should consider the following:

- *Public networks differ significantly in their capabilities to handle various types of data.* In deciding on the delivery of various forms of content, including simple data transfers, organizations should consider the bandwidth of the various public networks that are to carry the specific content required to support an e-commerce system. Organizations should also consider accessibility of those networks by users, as not all users have access to all types of networks.

- *Certain types of traffic, especially related to video, audio, and other multimedia content are not suitable for all public networks.* We observed in the earlier sections that cellular wireless networks and paging networks are limited in their abilities to transport rich content to the end user. Although technologies are advancing at a rapid pace, as of this writing it is too early to plan for e-commerce systems that will deliver rich content such as streaming audio/video news clips to a user's wireless handset.

- *Technologies have advanced significantly to enable integration of traffic.* Enterprises have always maintained separate networks to transport voice, data, and other specialized content. Organizations have used the PSTN for their voice needs and built internal LANs and WANs to cater to various forms of data traffic. With progressing requirements to support video and audio through computer networks, enterprises have also deployed islands of networking segments to accommodate those needs. As technologies and solutions for integrating voice, data, and audio are proliferating, enterprises should consider deploying such services to reduce costs and optimize operations. By leasing Sprint's ION, for example, organizations can also alleviate their implementation costs.

- *Networks that have the potential of providing appropriate quality may require special network technologies and/or configurations to guarantee appropriate quality of service (QoS).* An enterprise should plan for QoS even for network backbones that boast high capacity. For example, although most of the Internet backbone employs high-performance networking segments (e.g., SONET rings and the use of ATM switching technology), QoS cannot be guaranteed without appropriate planning and the leasing of specific QoS services from ISPs. QoS is a vital parameter for delivering high performance e-commerce systems. The following section presents more details on this issue.

Quality of Service (QoS)   QoS is the guarantee for delivering specific bandwidth/throughput to the appropriate applications. For example, if a videoconferencing application requires a minimum bandwidth of 2 Mbps, QoS will ensure delivery of the appropriate throughput to the application. The use of shared network links poses numerous QoS issues, as other applications simultaneously share the network's bandwidth and achieving appropriate QoS may not always be possible. QoS poses even greater challenges when applications run on public networks such as the Internet. This is because data packets can traverse multiple nodes and NAPs before arriving at their destinations, and the Internet does not provide end-to-end network control for traffic, unless an ISP can specifically route traffic over the segments that it controls.

Various technologies have surfaced in the past to tackle QoS issues. Those technologies range from providing high-capacity cables, hardware that incorporates new technologies, and sophisticated software that runs on the networking hardware. However, successful implementation of those solutions depends upon factors such as the nature of the application, type of traffic, and the specific network that implements the appropriate QoS solution.

Resource Reservation Protocol (known as RSVP) and Multi-Protocol Label switching (MPLS) are the two technologies that enable applications to request allocation of a specific bandwidth on the network. For RSVP to work, both routers at the source and destination have to implement the RSVP protocol. When an application requests the appropriate bandwidth on a network, the routers can grant or deny the request depending on the availability of network resources. The disadvantage to this mechanism is that the required bandwidth is not guaranteed for availability at all times.

MPLS works by allowing network managers to route different types of traffic on different networking paths. This is accomplished by inserting a label in each data packet and informing the networking devices of the significance of various labels. When the devices encounter these labels, they can route traffic according to the label information. Thus, traffic requiring high bandwidth may be labeled uniquely to enable the networking device to route that traffic on an appropriate path. Other traffic may be forwarded on a different path, thus alleviating the load of the network and guaranteeing appropriate levels of service. Alternatively, appropriately labeled traffic may be routed onto a network (e.g., ATM) that will provide the suitable QoS levels for that traffic.

Another means of providing appropriate levels of QoS is to leverage the networking infrastructure of the appropriate ISP. An ISP may offer multiple POPs in geographic locations that can accommodate the organization's source and destination sources. Since in this case the ISP has full control over their network (as traffic does not route through uncontrolled nodes and is based on the ISP's backbone), the ISP can provide appropriate QoS guarantees to the end user. However, if the source and destination are not on one ISP's backbone, this may not work effectively.

Not all applications require QoS guarantees. Applications such as Internet telephony and videoconferencing require appropriate QoS levels as the various elements of the data transmission have to reach the destination in a specific time period in order to enable practical and effective use of these technologies. For example, an Internet telephony session with a bandwidth of less than 28 Kbps will not be of high enough quality to be audible at the end user's destination. Therefore, depending on the appropriate levels of QoS, the enterprise has the following options:

1. Lease appropriate QoS levels from ISPs;
2. Self-deploy a network architecture that provides sufficient capacity to ensure delivery of appropriate QoS guarantees to the application or incorporates QoS technologies (e.g., RSVP, MPLS) into its networking topology.

Traffic Utilization

Deploying e-commerce systems also requires analysis of traffic utilization. For example, an organization may determine that it needs T3 lines to handle its Web site loads. However, leasing a dedicated connection from a NSP may not always be cost efficient if such requirements hold true only for peak hours. An organization can thus request special services from NSPs in order to accommodate such demands at reduced rates.

3.6.1.3   Assess Network Stability

The criticality of an organization's e-commerce systems can gravely influence the enterprise's virtual network architecture. With the availability of multiple ISPs and network services, an organization has to evaluate services and the appropriate NAP based upon the criticality of the e-commerce systems that it plans to deploy for its internal and external customers. A bank, for example, should perform a thorough network stability analysis of the ISP and its services before deploying a banking e-commerce system on the ISP's backbone.

The following four factors determine a network's stability. First, the physical network backbone should include appropriate controls to provide maximum availability. Second, the NSP should provide adequate site contingency to accommodate any natural or other disasters. Third, the NSP should employ a reliable network management infrastructure to provide appropriate statistics on network usage, bottlenecks, and other parameters that may affect availability. Finally, an enterprise should assess the NSP's security to prevent misuse of an enterprise's data.

Network Connection Redundancy

In deploying networks, an organization should appropriately plan for the contingency of its network cables as well. Numerous cases exist where incidents related to cable cuts and other natural disasters have rendered networks unavailable. Many ISPs, for example, deploy multiple SONET rings to provide redundancy of their network connections. Depending on the criticality of the e-commerce systems, the organization should either deploy appropriate contingency measures or assess an NSP's network redundancies. For example, to provide backbone redundancy, some ISPs devise alternate paths by ensuring that redundant cables are not included in the same physical pipe. Inclusion of alternate cables in the same pipe may provide network redundancy at a network level but it does not address natural or man-made disasters such as cable cuts during construction in the area.

For connections from the enterprise to the NSP that provide backbone services, an enterprise can request appropriate backup services. Usually, organizations access the NSP facilities through local loops. As a contingency measure, enterprises can purchase local loop services from multiple carriers (e.g., CLECS and LECs) to ensure availability of local loop connections. NSPs also provide automatic network fail-over mechanisms, in which traffic switches to an alternate path if the primary link experiences problems.

NSP Site Contingency

An organization should adequately analyze the contingency plans of its NSPs, especially if the enterprise intends to use the NSPs to deploy their mission-critical e-commerce systems. Not all NSPs provide adequate contingency plans. In assessing site contingencies, organizations should check for the following:

- Does the NSP have contingency sites?
- How transparent is the process to switch operations to the contingency site? What are the various fail-over mechanisms?
- How long is the unavailability period before a site is fully operational on the contingency site?
- What loss of services or quality should the customer organization expect when the NSP switches to the contingency site?
- Does the NSP periodically test its contingency mechanisms?

An organization should compare the answers to these questions against its requirements. Consider the case of a shopping Web site that attracts millions of users and leases networking services from an NSP for the hosting of its content. An unsatisfactory response to the above questions should prompt the organization to either switch to another NSP or plan for alternate contingency measures.

Some organizations pay little attention to the NSPs' network operations and do not cater to alternate plans to react to such outages. In mid-1998, a satellite lost service, and the operations of numerous network services including those of paging companies, TV content, and others were directly affected. Organizations should not overlook the possibility of such contingencies and devise plans to react to such cases. These plans may include devising contingency business processes or devising processes to inform customers of such outages.

Network Control Required to Assess Stability

Network management deals with the overall well-being of the network. Specifically, it addresses issues related to the availability and performance of the network. Network management deals with all issues related to networks that include cables, networking software, hardware, protocols, and traffic flows. The stability of the network therefore requires an appropriate network management infrastructure to respond proactively to any outages. An organization should either deploy its own network management infrastructure or collaborate with the NSP in providing overall control of the network that may impact the availability of its business. Assessing the granularity of network management controls that an NSP provides is equally vital. Chapter 6 discusses network management in further detail.

Table 3.2
NSP/ISP Selection Criteria

Subject Area	Selection Criteria
Network services	Assess the breadth of network services required to support an enterprise's e-commerce systems requirements
Network reach	Identify points of presence (POPs)
	Assess breadth and depth of networking options required to meet an enterprise's network reach
	Assess geographical coverage for wireless networks
Network capacity	Assess networking speeds for various connections
	Assess custom traffic utilization services
	Integration of various forms of traffic
	Assess wireless network's bandwidth
	Assess the breadth of broadband network services
	Assess network backbone's capacities
	Assess QoS services
Network stability	Assess network management infrastructure
	Discuss granularity of network management controls
	Assess network backbone's contingency
	Assess Network Operations Center's (NOC's) contingency
	Assess contingency capabilities
	Assess test results of ISP's contingency plans
Network security	Assess internal security procedures
	Assess enterprise's network security services
Customer support	Assess customer support infrastructure (expediency in reacting to problems, ability to dispatch field staff to rectify problems, etc.)

3.6.1.4   Assess NSP's Security Controls

Assessment of NSP's security controls can be very crucial for organizations that transmit sensitive information over such networks without the use of an appropriate network security solution. Besides ensuring the network security (e.g., deploying VPNs) of the e-commerce system, numerous other security issues exist in leasing network services from NSPs. For example, an enterprise should assess procedural security controls of the NSP that hosts an enterprise's content. If an NSP hosts an enterprise's content, it may provide backup services for its Web site. Accordingly, the NSP may store backup tapes at other locations. In such cases, the enterprise should assess the security controls in order to ensure that tapes are not stored in an insecure location thus exposing the enterprise to a breach in information confidentiality. Addressing such issues should be part of an enterprise's plan to address end-to-end security of its e-commerce systems. Chapter 8 thoroughly discusses such security issues.

3.6.2   NSP/ISP Selection Criteria

Table 3.2 provides a summary of the points covered in this section. Enterprises can employ this checklist to determine the choice of appropriate NSP/ISP. Accordingly, organizations should formulate appropriate SLAs with appropriate NSPs in order to receive service delivery guarantees as well. Chapter 9 further elaborates on the topic of SLAs.

Notes and Web Sites

[1]   www.japan-us.org

[2]   www.Internet2.edu, www.ngi.gov, www.ucaid.edu/abilene

[3]   The reader should note that ISPs offer these connectivity mechanisms. Other networks (wireless, cable, etc.) also provide connectivity to the Internet backbone. Later sections will elaborate on those issues.

[4]   The numbers next to the standards designate the frequencies at which these standards operate.

[5]   Heine, Gunnar. *GSM Networks: Protocols, Terminology, and Implementation*, Boston: Artech House, 1999.

[6]   www.att.com

[7]   www.wirelessknowledge.com

Additional Web Sites

www.cdpd.com
www.motorola.com
www.pagemart.com
www.sprint.com

4

E-Commerce Services and Application Repositories

The previous two chapters highlighted innovations in information appliances and networks that enable organizations' users and customers to access an enterprise's IT-enabled services. Organizations provide these services through a mix of legacy applications and new e-commerce–enabled solutions. These applications automate a firm's business processes, empower customers to carry out self-service transactions, and provide appropriate customer support services to an organization's customers. This chapter focuses on enterprise applications and the manner in which they provide such services to its users. An organization can engineer these e-commerce services to enable its customers, employees, and strategic partners to access enterprise services through various means.

An organization's e-commerce services could be a mix of plain information access, live consultation with an organization's agents/representatives, and access to complex backend transaction-processing applications. Applications that provide these services therefore incorporate three distinct session types to provide these services. These sessions consist of inquiry, transactional, and consultative session types. An inquiry session involves database access for inquiry purposes only. The significance of these sessions in the e-commerce computing paradigm has increased manyfold as inquiry sessions against information, and knowledge repositories are the cornerstone of an enterprise's knowledge management infrastructure.

Transactional sessions involve committing database state changes. An enterprise's employee entering orders, a system updating order statuses, a business partner accessing an enterprise's systems to update inventory information, and a customer accessing the system to enter a stock trade are all examples of transactional sessions.

Consultative sessions refer to interactive and noninteractive communication sessions. Communication in interactive mode involves a multimedia (video, audio, and/or interactive text chat) interface with a live person or an interactive response system such as an IVR. E-mail and pager messages are examples of noninteractive communications.

Enterprise applications have evolved through three distinct phases in their support of the various session types for the three business domains. The first phase primarily consisted of applications that constitute today's legacy and mainframe applications. Since the early days of computers, most of these applications have existed on enterprises' mainframe systems. These applications were developed in traditional programming languages such as COBOL and resided on application server environments powered by transaction processors such as CICS and others. Network and hierarchical databases (e.g., IDMS, IMS) stored information related to these applications. These systems are still operational within enterprises and in some cases still form the core of their business process operations. Table 4.1 illustrates this first application phase within the enterprise and its support for the three session types. The primary characteristics of first-phase applications are as follows:

- These applications automated the core business processes of an enterprise, including internal financial applications (e.g., accounting, payroll), human resources applications, and other vertical applications developed to match an enterprise's business processes.

- The span of these applications was primarily confined within the enterprise. These applications had no connectivity to consumers.

Table 4.1
Mainframe Era Support for Different Types of
Application Models for Various E-Commerce Domains

	Transactional Applications	Inquiry Applications	Consultative Applications
Business-to-business	Limited (Dumb terminal applications for EDI)	Limited (Dumb terminal applications for EDI)	✗
Business-to-consumer	✗	✗	Limited (call-centers)
Intrabusiness	✓	✓	Limited (e-mail applications)

Larger enterprises were among the first to develop EDI applications for business-to-business transactions.

- These applications merely focused on automating business processes and had limited contributions in the nurturing of a knowledge-based culture within the enterprise.
- These applications ran only through dumb terminal interfaces.
- Consumer-oriented businesses offered services through brick-and-mortar branches, where customer service representatives served customers by accessing dumb terminal applications. A few enterprises enabled customers to call enterprises' call centers through the PSTN channel to request basic services.

During the second phase, enterprises wanted to integrate their business processes, enable speedier and more effective information flows between business processes, and overhaul their internal and external linkages. The proprietary systems of the earlier phase did not meet requirements that had surfaced to accommodate the new business challenges. Organizations looked for alternatives to effectively manage their supply chains, improve cycle times, and optimize internal operations, and they required better planning and forecasting processes. This triggered the emergence of client/server applications and legacy application migration toward enterprise resource planning (ERP) systems. Client/server computing enabled enterprises to automate back-office and non-transactional business processes, whereas ERP systems, coupled with innovations of client/server systems, offered business integration and streamlining opportunities to enterprises. Table 4.2 illustrates the second era of enterprise

Table 4.2
Client/Server Era Support for Different Types of
Application Models for Various E-Commerce Domains

	Transactional Applications	Inquiry Applications	Consultative Applications
Business-to-business	Limited (dedicated connections)	Limited (dedicated connections)	✗
Business-to-consumer	✗	✗	✓ (enhanced call-centers)
Intra-business	✓	✓ (data warehouses)	✓ (enabled by client/server groupware)

computing [1] and its support for the three session types. The primary characteristics of this era are as follows:

- Client/server applications enabled enterprises to strengthen linkages with established business partners and encouraged them to explore other strategic linkages as well. The primary reason for this was the decreased cost associated with the development and deployment of client/server applications and flexibility in their design.

- Client/server applications enabled enterprises to tap enterprise databases to use their data stores effectively. Deployment of data warehouses to analyze enterprise data enabled enterprises to appreciate the value of this data and pushed them into the information age.

- Client/server applications of this era still had limited accessibility to the consumer market. The primary reason for this was the limited empowerment of the consumer market with information appliances. Limited proliferation of public computing networks was another obstacle.

- The emergence of client/server computing enabled numerous innovations in the computer telephony integration (CTI) arena thus facilitating the deployment of sophisticated call centers. This enabled enterprises to build stronger customer relationships.

Another primary development of this era was the emergence of ERP applications. ERP applications comprise a set of software modules that correspond to the various business functions and processes of an enterprise. These business processes include inventory control and management, accounts receivable, shipping and logistics, human resources, production planning and control, and others. ERP applications provide numerous benefits to the enterprise. ERP software presents opportunities to standardize various business processes and thus optimize operations across the enterprise. Integration of various business processes facilitates streamlined operations. Besides providing appropriate tools, ERP software provides opportunities for streamlining various activities within the business processes.

ERP application providers include SAP AG, Oracle, Peoplesoft, Baan, and others. ERP systems enable the integration of various business processes, which was not feasible in the legacy application environment. Organizations can integrate various business functions under one umbrella. For example, SAP offers various business functions related to the travel industry. SAP's Travel Planning software integrates business functions that include air, car, and hotel reservations; travel expense tracking; and accounting management. This ability to integrate a firm's business processes provides a seamless view as well as a flow of information across various interrelated parts of the firm's value chain. This enables information sharing among various business departments by integrating various business functions at the application level. Besides, Unified Data also provides opportunities for intelligent forecasting and other types of analysis.

ERP systems also provide better opportunities for linking with e-commerce applications. This is because legacy systems suffer from data update and access limitations since they build on older technologies and do not scale appropriately to meet e-commerce requirements. Most ERP solutions incorporate high availability and scalability features that suit the e-commerce arena, in which customer volume is continuously increasing by enormous proportions. In line with the latest application architecture trends, ERP systems build on object technologies that allow them to offer component solutions and thus are extensible to plug in with other key business software, thereby extending the business processes.

The client/server era and parallel deployment of ERP applications did not effectively resolve the problem of appropriately managing external linkages with customers and business partners. Organizations relied on EDI technologies and techniques for such issues. EDI, however, was feasible only for larger organizations that could afford the costs associated with implementing it. Smaller organizations could not benefit from offering business applications to parallel the reach and linkages of the larger organizations. This pushed the industry into the current and third phase that marks the era of e-commerce. In this era, enterprises are deploying special e-commerce applications and hooking them to the backend ERP and legacy systems. Table 4.3 illustrates the e-commerce era and its support for the three session types. The primary characteristics of applications in the e-commerce era are as follows:

- E-commerce applications empower enterprises to exploit business linkages with all types of business partners. Internet-based computing enables small and large organizations to engage in business transactions (unlike the earlier era that provided such benefits to larger enterprises that could afford the costs of EDI).

- Internet-based computing and applications enable a vast consumer market to touch enterprise services.

- Customers are empowered to access enterprise applications through various electronic channels (e.g., PCs, PDAs, wireless telephones). Chapter 2 extensively discussed these issues.

Table 4.3
E-Commerce Computing Era Support for Different Types of
Application Models for Various E-Commerce Domains

	Transactional Applications	**Inquiry Applications**	**Consultative Applications**
Business-to-business	✓ (e.g., Internet EDI, OBI protocol)	✓	✓ (Internet groupware)
Business-to-consumer	✓ (Available through Web, IVRs, wireless telephones)	✓ (Available through Web, IVRs, wireless telephones)	✓ (Available through Internet, pagers, wireless telephones, Web-enabled call-centers, etc.)
Intrabusiness	✓	✓ (Enabled by knowledge repositories)	✓

- The various electronic channels facilitated the emergence of virtual enterprises. Virtual enterprises have minimal to no brick-and-mortar presence. E*Trade Securities, Inc.; eBay, Inc.; Amazon.com; and eToys are a few examples that have established a strong presence in their respective fields.

- E-commerce computing and applications enabled enterprises to truly push data and information in the realm of knowledge. Availability of information sources on enterprise intranets and the Internet coupled with groupware tools for internal collaboration–enabled opportunities for massaging information into knowledge.

The remainder of the chapter focuses on enterprise applications, their architecture, and their role in providing various inquiry, transactional, and consultative services to enterprise users. The chapter also elaborate the roles of application access gateways in connecting users equipped with various information appliances to backend enterprise applications. Figure 4.1 illustrates this architecture. Readers should note that this chapter does not elaborate on the ancillary technology elements (middleware, directory services, etc.) that make up these applications. Chapter 6 discusses those components in detail.

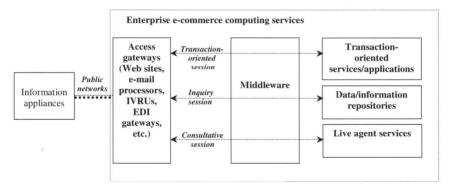

Figure 4.1   E-commerce application architecture.

4.1   Application Access Gateways

Access gateways provide frontend services for users and systems requiring access to enterprises' backend computing resources and services. Access gateways form the interface layer between information appliances and the backend informational, transactional, or live agent services. These gateways interact with information appliances to handle their requests and route them to backend service information and application repositories for fulfillment of those requests. Accordingly, these gateways route the response from the backend repositories back to the information appliances.

The specific nature of the information appliance and services determine the choice of appropriate gateways. For example, Web servers act as access gateways to interact with Web browsers and handle users' requests by interacting with an appropriate backend service. Telephony gateways are another form of gateway that provide the necessary software and hardware to handle IP-based voice from the user's information appliance and map it to regular circuit-switched data streams (PSTN format).

Figure 4.2 illustrates the popular access gateways for various information appliances. The focus of this section will be on the inherent characteristics of access gateways. Later sections will elaborate the role of these gateways in providing end-to-end e-commerce services.

4.1.1   Web Servers

Web servers are the most popular types of access gateways that enable users with browsers to access content that ranges from HTML pages to high-end

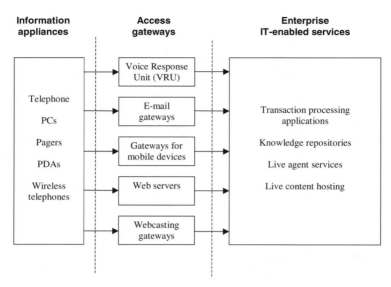

Figure 4.2   Application access gateways.

mission-critical applications. Web servers interact with browsers using the HTTP protocol and provide the necessary underlying infrastructure to enable enterprises to bring e-commerce services to the Internet.

As Web servers host connections from a wide range of users on the Internet, the Web server platform provides the necessary technology components to handle high Internet traffic and transaction volumes. Web server software is designed to optimally exploit the underlying operating system and hardware technologies to offer multiprocessing, fault-tolerance, and performance to cater to such high traffic. These servers also incorporate the necessary technologies to interface with enterprise's backend services and applications and route user requests from browsers to the backend applications and vice versa.

Many Web servers are available for different operating systems. For example, Sun's and Netscape's iPlant Web server is available for both UNIX and NT systems. Microsoft's Internet Information Server (IIS) runs on Windows servers, and Apache Web server is available for a host of systems, including the Mac OS X server. Various Web servers are also available for the Linux operating system including the Zeus Web server, iMatix Corporation's Xitami Web server, and others.

4.1.2   Interactive Voice Response (IVR) Gateways

An Interactive Voice Response (IVR) or Voice Response Unit (VRU) handles customer calls through the PSTN and provides the caller a menu (also referred to as the *call tree*) of choices based on a predefined and preprogrammed script. IVRs then can execute an appropriate function based on the customer's input. The input can be natural language voice prompts or telephone touch-tone input. For example, a user may call a stock trading institution and be greeted by an IVR that in turn prompts the caller with various inputs (e.g., press 1 to enter an order or press 0 to speak to an agent).

IVRs are special software programs that require a hardware card and run on popular operating systems (e.g., NT and UNIX). Installation of IVRs alleviates the call-center agents from handling the flood of customer calls by providing a self-service technology infrastructure. Enterprises can build various functions within the IVR menus and link those functions with backend enterprise systems. For example, the IVR can act on the user's natural language voice prompt to trigger a backend transaction-processing function. If the caller does not want to engage in a self-service transaction, the IVR can be set up to route the call to an appropriate service representative (call-center agent) to handle the caller's requests.

IVRs are capable of receiving user input either through DTMF tones or through natural language speech. An IVR recognizes the user's DTMF inputs and translates them to appropriate messages for the backend systems. In the speech mode, the IVR translates caller's voice input through appropriate speech-to-text technologies and formulates an appropriate message for the backend systems. Interface to the backend systems is thus a crucial feature of the IVR systems. Many IVR vendors provide prebuilt plugs and interfaces to popular backend systems. The returned response from the backend systems is converted using a text-to-speech technology (TTS) that reads the response back to the caller.

The use of IVRs is on the rise because IVRs handle customer's requests through the telephone, which is the most popular and ubiquitous information appliance. Almost all enterprises that offer e-commerce services through the Internet or other electronic channels also provide IVR-based services. InterVoice, Inc. offers IVR and other types of solutions to facilitate building of such e-commerce services. For example, InterVoice's OneVoice is an IVR platform and SpeechAccess is a speech recognition solution [2].

4.1.3   Internet Telephony Gateways

As mentioned in Chapter 3, the increasing popularity of the Internet and availability of high-bandwidth networks is fueling the use of Internet telephony. Enterprises are therefore rushing to use this channel to handle customer calls and for internal uses as well. Popular implementations involve using the Internet telephony channel to route customer calls to call-center agents.

Internet telephony is packet-based (it uses the H.323 standard to transport voice over the Internet). On the other hand, traditional telephony channels build on circuit-switched technologies. Handling of Internet telephony calls therefore requires the installation of appropriate gateways to translate between packet-based and circuit-switched telephony. These gateways employ appropriate hardware cards to perform the required mapping. The gateway receives packet-based calls through the Internet (or an intranet), performs the required translation, and launches a circuit-switched call on an appropriate voice trunk. This allows for leveraging the legacy infrastructure to handle Internet telephony calls. The voice trunk from the telephony groupware, for example, may terminate at a PBX, which then routes the call to the traditional circuit-switched half of the network to an appropriate person that answers the call. For example, VocalTec Communications Ltd. offers VocalTec Telephony Gateway Series products that provide IP telephony solutions [3]. ECI Telecom Ltd. is another vendor that offers an Internet Telephony Expander (ITX) series of products that provide IP/PSTN services [4].

4.1.4   Wireless Gateways for Web Access

Chapter 2 provided various ways that enable users to use their mobile information appliances to interface to the Internet and the Web. The primary issue in accessing the Internet through mobile appliances is the small display of the mobile appliances coupled with the limited bandwidth of the wireless network that inhibits the transport of rich content. The WML surfaced to resolve some of these issues. WML technology allows only the text portion of the rich Internet pages to be displayed on the appliance's small display. In this mode, a gateway at an appropriate NSP translates the Internet content and transmits it to the user's handset.

Multiple vendors offer such gateways to enable mobile users to browse Internet and Web content. Some of the popular solutions are as follows:

- Phone.com (formerly known as Unwired Planet) provides a suite of software gateways to bring data services to wireless information appliances. The organization offers the UP.Link server that translates Internet content and delivers content suited to the user's mobile information appliance. Users can use the UP.Browser on the information appliance to browse Internet content.

- Spyglass's prism solutions translate rich Web content into formats suitable for information appliances.

- Oracle has introduced a new technology called Portal-to-Go, the objective of which is to bring Internet and database content to users' wireless devices without the need to modify it. This project is a server that performs the appropriate translation [5].

4.1.5   E-Mail Gateways

Internet e-mail is one of the original Internet services and has evolved to become one of the primary means of communication among Internet users. Accordingly, e-mail is also becoming a primary communication channel within the e-commerce arena, and enterprises are installing appropriate systems to handle e-mail–enabled customer communication. Earlier implementations of e-mail systems for customer communication leveraged the traditional approach that involved routing all customer e-mail to a specific recipient address. Depending upon the volume of e-mail received, various service agents handled individual e-mail requests.

This implementation works adequately for handling a small volume of customer e-mail. However, a high volume of customers coupled with the requirement to open multiple e-mail access channels to handle requests for services such as customer inquiries, comments, and other types of requests necessitates automated and efficient methods for handling such e-mails. This is because a high volume of customer e-mails destined for multiple departments makes the traditional approach for handling e-mail burdensome and in some cases unmanageable. Deployment of intelligent e-mail gateways resolves these issues by offering some of the following features:

- Identifying the intent of the e-mail message by parsing the subject header of e-mail;

- Routing the e-mail to the department or service representative best qualified to handle the specific e-mail (request);

- Identifying the message type by an intelligent parsing of the e-mail message (body) and automatically triggering an appropriate predefined (preprogrammed) response;

- Routing the results of the e-mail request back to the customer;
- Tracking e-mail interactions in a repository.

Various vendors offer e-mail-handling solutions that blend with enterprise call centers. For example, Lucent offers CentreVu Internet Solutions, which manages the delivery and response of all e-mail messages and blends this solution with Internet-enabled call-center solutions.

4.1.6 Webcasting

Webcasting is a set of technologies that facilitates delivery of multimedia and video/audio content through the Internet/Web. Webcasting's initial use was for the media and entertainment organizations that delivered specialized content to its viewers and users. However, Webcasting has found numerous applications in all facets of e-commerce. Some of the popular applications of Webcasting are the following:

- Playing live news feeds. Media organizations can offer this service to their consumers.
- Conducting live training sessions or conference presentations over the Web (distance learning). Enterprises can offer these solutions to their internal employees as part of their intrabusiness initiatives.
- Company promotions on products and services to enterprise and consumer market customers.
- Audio and video entertainment—primarily for the consumer market.
- Captivating Web sites pull customers toward the business

Figure 4.3 illustrates two typical Webcasting configurations. The first configuration allows for playing of live audio or video streams to the customers. In this configuration, a regular video/audio source (e.g., a camcorder) records a live event, such as a conference presentation. The camcorder in turn interfaces to a video/audio capture card, which digitizes and encodes signals into appropriate formats. The video/audio signals are then transmitted to a distribution server, which interfaces with client's browsers to receive live video/audio streams.

The second configuration facilitates the streaming of stored content to customers. For example, customers may launch an on-demand request for the playback of a recorded event (e.g., a sports clip). In this configuration, content is stored on an appropriate multimedia storage disk. Upon retrieval of the

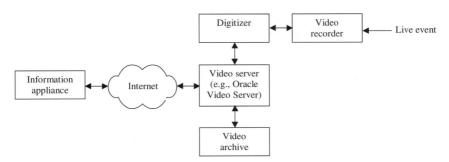

Figure 4.3 Webcasting scenarios.

appropriate content, the distribution server streams the content to the appropriate users on the Internet/intranet.

Unicast and multicast are the two primary methods for streaming multimedia content. Unicast is the more popular of the two and involves sending a separate media stream from the media server to every client that requests the appropriate media. In multicast, the media server transmits only one stream that is picked up by all the clients on the network. Multicasting therefore has less stringent network bandwidth demands but requires complex networking. Multicasting is generally popular for implementation on enterprise intranets, especially for applications such as enterprise training where multiple users can simultaneously attend training sessions. All users thus can pick up the one media stream. On the other hand, for distributing media over the Internet, organizations use unicast techniques as users' requests over the Internet may be at different times and users' locations are distributed remotely and over different and unknown network segments.

4.2   E-Commerce Solutions

This section focuses on the business applications that automate the core business processes of an enterprise spanning all three e-commerce business domains (business-to-business, business-to-consumer, and intrabusiness). These applications enable enterprises to run their supply chains, automate various financial processes, let consumers carry out self-service transactions, enable enterprises to transact with business partners, and so on.

As mentioned earlier, an enterprise's application portfolio is primarily comprised of legacy and ERP applications to which enterprises have begun to add the new arsenal of specialized e-commerce applications. E-commerce applications have enabled enterprises to enter into the e-commerce arena. In parallel, enterprises are also opening their legacy and ERP systems to e-commerce transactions and integrating them with the specialized e-commerce applications. This is facilitating enterprises to integrate all business processes over virtual channels such as the Internet. For example, in a stock trading system, customers execute trade orders through a transaction-oriented system, and the data is updated real time in the backend legacy databases.

E-commerce applications enable all types of businesses (smaller and large) to get on the bandwagon of e-commerce. For larger enterprises these systems link to backend systems and applications. This section first focuses on the e-commerce applications followed by a discussion of the trends that are opening the legacy and ERP applications to e-commerce.

4.2.1   E-Commerce Applications

E-commerce applications build on Internet-based technologies, varied computing networks, and Internet-based business protocols. Popular Internet-based technologies include HTML and HTTP, whereas Internet, pager networks, and others reviewed in the previous chapter form the various computing networks. Internet-based business protocols include Open Business on the Internet (OBI), Open Financial Exchange (OFX), and EDI. These applications incorporate capabilities, such as Web-enabled user interfaces and forms and coupling of workflow with intelligent collaboration technologies (e.g., e-mail), and they facilitate an inexpensive means for developing transactional and nontransactional enterprise systems.

Numerous e-commerce application suites have surfaced that enable deployment of effective e-commerce solutions. These application suites facilitate deployment of Web-based storefronts, Internet-based procurement solutions, Web-based advertising, and more. Both small and large enterprises can benefit from these applications. These applications provide out-of-the-box functionality for smaller organizations, whereas larger enterprises can integrate these applications with backend ERP and legacy systems and thus automate all legs of the Internet-based supply-chain business processes.

E-commerce applications differ from the ERP and legacy applications in that they build on Internet-based standards and technologies, thus enabling enterprises to immediately reap the benefits of e-commerce. For example, using various e-commerce application suites from vendors such as Microsoft and Netscape, organizations can set up a Web site and start selling products in little time. To erect a similar application based on legacy systems would require a significant amount of investment and time, which in most cases may seem impractical.

E-commerce applications and packages provide various features that enable organizations to engage in e-commerce activities. Some of the primary features and characteristics of e-commerce applications follow.

4.2.1.1 Streamline Business Processes

E-commerce applications streamline and automate business processes that otherwise proved challenging in the legacy and ERP application domains. For example, earlier approval process for procurement required approvers to sign on to specific business applications to approve orders. E-commerce applications enable solutions that merge the approval process with the e-mail infrastructure. The order routes to the approvers part of their regular e-mail, and approvers can simply approve orders by opening their e-mails, checking the order details and pushing a button to signal approval or disapproval of the order. Similarly, stores and branches of a large organization can deploy e-commerce applications that would directly connect to the manufacturer over the Internet to order goods. Earlier solutions forced them to request these goods through the main office, which would then connect to the supplier and then to the manufacturer for the ordering of goods.

4.2.1.2 Lower Cost of Deploying Internet-Enabled Business Processes

E-commerce application suites provide functionality that enables rapid deployment of business processes over the Internet and lowers cost of entry for enterprises. Built-in forms and prepackaged business-process–specific templates enable organizations to establish presence over the Internet without significant investment in time and money. For example, iPlanet/Netscape's BuyerXpert solution provides bundled functionality that enables small and large organizations to engage in enterprise procurement transactions over the Internet. Similarly, Netscape's SellerXpert enables organizations to sell products and services over the Internet.

4.2.1.3 Specialized E-Commerce Functionality

E-commerce application suites solve problems that were nonexistent in the legacy mainframe era. Selling products and services over the Internet, empowering customers to enter self-service transactions, and other such features require specialized applications and packages that include functions specific to Internet-based commerce. The following include some of the more popular functions that e-commerce packages incorporate in their core.

- *Shopping catalogs:* For procurement solutions, catalog software allows organizations to incorporate catalog functionality in their applications.

Products such as Live-commerce from Open Market allow organizations to store a large number of items in catalogs and provide features for searching and retrieving catalog items. IBM offers a comparable product called Catalog Architect that works with IBM's merchant servers and facilitates the building of catalogs and importing data from other catalogs.

- *Shopping software and agents:* Shopping cart software allows users to pick products as they shop at a Web site and place them in an electronic shopping cart. Users have the option to view, add, or delete items from the shopping cart. Other versions of shopping software have built-in intelligence. These agents surf the Web and identify merchants and products that suit user preferences. The mySimon.com portal is an example of an intelligent shopping experience. mySimon offers the Simon shopping agent that collects numerous information necessary for influencing a shopper's decision by traversing through thousands of retailers' sites.

- *User-interaction tracking modules:* Tracking user actions enables enterprises to offer focused solutions to the customers. By tracking and maintaining a history of user interactions, organizations can predict preferences of their visitors and offer them appropriate offerings and services. If a customer interacts with a Web site and searches for stereo players, the organization can offer discounts on stereos from various participating organizations. For example, Andormedia's ARIA and LikeMinds suite of tools enable organizations to track user actions in real time. These tools provide rich reporting and trend analysis mechanisms to marketers to enable them better focus their offerings.

- *Customization and personalization software:* Using one-to-one relationship tools, organizations can offer personalized solutions to their customers. By remembering customer needs and preferences, organizations can offer tailored user interfaces, functions, and cross-selling and up-selling opportunities to their customers. For example, Microsoft's Site Server solution offers a Direct Mailer function that creates a personalized e-mail marketing program based on Web visitors' preferences and profile information.

- *Ad servers:* Web-based advertisements are a popular means of generating revenue and attracting potential customers. Ad servers incorporate various functions such as creating ads, managing and scheduling ads, and tracking revenue generation. Microsoft's Ad server is an example of a product that provides such functionality. Chapter 9 provides more details on the process of Internet advertising.

4.2.1.4 Integration With Legacy and ERP Systems

Integration of e-commerce systems with the primary fabric of an enterprise's core business processes requires a seamless integration of these systems with legacy and ERP systems, as legacy databases and ERP systems form the central nervous system of an enterprise's business processes. It is therefore quite common to see that e-commerce package vendors are incorporating more backend system integration features besides merely enhancing frontend functionality for Internet access. Legacy systems and ERP vendors are also aiding in this initiative by opening their systems to e-commerce packages. Such backend integration features are required to achieve end-to-end integration of business processes. For example, a common need for organizations is to tie their Web frontends to the backend order processing, accounting, and distribution systems that may be deployed on either legacy or ERP systems.

These backend integration features enable organizations to integrate e-commerce packages with legacy and ERP systems. For example, BuyerXpert can generate EDI-based orders and transmit these orders to business partners that can only handle EDI-based transactions through their legacy platforms. Other features, such as the ability to access multiple data sources through standard database interfaces (e.g., ODBC); the ability to handle message-oriented and batch data formats; support for newer data formats, such as Extensible Markup Language (XML); development of application adapters to hook e-commerce packages into backend systems; and specialized business process workflow engines enable the integration of e-commerce packages with backend systems.

4.2.1.5 Facilitate Integration With Other E-Commerce Applications

The e-commerce application suites have built-in integration with the Internet, Web, and other technologies thus enabling easier integration and building of end-to-end e-commerce business processes. These application suites, for example, provide Web-based interfaces and tightly integrate with Web server software (e.g., Microsoft's IIS) to bring such business processes on the Web. These applications also integrate with certificate authorities to provide appropriate security features that are required for e-commerce transactions. Similarly, these applications plug into e-commerce middleware thus extending the reach and functionality of these applications.

4.2.1.6   Integration With Groupware and Business Processes

E-commerce applications integrate with groupware functions to automate all legs of an Internet-based business process. This reduces the cycle time for such processes. For example, incorporating groupware functionality (e.g., e-mail, Internet fax, reminders) eliminates the manual processing of certain business process activities such as paper-routing, manual sending and receiving of faxes, making telephone calls, and so on.

4.2.1.7   Extend Business Processes Remotely

E-commerce application suites provide organizations with an inexpensive way to extend their business processes to internal employees, business partners, and customers. For example, all internal employees can enter orders through their browser interfaces supported by multiple information appliances (PDAs, PCs, wireless telephones, etc.) instead of requiring access to legacy and mainframe applications that were difficult to set up for all employees. Organizations can access supplier's catalogs over the Internet using the OBI standard, thus eliminating bulk transfers and loading of catalogs in internal systems. This also enables suppliers to self-manage their catalogs and other business information on the buying organization's servers. Similarly, customers can interact with an organization's shopping Web site to browse products and services portfolio.

4.2.2   E-Commerce Business Standards

Internet-based e-commerce also triggered the need for newer business standards and specifications. E-commerce applications incorporate those standards to enable organizations to conduct business over the Web and interoperate with other enterprise systems without the need to build customized interfaces. Numerous standards have surfaced to facilitate e-commerce among organizations, customers, and business partners. The following discusses the primary standards and frameworks prevalent in the e-commerce arena today.

4.2.2.1   Open Buying on the Internet (OBI)

The OBI consortium drives the OBI standard. OBI consortium is a nonprofit organization managed by CommerceNet. Membership in this organization is open to buying and selling organizations. OBI specifies the technical and nontechnical specifications for enabling buying and selling of low-cost goods over the Internet. OBI primarily targets buying and selling in the business-to-business arena. These transactions usually account for most of an enterprise's purchasing transactions. In many cases, the cost of processing these transactions is higher than the value of the purchased goods. However, by introducing a standard based on the Internet, these costs are greatly reduced.

OBI is an alternative to EDI systems, although both standards can work in concert with each other. The primary drawback of EDI systems was their difficult setup, integration with backend systems, and VAN cost issues. Most EDI users in the industry are large corporations that benefit from implementing such systems. OBI, on the other hand, provides benefits to both small and large corporations by offering a standard for buying and selling over the Internet. However, organizations usually use EDI to buy and sell all types of goods, whereas OBI focuses solely on the acquisition of indirect materials. The OBI specification also accommodates the necessary workflow required in business-to-business procurement processes. These include appropriate approval processes, routing transactions to relevant parties, and checking for dollar limits, and so on.

4.2.2.2   Open Financial Exchange (OFX)

OFX is specification for the exchange of financial information over the Internet. The OFX specification was created by Microsoft, CheckFree, and Intuit in early 1997 and supports various financial activities such as bill payment, funds transfer, bill presentment, and investment. OFX standardizes information exchange among financial institutions, customers, and other financial processors. OFX provides the mechanisms for a standard interface to various financial institutions through client software (e.g., Microsoft Money or Quicken), other channels (e.g., IVRs or ATMs) and for inter-organization financial information exchanges. This eliminates the need for financial institutions to develop custom interfaces for various electronic channels. OFX defines the common set of operations that all parties can support in their offerings, thus providing interoperability.

OFX enables organizations to engage in online financial transactions with their customers as it supports Internet-based transactions, thin clients, and personal financial software that resides on various information appliances. Various software vendors provide servers that conform to OFX specifications. For example, Sybase Financial Server (SFS) incorporates OFX to enable financial organizations to connect to their customers and other financial business partners to exchange financial information and transactions over the Internet.

4.2.2.3   Microsoft's BizTalk

BizTalk is Microsoft's e-commerce strategy for providing a common framework that integrates the data and protocols required for integrating business processes. BizTalk is based on XML and enables organizations to exchange information in a standard format. According to Microsoft, BizTalk provides the glue that enables enterprises to integrate into one framework their various business processes that build on different protocols. For example, organizations engaged in EDI transactions can potentially use BizTalk to migrate to XML-based data interchange. Microsoft's BizTalk server offers translation services thus enabling two organizations using different business protocols to transact business services over the Internet.

4.2.2.4   CommerceNet's eCo Framework

CommerceNet has chartered the eCo project, which is sponsored by Veo Systems, Inc. As the past couple of years has witnessed a tremendous onslaught of e-commerce based standards, eCo framework promises to provide a common framework for interoperability among various standards. These standards include OBI, OFX, EDI, BizTalk, and many others [6]. This facilitates interoperability between vendors engaged in various e-commerce activities. The framework will build on XML as one of the standards for information exchange. The eCo specification defines an e-commerce system into seven categories. This enables other e-commerce business systems to identify, discover, and interact with other e-commerce systems and ask questions such as [7]:

- What other businesses can I find?
- What services do they offer?
- What kinds of interactions do they expect?
- What protocols do they follow?
- Can our systems communicate?
- What application interfaces do they provide?
- Are our interfaces compatible?
- What information must we exchange?

4.3   Enabling ERP Systems and Legacy Applications for E-Commerce

E-commerce–enabling business processes require appropriate modifications to legacy/ERP applications and systems. E-commerce–enabling systems allow enterprises to offer self-service applications and systems to customers and business partners. Enterprises, for example, can open up their backend ERP and legacy transaction systems to enable customers to perform money transfer transactions and to order stocks through various information appliances. These information appliances access the backend systems via appropriate gateways. For example, a user calling through a telephone can be greeted by an IVR system that in turn can format appropriate transactions and route to the backends for execution.

E-commerce–enabled backends also streamline business processes and alter their workflow to enable corporate users to access applications and systems remotely when they are mobile. For example, sales staff in the retail industry can access and update corporate inventory systems when they are on the road. They can also access suppliers' backend systems to order items instead of going back to their offices to access those systems. Similarly, organizations can streamline business processes by delegating relevant activities to business partners. For example, organizations should not have the onus of maintaining external supplier's catalogs on their systems. This burden can be relegated to the supplier organization, which can access buyer organization's systems and maintain the catalogs.

Enterprises can e-commerce–enable backend systems by building appropriate Web-based interfaces. For example, users can use browsers to directly access a CICS-based application. Alternately, enterprises can build other interfaces to provide such an access. Enterprises, for example, can use an IVR-based interface to enable users to access backed applications (e.g., checking of order status). This section explores both options in further detail.

4.3.1   Web-Enabling Backend Systems

Providing appropriate Web interfaces is the most popular method for opening backend applications and systems for e-commerce. Enterprises have multiple options to Web-enable those applications. These include porting of Web servers on proprietary platforms, developing middle-tier functionality to map between browser and backend protocols, and interfacing e-commerce applications that have standard browser interfaces with backend applications. The following further explains these options.

4.3.1.1   Porting of Web Servers on Legacy Platforms

Figure 4.4 illustrates the scenario in which vendors Web-enable their proprietary applications by porting Web servers to run on the native backend application's operating system. This scenario therefore does not require a separate server in the middle tier. Instead, the Web interface on the proprietary system supports HTTP connections. For example, IBM provides a CICS Web Inter-

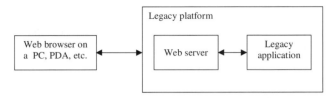

Figure 4.4   Porting of Web servers on legacy platforms.

face (CWI) that enables a browser to connect directly to a CICS program running on the OS/390 platform.

4.3.1.2   Vendor Interface Modules for Web-Enabling Legacy Platforms

In this scenario, vendors provide interfaces such as Java modules or dedicated servers that interface to their proprietary backend systems. This enables organizations to offer browser-based solutions without modifying backend applications. Figure 4.5 illustrates this scenario. One method involves mapping browser input to backend TP monitors' data and communication streams (e.g., SNA or TCP32) through an application server. These gateways thus mask the browser input and other Internet-based technologies from the legacy and ERP systems. For example, the CICS Transaction Gateway provides a terminal servlet that provides turnkey browser access to 3270 screens thus masking the details of the browser and Internet environment from the backend applications. Similarly, ERP vendors provide Web interfaces for their systems enabling users to access ERP application functions through Web browsers.

4.3.1.3   Interfacing E-Commerce Applications With Legacy Platforms

Enabling backend applications does not necessarily require providing a direct connection to the users. At times, e-commerce applications solicit user inputs, provide various business functions, and then interface with the backend sys-

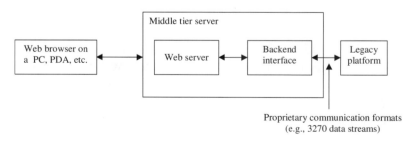

Figure 4.5   Interfacing with backends using middletier servers.

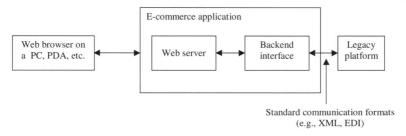

Figure 4.6   Interfacing with backends using e-commerce applications.

tems for specialized activities. For example, an organization may set up a selling solution, based on Netscape's SellerXpert project, that provides selling functions over the Internet. However, the enterprise may require interfacing it with backend systems to complete the selling loop. This therefore necessitates interfacing the e-commerce application to the backend systems. However, the user does not have to connect to the backend application directly. Rather the user interfaces to the backend application through the e-commerce application. For example, IBM's Commerce Integrator integrates IBM's e-commerce application suites with backend ERP and legacy systems. IBM also offers appropriate plugs (e.g., SAP ERP plug) for each of the systems that need to be interfaced with the frontend e-commerce solutions. Figure 4.6 illustrates this scenario.

4.3.2   Enabling Backend Systems for Nonbrowser-Based Access

Earlier solutions discussed the possibility of providing Web-based interfaces to backend systems. However, sometimes organizations empower customers to access backend systems using non-browser-based appliances and applications. The following discusses scenarios for opening backend systems for access through IVR, e-mail, and GSM telephones.

4.3.2.1   Interactive Voice Response (IVR)

IVR systems function by receiving user input either through the telephone's touch-tone keypad or user's speech input. IVR systems have preprogrammed scripts that guide users through a menu of choices. A typical flow to process a user request for inquiring an account balance would be as follows:

- The IVR collects user input. The IVR can collect user input by prompting the customer to enter their account number on the telephone's keypad or to "read" the account number. If the customer reads the account number, the IVR uses a voice processing system to convert the customer's spoken words.

- IVR forwards the input to a gateway that uses the customer input to formulate the backend transaction in an appropriate format.

- The gateway interfaces with the backend system using the backend system's communication protocol and submits the transaction.

- The gateway receives the input from the backend system and parses the input for the IVR.

- After receiving the input from the gateway, the IVR uses Text-To-Speech (TTS) technology to read the result to the user.

The design of such a system requires the following steps:

- Acquisition of an appropriate IVR system (or leasing IVR services through an NSP) that suits business requirements. For example, IVRs can process touch-tone input as well as users' speech input through appropriate speech recognition software.

- Developing business-specific scripts for the IVR. This involves developing menus and linking backend functions to menu selections.

- Interfacing the IVR system to the call center's PBX system for routing customer calls to call-center agents if the customer asks to be transferred to an agent.

- Interfacing the IVR with the gateway and developing custom program logic to interface with the backend systems for self-service transactions. For example, on detecting a user request for a money transfer, the IVR invokes the appropriate application on the backend legacy systems.

4.3.2.2   Intelligent E-Mail Processing

Similarly, an organization can deploy an intelligent e-mail gateway to process a user's request. An intelligent e-mail engine parses the content of the e-mail message and triggers a predefined response to the customer. In some cases, additional processing (or routing) of the e-mail message follows as well.

The following depicts the scenario for intelligent processing of e-mail. This scenario can apply to e-commerce transactions as well:

- A customer sends e-mail to a company (e.g., a customer requests mortgage information).

- An e-mail server receives the e-mail message.
- The e-mail server forwards the message to an intelligent e-mail gateway.
- The e-mail gateway parses the message and applies appropriate business rules (e.g., the e-mail gateway recognizes the intent of the message by parsing the subject title that includes the word *mortgage*).
- The e-mail gateway triggers an acknowledgment of the message and instructs the customer to wait for the necessary information.
- The e-mail gateway through its business rules identifies the appropriate server that stores the appropriate information and requests that information from the server.
- The server forwards the appropriate information to the e-mail gateway.
- The e-mail gateway composes a message with the relevant information and sends an e-mail to the customer.
- If the e-mail gateway does not identify an appropriate response for the customer, the request routes to an appropriate PBX to be forwarded to a live agent that handles the customer request.

4.3.2.3 Wireless Telephones

Chapters 2 and 3 highlighted the various characteristics of wireless telephones to handle data-related functions. For example, GSM telephones can send and receive SMS messages. Enterprises can exploit this feature to offer various e-commerce services for self-service transactions. The following depicts a typical process to handle such self-service transactions:

- A customer through a loaded application on the GSM handset sends a request (e.g., an account balance request) to a financial institution. For GSM handsets, the application is loaded on the SIM card. The application formulates an SMS message and sends it for processing.
- The GSM service provider receives this SMS message and routes this message to a special gateway that the organization has set up in cooperation with the GSM service provider.
- The gateway translates the SMS request into a transaction format and routes this transaction to the backend systems for processing.
- On receiving the response from the enterprise's backend system, the gateway maps the input to an SMS format and sends it to the user's handset through the GSM service provider's SMS infrastructure.

The design of this solution involves the following steps:

- Design of the application that is loaded on the SIM. Enterprises can develop this application using various SIM toolkits that enable development of such applications.
- Development of the gateway software that maps between the SMS messages and the enterprise's backend systems.
- Formulating mutual agreements with the GSM service provider to collocate the gateway at the GSM service provider's premises.
- Establishing appropriate network connections between the GSM service provider and the enterprise.

4.4   Knowledge Repositories

Knowledge management is an emerging discipline that focuses on leveraging an organization's information assets and intellectual capital to enable organizations to operate more effectively and differentiate themselves in the knowledge-based economy.

Data warehousing solutions provided enterprises with the first primary push toward the knowledge management paradigm. However, whether data warehouses provided knowledge or merely reflected various business process–related trends is a subject of extensive debate. Knowledge depends on intellectual capital that resides within the minds of people who run the organization. Most of the intellectual capital–related "data" flows through e-mail systems and other collaboration channels, which provide a medium for problem discussions

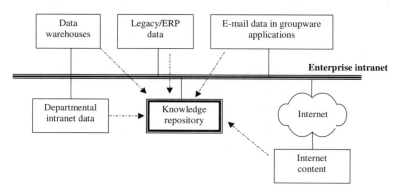

Figure 4.7   Building knowledge repositories.

and facilitate communication of solutions and other information. Furthermore, organizations have recognized that data warehouses that store operational data have to be looked at in the context of external data sources that include news, market analysis, and other relevant information to reap the real business insights that form the basis for true knowledge. The goal of the knowledge management thus transcends beyond simply integrating all knowledge bases and providing their access to knowledge workers. As *Harvard Business Review on Knowledge Management* [8] mentions, "creating new knowledge is not simply a matter of 'processing' objective information. Rather, it depends on tapping the tacit and often highly subjective insights, intuitions, and hunches of individual employees and making those insights available for testing and use by the company as a whole."

The past few years have seen various organizations establish an e-commerce presence to differentiate themselves in the market place. However, these trends (e.g., setting up a Web site to sell products or services) have shifted from providing competitive advantage to becoming a competitive necessity. The challenge of the new millennium lies in strengthening business and customer relationships through e-commerce computing by offering personalized solutions and services. A key ingredient of this strategy will be to intelligently analyze and harness various types of information that lies within the organization to offer better and innovative solutions to customers. This applies to intrabusiness and business-to-business paradigms, too, as enterprises will have to streamline their internal processes and establish better linkages and relationships with their business partners, respectively. Intelligent harnessing of organizational knowledge and the establishing of an appropriate infrastructure are thus vital to react to business pressures and external market forces.

An appropriate knowledge management infrastructure enables the sharing of organizational knowledge. Unfortunately, most of this knowledge resides in places that are difficult to access by knowledge workers. For example, a networking specialist may send e-mail to a colleague about the special tricks involved in the configuration of a certain software package. This information resides on a server or user's mail box and other users attempting to resolve such issues will waste enormous time (e.g., making telephone calls) to resolve this problem. The availability of a knowledge repository that indexes this topic for retrieval would benefit everyone within the organization. The establishment of a knowledge management infrastructure empowers thus knowledge workers in gaining insights for reshaping business processes and taking effective and timely decisions. Sharing of ideas in real time leads to making vital decisions and devising innovative solutions

Establishing an effective knowledge management infrastructure requires the following steps:

- Identifying sources of information within the organization;
- Capturing information from those sources in knowledge repositories and continually updating that information;
- Organizing information to enable meaningful and faster access;
- Providing the tools to access that information.

The remainder of this section discusses these steps in further detail.

4.4.1   Identification of Knowledge and Content Sources

The first step in establishing intelligent knowledge repositories is to identify all the sources of information internal and external to the enterprise. Figure 4.7 illustrates the various sources of information that enterprises can cull to build appropriate knowledge repositories within the enterprise. The following describes those information sources and the value that they hold within the organization.

4.4.1.1   Operational Data Stored in Legacy and ERP Systems

Legacy and ERP systems hold operational information associated with core business processes of the enterprise. Within the knowledge management context, ERP data is vital for analyzing trends, identifying patterns, forecasting, and identifying areas for improvement. Within the e-commerce arena, this information can be vital in developing new solutions. For example, ERP data can be tapped to study specific customer demographics that can enable an enterprise to better position its product and service offerings to its customers.

4.4.1.2   E-Commerce Sites

Responding to the skyrocketing demands of e-commerce, enterprises are deploying a large number of commerce sites. In most cases, especially in larger enterprises, separate departments deploy these solutions and collect enormous data on customers, including demographic information and customer interaction behaviors. Collectively, this data helps enterprises offer better e-commerce solutions to their customers.

4.4.1.3   Internet and Web Information

The Internet and the Web are a vast repository of information. Collectively, this information can be extremely useful in enabling a knowledge worker to make effective decisions, reshape new business processes by accessing news, analyze trends, and access relevant information.

4.4.1.4 Communication and Collaboration Data

Communication and collaboration data provides people with insights and judgments necessary to solve problems and issues. For example, e-mail is a vital element in the building of knowledge bases. The primary reason is that e-mail information reflects human insights and experiences, the basic tenets of a knowledge management strategy. Organized more effectively, e-mail information thus can be a valuable source for the knowledge worker.

4.4.1.5 Data Stored in Isolated Systems

The ease of installation of client/server systems has enabled various departments within organizations to deploy isolated systems on their intranets. This data can prove to be useful in studying business trends.

4.4.2 Knowledge Capture

Subsequent to the identification of all data sources within the enterprise, the next step is to extract data from sources that store information in multiple formats. Data extraction poses numerous challenges as information stored within the organization is available in various formats that include the following:

- Nonrelational data stored in legacy files (e.g., VSAM files, IMS, and IDMS databases);
- Relational data stored in an organization's databases (e.g., DB2, Oracle, and Sybase);
- Various office-document formats (e.g., Microsoft Word, Acrobat files, and other popular office software).

Various means exist to extract these data sources, and a few are discussed in the following sections:

4.4.2.1 Data Extraction From Standards-Compliant Databases

ODBC and JDBC are some of the database interface standards that enable data extraction from popular databases such as DB2, Sybase, and Oracle. Extracted data or respective information links can be stored in the knowledge repositories. For example, Oracle offers data adapters that extract data from ODBC data sources and ERP systems.

4.4.2.2   Data Extraction From Legacy Databases

A large installed base of legacy databases that include file formats such as VSAM, IMS, and IDMS still exists within organizations. Various vendors have developed tools that understand the native data formats of these nonrelational and legacy data sources and extract data and import them into relational or newer format databases. For example, tools by CrossAccess enable users to run SQL queries and extract data from nonrelational databases and import it to RDBMSs such as DB2, Informix, Sybase, and Oracle. Platinum's InfoTransport is another tool that retrieves data from MVS/DB2 systems in an ASCII format. This data can then be exported to any database that can import ASCII data.

4.4.2.3   Converting Data in Standards-Based Formats

With XML becoming a popular industry standard, numerous vendors are offering tools that convert information stored in various formats to the XML format. This facilitates effective querying, retrieval, and distribution of information. Various vendors are also XML-enabling backend ERP systems and applications. This facilitates end-to-end standardization of information flow and better integrates various steps of the business processes. For example, StreamServe, Inc. offers solutions that facilitate receiving data feeds from ERP systems (e.g., invoices or statements) and maps them into appropriate XML formats.

4.4.2.4   Data Extraction From Structured Documents

Instead of extracting data from one data source at a time, organizations can also explore tools that extract data from multiple data sources. For example, Knowledge Server from Autonomy, Inc. offers a knowledge management solution in which information can be loaded into the server from multiple sources. The company offers modules that extract data from multiple sources such as Notes databases, Microsoft Exchange (e-mail), and others. The organization also provides the necessary tools to build modules to enable data extraction from various other sources.

4.4.2.5   Focused Data Extraction From the Internet and the Web

The Internet/Web is laden with tremendously useful information. Corporate users can search the Internet to find relevant information. However, this method can be very ineffective as public search engines return enormous results and the content may not be relevant. In one popular solution, site administrators establish links of valid sites in a directory and enable workers to wade through those sites. Various products in the market extend this concept by automatically maintaining these directories of useful information links and sites. For example, Inktomi Corporation's Directory Engine enables administrators to build customized directories of information. The Directory Engine uses Concept Induction technology to automatically analyze information stored in various sources specified by the administrators and accordingly updates and categorizes the information for future retrieval.

4.4.2.6 Capturing Paper-Based Assets Into Knowledge Repositories

Paper-based assets form a major source of knowledge within organization. This includes analysis reports, solutions, and other vital information. Various technologies such as digital scanning enable the capture of such paper-based assets into standard formats. Advances in scanning technologies (e.g., optical character recognition, or OCR) allows the recognition of text, spreadsheets, tables, and figures and allows their storage in popular file formats, which then facilitates editing and storing of those documents in knowledge repositories. For example, OmniPage Pro is a popular OCR software that facilitates such functions. However, OCR requires extensive error checking because it involves multiple steps to scan documents that include character recognition and analysis of the entire image. The final output of the document is therefore not completely error-free and requires editing and error correction.

4.4.3 Knowledge Organization and Access

A critical element for establishing an appropriate knowledge management infrastructure is to populate knowledge repositories with information that is relevant for various business conditions. For example, an organization may build a knowledge repository for its research scientists that contains up-to-date information from various research sites. Similarly, a shopping Web site may provide its customers with the ability to access detailed information on certain products by enabling them to search through various categories of information that the enterprise catalogs in its knowledge repositories.

Similar to other applications, organizations can develop knowledge repositories and knowledge management solutions or acquire external packages. The specific need depends upon the business requirements, the volume of data, customer demographics (internal or external), and so on. In any case, selection of the appropriate database platform to store knowledge, data interchange format, and query mechanisms for retrieving data are the critical elements of any knowledge management solution. Subsequent sections review these three elements and discuss various related issues.

4.4.3.1   Extensible Markup Language (XML)

The underlying challenge behind the building of a knowledge management solution is the ability to represent data in standard formats that can then be queried and presented to users. Depending on the data types, diverse data sources have their own way of storing and representing information and presenting it to the user. For example, RDBMS data can be queried by the user, whereas content stored in a Microsoft Word format file cannot be easily queried or retrieved by the user. Therefore, as knowledge management involves unifying various data sources into one repository to enable widespread access, the need for a standardized information exchange language becomes evident. The XML specification holds that potential in devising effective knowledge management solutions and is being widely adopted as a standard for information representation and exchange. XML has enormous benefits, some of which are as follows:

- XML separates content from styling information, unlike HTML, which blends both in its syntax. XML specification highlights the content of the document as opposed to HTML, which focuses on displaying the document. Documents can be indexed and thus can be searched. For example, supplier catalogs in an XML format allow buyers to search against the catalog entries, as opposed to just viewing the catalog. Thus, decoupling of content and styling information provides the ability to search all information stored within the XML document.

- Information from various sources (e.g., databases, ERP systems, legacy applications) can be described using XML and retrieved based on the XML syntax.

- Various applications and systems can communicate using XML. Initiatives are already underway to use XML-based information exchanges for business transactions instead of using EDI formats.

- XML highlights the content in databases and documents. Users and applications can selectively update content without updating the entire document.

- XML thus allows sharing of information within the organization and across value chains.

4.4.3.2   Database Platform

Choosing the right database platform is crucial for building knowledge repositories. Databases from various vendors employ different database technologies for the storage of data. Popular databases include flat file, network and hierarchical databases, relational databases, object-relational databases (ORDBMS) and object-oriented databases (OODBMS). Each type of database suits different applications that may vary in their data storage requirements, data replication requirements, and query characteristics. The choice of a particular database should therefore consider an enterprise's knowledge management requirements. The following presents various considerations relevant to the choice of appropriate database technology for an enterprise's knowledge management solutions.

Flat files were among the first ways to store an enterprise's data. However, flat-file databases lack the security and transactional reliability features that mission-critical applications require. Hierarchical and network databases, such as IDMS and IMS, emerged to handle these requirements and still enjoy a large installed base within enterprises. Later, RDBMSes surfaced and became very popular due to their simplicity in representing data in a tabular fashion and due to their strong query and retrieval features. RDBMSes therefore enjoy the largest installed base of repositories and continue to enjoy large user support. RDBMSes support various data types including character-string data types, date formats, and floating-point data. Examples of popular RDBMSes are Oracle 7.x systems, Sybase system 10/11, and IBM's DB2.

RDBMSes do not inherently support data types associated with richer data content such as multimedia data types (e.g., video, audio, images). E-commerce applications, on the other hand, require the new multimedia data types to offer rich and robust applications and content to their users. For example, vendors can offer captivating product demonstrations and enable users to customize products by interacting with product images. Such applications require robust database technologies and object-oriented application frameworks that support multimedia data storage, query, and retrieval.

ORDBMSes surfaced to resolve issues associated with storing of complex data types. However, as the core of an ORDBMS still consists of a relational model, vendors employed several translation techniques to store multimedia and object data types within the relational model. Although this enables the storage of such complex data types indirectly, it gravely impacts performance for large installations that require superior data retrieval capabilities. Examples of ORDBMSes include Oracle 8.x systems and IBM's Universal database.

Object-oriented DBMSes (OODBMS) are the latest trend in database technologies and directly resolve issues related to representing multimedia objects and storage of complex data types. OODBMSes enable storage of multimedia objects and provide the best performance for the storage and retrieval of multimedia objects. With the increasing popularity of object-oriented programming languages and resultant applications, OODBMSes provide the ideal foundation and prove to be the best fit for data requirements of such applications as OODBMSes provide results of queries in object formats. This is because object-oriented applications do not have to perform complex operations between object-oriented data that they handle and the tabular data represented by the RDBMS. An OODBMS coupled with object-oriented middleware and applications can provide the new foundation needed to support scalability and flexibility demanded by the new generation e-commerce systems. Popular examples of OODBMSs include Gemstone, Jasmine, O2, Objectivity/DB, and Versant ODBMS.

Based on their ability to optimally store various data types and other related features, it is quite common to see a broadening support of OODBMSes for the new knowledge management solutions. For example, Glyphica, a vendor that delivers portal solutions, builds on Versant's OODBMS technology to enable handling of complex data structures (e.g., XML data), enhanced navigational features, and large volumes of data required for e-commerce applications [8]. Enterprises, therefore, should consider OODBMSes for building customized knowledge management solutions.

4.4.4   Using Search Engines for Knowledge Retrieval

Search engines are vital technology tools in the implementation of knowledge management solutions within the enterprise. Enhanced search capabilities allow users to tap into the vast online and offline information sources and knowledge repositories. Enhanced and robust retrieval capabilities coupled with intelligent categorization of information provides the users with the necessary information, knowledge, and insights required to further various business processes.

Within the context of e-commerce, search engines established their importance within the realm of public portals, which provided users with the ability to scour through the vast information base of the Web. However, with the emergence of enterprise portals, organizations found an equal need to maximize the potential of the stored information repositories and provide their users with the necessary tools to transform those repositories into knowledge bases. (Chapter 5 discusses portals in more detail.) Search engines provide timely access to information, facilitate enhanced decision making capabilities, and link users to various information sources and/or populated knowledge repositories.

Initially, search engines were simple and provided the ability to search the text and HTML files published on various Web sites. However, numerous nodes of the Internet, the Web, and organization intranets populated with a wide array of content have triggered the emergence of sophisticated search engines that provide the ability to search through all types of information. For example, users can use search engines to peruse news, stock market analyses, and other useful information. Almost all public-portal sites offer search engines of varying capabilities to their visitors.

Search engines are also extremely useful on Web sites that provide their visitors with the ability to search for information that would otherwise be cumbersome and, in certain instances, impossible to retrieve. Providing access to sophisticated search engines maximizes the value of information and empowers an organization's users to obtain valuable information.

The following delineates various features of search engines and the value they contribute to the building of inquiry-type enterprise applications.

- *Search capabilities through various sources:* The search or retrieval systems should be able to connect to diverse sources of information to retrieve the desired results. The various sources include the Internet; knowledge repositories on intranets, extranets, legacy databases, and digitally scanned databases; and others.

- *Search capabilities of various types of content:* The search engine should extend beyond text-based systems to incorporate searching of various types of content that include video, images, audio, and documents of popular word processing software packages. Although there are numerous challenges in the search and retrieval of these types of content, advancement in technologies, databases, and algorithms facilitate such search requests. For example, search engines from Excalibur Technologies Corporation's RetrievalWare product family [9] enable searching of graphics and illustrations, full animation, and motion video content. The search criteria for these content formats is not limited to a description of the content but allows searching based on color, texture, and shape of images. Various algorithms for pattern recognition and image analysis facilitate these searches. Popular applications of these search features enable users to submit images as search queries and retrieve images that closely match the original image based on various characteristics of the image. Besides, Verity, Inc.'s search engine called Verity Spider searches through information that is stored in multiple file formats, including PDF, Microsoft Office, Frame-Maker, and other document formats.

- *Flexible input handling:* Search systems should be able to collect a variety of input from users to facilitate more focused searches. For example, the systems should be able to specify the source of data (e.g., specific site search or an entire Web search), date requirements, relevance of document, desired content types (e.g., search only Microsoft Word documents), and so on.

- *Intelligent categorization of search results:* Categorization refers to the grouping of appropriate subject matter into groups that facilitate human understanding of the retrieved content. The search engine, for example, should be able to provide intelligent categorization of information based on the source of information, such as company databases, Internet, business partner's intranet, vendor's sites, and so on. Also, the search engine should have the capabilities to present a sorted view of information based on date, relevance, and other criteria. For example, Inktomi Corporation has introduced a new technology called Concept of Induction that closely mimics the human model of understanding of information. The powerful algorithms behind this technology facilitate analysis of vast amounts of information on the Internet and intranets and presentation of pertinent documents in an intelligible format.

- *Support for enhanced user input:* This refers to the capability of the search engine to go beyond searching for simple words to varying degrees of sophisticated search abilities including search on phrases, questions, the use of rules for filtering (Boolean), pattern and concept matching, and so on.

- *Ability to function appropriately on high-traffic sites:* For incorporating search functionality on intranets, organizations need to ensure that their search engines can handle the user load. Assessing this capability is more vital if the search capability is part of the staff's portals for mission-critical applications and not merely to enhance the knowledge management culture of the organization.

- *Focused searches:* Organizations that need to do research to support decision making cannot afford to flood their knowledge workers with results. To enable such organizations to find appropriate results, various vendors have developed products that enable searches against prequalified sites. These search engines then search only those sources of information and present information to their users in an intelligible format. For example, healthcare organizations can offer such solutions to their workers to enable them to search health sites. IBM's Intelligent Miner for Text is one tool that enables organizations to selectively index sites and build vertical search solutions.

4.5   Live Agent Services

Earlier sections discussed the various ways in which enterprises are making their backend systems accessible to their customers and business partners. This enables enterprises to reduce the number of calls from their customers to the enterprise's call centers. Customers, for example, can execute self-service transactions such as stock ordering, money transfers, bill payment, and other similar transactions. Furthermore, the increasing emphasis on customer service is driving organizations to build better relationships with customers through different information appliances using multimedia channels. Organizations are building multimedia call centers to address issues with customer relationship management (CRM) processes and to differentiate themselves in providing state-of-the-art customer services to their customers. For example, an organization may provide its customers live access to its agents through telephones, Internet voice, Internet chat, and video sessions. Establishing an infrastructure to support multiple information appliances using diverse media types involves many issues. This section discusses relevant issues for the establishment of such an infrastructure.

Enterprises establish call centers for many reasons, some of which are as follows:

- *Handle requests that are not possible to process through a self-service infrastructure:* Enterprises are undergoing a transition period to e-commerce–enable backend systems for self-service, and all customer requests are not feasible for fulfillment through a self-service infrastructure. Therefore, enterprises have to establish call centers to host calls from their customers to fulfill certain request—for example, not all banks offer Internet banking services. However, most banks offer common banking services such as automated account balance inquiries either through telephone channels or by enabling customers to call an enterprise's call center to inquire about such information. Not all customers have equal access to electronic channels to enable them execute self-service transactions. For example, although most countries have access to the Internet, a large segment of users still use the telephone to reach an enterprise's call center for the fulfillment of various requests. Enterprises therefore have to establish call centers to cater to these requests.

- *Transaction sensitivity:* Certain business services require higher levels of customer authentication, thus forcing enterprises to require their customers to call an enterprise's service representatives before requesting such services. For example, entering stock orders through e-mail is unreliable due to technology factors. Similarly, using the telephone channel to automatically enter stock orders is risky due to security issues inherent in the telephone channel. Therefore, most financial institutions that do not offer stock ordering services through the Internet require customers to speak to a service representative to enter orders. Service representatives then authenticate the customer by asking certain questions and execute the stock order on the customer's behalf.

- *Heeding to customer preferences:* Although organizations are establishing various electronic channels (e.g., IVR, e-mail) to receive and process customer input, certain customers still feel more comfortable interacting with live service representatives.

- *Complexity of systems and transactions:* Fulfillment of certain customer requests necessitates accessing numerous backend applications and systems. This could be due to the complexity of the transactions or the specific implementation of business processes within the enterprise. Opening those systems for self-service through various electronic channels could be quite challenging. This therefore requires that customers interact directly with the service representatives to request fulfillment of such transactions. For example, an agent's live intervention is usually necessary to review a credit increase request for a customer who has a controversial credit history.

Establishing call centers to host calls through various electronic channels involves three steps. The first step entails establishing the necessary technology infrastructure to receive customer calls through various channels (e.g., Web-enabling a call center, receiving calls through video kiosks, Internet chat calls). Second, enterprises have to establish a proper call routing infrastructure that intelligently routes calls to service representatives. Finally, enterprises have to empower service representatives with appropriate tools and information to enable them to process and fulfill customer requests. This section discusses the three steps in further detail.

4.5.1   Enabling Call Centers for Various Electronic Channels

This section discusses the steps required to enable call centers to handle customer calls through various channels. The section focuses on the popular electronic channels.

4.5.1.1 IVR

Interactive voice response (IVR) systems receive user input either through the telephone's touch-tone keypad or speech input. IVR systems have preprogrammed scripts that guide users through a menu of choices. A typical flow to transfer a user request for inquiring about an account balance would be as follows:

- The IVR collects user input, which indicates the customer's desire to speak to a live agent.
- IVR transfers the call to the call center's PBX unit, which selects an agent to handle the call.
- The IVR transfers the user-input data (e.g., account number) to a CTI middleware application (e.g., Quintus CTI), which uses the data to retrieve the appropriate user's profile from the backend databases.
- The PBX rings the agent's extension. On answering the telephone call, the CTI software sends the customer profile information in the form of a screen pop-up on the agent's screen.

The design of such a system requires the following steps:

- Acquisition of an appropriate IVR system (or leasing IVR services through NSPs) that suits business requirements. For example, IVRs can process touch-tone input as well as speech input through appropriate speech recognition software.
- Developing business-specific scripts for the IVR. This involves developing menus and incorporating various functions within the menus.
- Interfacing the IVR system to the call center's PBX system for routing customer calls to call-center agents if the customer requested to be transferred to an agent.
- Interfacing the IVR with the backend systems for self-service transactions. For example, on detecting a user request for money transfer, the IVR should invoke the appropriate application on the backend legacy systems.

4.5.1.2 Internet Telephony

With advances in Internet telephony, enterprises are enabling their customers to speak to call center agents through the Internet telephony channel. For example, customers that have access to multimedia information appliances, such as PCs, can click a button on the Web page that appears at the organization's e-commerce site and speak to the call center agent without leaving the Web page. The typical process to establish the Internet telephony channel includes the following steps:

- The user browses the organization's e-commerce site and elects to speak to a customer service representative by clicking on a button that appears on the Web page.
- The button click invokes the appropriate software (e.g., NetMeeting) on the user's PC.
- The button click invokes a parallel process by downloading an applet from the organization's Web site.
- The applet connects to a backend process thus informing appropriate CTI processes about the incoming telephony call.
- The backend process routes the call to the PBX requesting the availability of a call-center agent. The PBX queues the call for the next available agent.
- The NetMeeting software establishes a packet-switched call from the user's PC to an appropriate access gateway (e.g., Internet telephony gateway).
- The Internet telephony gateway opens a connection to the PBX.
- On routing the call to the agent, the voice call flows from the user's PC to the call-center agent. The appropriate access gateway translates the calls between the packet-based call (from the user's PC to the access gateway) and the circuit-switched call (between the access gateway and the PBX connected to the call-center agent).

The design of an Internet telephony channel requires the following steps:

- Acquisition of an appropriate access gateway (e.g., Internet telephony gateway).
- Development of the necessary middleware software that routes the calls from the user to the right agent.
- Design of the network to accommodate complications surfacing from the placement of the access gateway, the enterprise's firewall, and other routers and switches.
- Downloading of the appropriate telephony software at the user's PC if the user does not have it installed.

- Development of software that embeds the necessary controls (e.g., push-button) on the organization's Web site.

4.5.1.3   Internet Chat

Internet chat or Internet Relay Chat (IRC) refers to a communication method in which two parties communicate by entering text on a small screen. This text is then transmitted to the other party through the Internet. The process flow for an IRC is as follows:

- The user browses through the organization's e-commerce site and elects to engage in a chat session with a customer service representative by clicking on a button that appears on the Web page.
- The button click invokes a process by downloading a chat applet from the organization's e-commerce Web site that will take user input (chat) and relay to the agent.
- The button click also downloads an applet that connects to a backend process thus informing appropriate CTI processes about the incoming IRC call.
- The backend process routes the call to the PBX requesting the availability of a call-center agent. The PBX queues the call for the next available agent.
- The chat applet software establishes a packet-switched call from the user's PC to the call-center agent's browser through the Internet.
- On routing the call to the agent, the chat call (text) flows from the user's PC to the call-center agent.

4.5.1.4   Video Kiosks

Video kiosks enable customers to access an enterprise's call-center agents through a video session. Video kiosks transmit the agent's video image to the customer. Similarly, a video camera mounted on the kiosk transmits the customer's video image to the call-center agent. Organizations place video kiosks in public places and enable customers to touch an organization through video sessions. For example, various banks have set up video kiosks in public places such as shopping malls and enable customers to invoke a video session with an agent.

A typical video session involves the following steps:

- A customer approaches a video kiosk and invokes the video application installed on the video kiosk by touching the screen.

- If the video kiosk connects directly to the call center, the kiosk network software establishes a network connection between the video kiosk and the call-center agent's PC.
- If the video kiosk connects to the call center through routing software, the PBX queues the call for the next available agent.
- On connecting to the agent's station, the customer can interact with the agent through video and audio sessions.

The design of a video kiosk involves the following steps:

- Acquisition of video kiosks.
- Selection of networking connections that can handle video sessions between the agent and the customer. ISDN is a popular choice for such connections because it can handle video and audio calls.

4.5.1.5   E-Mail Processing

Enterprises set up e-mail solutions to enable their customers to send e-mail requests. Users can send e-mail to call centers to request information on products and services or account-specific responses. The following describes two different architectures for handling customer e-mail messages by installing appropriate e-mail gateways. Enterprises can merge the two architectures to deliver one robust solution for handling customer e-mail.

Simple E-Mail Routing

Simple e-mail routing involves receipt of the appropriate message by an e-mail server that simply stores the sender's e-mail message in a mailbox that the recipient retrieves by using an e-mail client application. To handle customer e-mail, for example, an agent can sign on to the server and respond to all messages sent to the customer-feedback@enterprise.com address.

Intelligent E-Mail Routing

Intelligent e-mail routing involves receiving a user's e-mail by an intelligent e-mail gateway, analyzing the subject header of the e-mail, and routing it appropriately to a resource that will appropriately fulfill the e-mail request. In typical cases, an intelligent e-mail gateway receives the recipient's e-mail and performs the following steps:

- A customer sends e-mail to a company (e.g., a customer requests mortgage information).

- An e-mail server receives the e-mail message.

- The e-mail server forwards the message to an intelligent e-mail gateway.

- The e-mail gateway parses the message and applies appropriate business rules (e.g., the e-mail gateway recognizes the intent of the message by parsing the subject title that includes the word *mortgage*).

- The e-mail gateway routes the e-mail message to an intelligent PBX (e.g., Lucent Technologies' DEFINITY product). The e-mail groupware software through its business rules identifies the department best qualified to handle mortgage specific requests and routes the message to the appropriate PBX.

- The PBX queues the e-mail call for the next available skilled resource (an agent best qualified to handle mortgage types of requests).

- The PBX routes the e-mail to the next available agent within that skill category.

- The service representative appropriately responds to the e-mail (the agent forwards the appropriate information to the customer and closes the open request).

4.5.2 Establish Call Routing Infrastructure

Earlier sections discussed methods for routing various types of calls to call-center agents. Routing calls involves matching customers with service representatives who can effectively fulfill customer requests. Customer matching involves various parameters, some of which are as follows:

- For an ordinary request, customer matching may depend on the specific function that the user wishes to execute. For example, a customer may choose a certain option to request entering of stock orders. In this case, the routing engine routes the call to the service representative that is skilled to handle stock orders.

- Customer matching may depend on the user's profile. For example, upon receiving a user's input for an account number, the CTI middleware may detect the customer to be a high-profile customer. In this case, the routing software will route the call to special agents skilled to handle special customers. To cite another example, the CTI middleware may detect that the customer has defaulted on his payment and therefore will route the call to the credit department.

- Customer matching may depend upon the specific electronic channel that the customer is using to contact the call center. For example, handling of Internet chat, telephony, and video kiosk calls requires agents to be skilled in handling such calls and requires their stations to be equipped with appropriate hardware and software to handle such calls.

Handling of such intelligent routing depends upon the CTI middleware and the PBX that routes the calls to the call center agents. The PBX can be set up by defining agents with appropriate skills, and incoming calls can be designated to be matched against specific agents. CTI middleware, on the other hand, receives the customer input (e.g., entered account number), retrieves the customer's profile from the backend database, and pops up a screen with customer information on the agent's screen as the PBX routes the call to the agent's workstation. This enables the call-center agent to recognize a customer and engage in a friendly interaction with the calling customer.

4.5.3   Empowering Service Representatives With Appropriate Tools and Information

To enable call-center agents to handle customer calls effectively requires empowerment of the call-center agent's station with appropriate hardware and software. The various components that are required on the agent's station to enable them to service calls from customers include the following:

- *Software to handle calls through various electronic channels:* To receive calls through various electronic channels such as e-mail, Internet telephony, and IRC, and so on, the agent's workstation should be installed with appropriate software and hardware. For example, for the electronic channels discussed earlier, the agent's workstation would require an Internet browser and an Internet telephone. Similarly, to handle calls through a video kiosk, the agent's station should have a mounted camera to relay the agent's video image to the customer's video kiosk.

- *Access to backend systems and databases:* Agents should have access to backend systems and databases to enable them to execute transactions on the customer's behalf. For example, agents of a shopping organization should have access to the appropriate order entry system to enter customer orders as customers call in to request orders.

- *Trouble ticketing software:* The agent should have appropriate software to log customer's requests that may not be fulfilled immediately and that require higher-level management's intervention or further research. The trouble-ticketing software logs customer issues, assigns them to appropriate agents, and on resolution interfaces with CTI software that triggers outgoing calls to customers indicating the status of resolutions.

Notes and Web Sites

[1] Although client/server computing still forms the foundation of e-commerce–based computing, the term here is used to signify the client/server era that predates e-commerce computing.

[2] www.intervoice.com

[3] www.vocaltec.com

[4] www.ecitele.com

[5] www.oracle.com

[6] www.commercenet.com

[7] eco.commerce.net/specs/index.cfm

[8] *Harvard Business Review on Knowledge Management.* HBS Press, 1998, page 24.

[9] www.versant.com

[10] www.excalib.com

Additional Web Sites

www.lucent.com
www.microsoft.com

5

Establishing E-Commerce Application Access Infrastructure

Chapter 4 discussed applications that enterprises develop or acquire to enable their core business processes. The chapter highlighted technical and business strategies that enterprises can pursue to enable e-commerce business processes. However, the ultimate objective of an enterprise's IT strategy is not the mere accumulation of a robust applications portfolio. Rather, the paramount goals are to maximize business value (Chapter 1 highlighted relevant indicators), minimize operational costs, and maximize availability of business operations.

The e-commerce computing paradigm holds enormous potential to help enterprises reap maximum value from their IT investments by facilitating exposure to relevant information and IT-enabled services. Portals, for example, help enterprises match user's (internal and external) personal and business needs with appropriate information and applications. Similarly, extranets facilitate maximum collaboration among internal and external users through various network segmentation techniques, while systems and network management strategies enable enterprises to minimize operational costs and help avert business unavailability.

The nature of e-commerce computing has sparked an entire industry of service providers who offer various IT services to enterprises, who then integrate them into their e-commerce systems. This presents enterprises with outsourcing opportunities and lets enterprises focus on their core competencies (business processes and content for e-commerce). These service providers offer numerous advantages for enterprises including reliability of services offered and infrastructure cost savings. The time has marched forward from the era when a firm could depend solely upon its internal processes and technological infrastructure for the deployment of end-to-end solutions. In the e-commerce era, enterprises collaborate with various entities to build cost effective and efficient e-commerce systems.

Developing end-to-end e-commerce systems thus requires intelligent synergies with these service providers. This chapter focuses on strategies and related issues in engaging with these service providers to deploy intelligent e-commerce systems.

5.1 Extranets

Extranets extend an organization's intranet to include strategic business partners in order to facilitate work collaboration and information sharing. Extranets enable the reengineering of business processes by bringing relevant entities together to collaborate and engage in business transactions more intelligently. An extranet application can also provide links to other related sites, thus providing its users with additional information. For example, banks have deployed extranets to engage in transactions with various consumers. Similarly, they deploy extranets to collaborate with the businesses they serve and share information with other banks. Manufacturing organizations deploy similar extranets to streamline their supply-chain processes and activities.

The concept of extranets surfaced based on the premise that an intranet is an internal network freely accessible to an enterprise's employees, and access to the intranet's resources by external users requires regulated access controls. An extranet, therefore, referred to the appropriate physical and virtual segmentation of the intranet (at a network and application level) required to achieve that objective (grant controlled access to external users while leaving free access to internal employees). This concept held true during the initial days of intranet implementation. However, as enterprises migrated mountains of data and applications over their intranets, they sensed an equal need for implementing such security controls for their internal employees as well. For example, the research department wanted to limit access to their data and findings to the engineers working on the project. Those engineers could access that data locally (when physically present within the confines of the enterprise), or remotely. Providing such access therefore requires implementation of extranet technologies similar to what is required for granting such access to external customers and partners.

This shows that although extranets find a majority of their references and use in the context of an enterprise's external relationships, their use is equally valid for internal implementations. An extranet therefore provides a specific group of users controlled access to an enterprise's applications and content.

Table 5.1

Comparing Intranets and Extranets

Characteristics	Intranets	Extranets
Users	Internal	Users belonging to an enterprise's entire value chain
Technologies	Internet-based technologies	Similar to intranets with implementation of security technologies such as VPNs.
Nature of content	Internal to the enterprise; all-purpose suiting organization's various departments	Mutually applicable to internal users and strategic partners and customers; focused and vertical, suiting the business processes fulfilled by the extranet
Content publishing environment	Usually uncontrolled; various departments on the intranet attach servers and publish desired content	Very controlled and regulated per preestablished guidelines
Security	Minimal	Stringent security controls
Administration	Performed through internally formulated procedures	Requires formal approved processes mutually agreed to by all parties

These applications or content could be of any form, as defined in earlier chapters. The group of users could include all users within the enterprise's value chain (e.g., an enterprise's internal employees, business partners, or external customers).

Intranets and extranets have varying characteristics. The different customers that extranets serve, extranet content, and regulated access differentiates an intranet implementation from an extranet. The following delineates the primary differences, which are also summarized in Table 5.1:

- Intranets are usually accessible to all of an organization's employees. Employees who have on-site or remote access to the intranet can use all available content and services, unless specifically restricted. On the other hand, users accessing extranets have restricted permissions to access its data and services.

- Extranets are similar to intranets in that they are built on technologies similar to the Internet. Extranet development, therefore, requires the same tools and technologies as intranet development. However, extranets employ additional security technologies—such as Virtual Private Networks (VPNs)—to regulate access to the network.

- Extranets are complex due to the granular security levels required for different security parameters, such as authentication, authorization, and so on, unlike intranets, which do not employ such stringent security controls.

- Extranets usually provide vertical content geared for a specific group of users. For example, a manufacturing extranet will host content (information and applications) relevant to the manufacturing industry.

- Extranets usually serve mission-critical business applications and require appropriate business and technology infrastructure to maintain availability and other service levels.

- Management of extranets requires predetermined and preapproved processes and guidelines.

Extranets are the primary enablers behind the success of business-to-business e-commerce. The following summarizes some of the advantages of deploying extranets:

- *Streamline operations and thus reduce cycle times:* Extranets streamline an organization's supply-chain processes. For example, suppliers can access a manufacturing organization's databases to get appropriate insights into the inventory levels, thus enabling them to deliver appropriate parts and supplies in due time. This can also enable the supplier to schedule its production adequately and thus reduce associated cycle times.

- *Reduction of costs:* Extranets facilitate reduction of costs for all parties. For example, certain suppliers spend enormous amounts of money to publish paper-based catalogs of their products. In many cases, these catalogs require frequent updates and thus incur huge publication and distribution expenses for the supplier. By deploying appropriate extranets and publishing information on their extranets, suppliers reduce such overheads. By further extending the extranet services to include ordering and bill payment, both suppliers and manufacturers reap the benefits of reduced operational and publication costs.

- *Increase in EDI and other business transactions:* As explained in earlier chapters, smaller organizations that could not benefit from the EDI infrastructure due to its higher costs can freely use the Internet for their business transactions by deploying appropriate extranets.

- *Secure transfer of transactions:* One of the primary advantages of VANs in an EDI setting was its ability to provide a secure mechanism for information exchange between partner organizations. However, the maturity of Internet security technologies and associated products enables a secure exchange and transmission of mission-critical and information-sensitive transactions over the Internet. The advancement of security technologies also enables the establishment of nonrepudiation measures that are usually involved in sensitive and high-dollar transactions. For example, members of an extranet organization may receive pricing information, or confidential information on a certain industry alert.

- *Interactive EDI transactions:* Traditional EDI transactions were batch oriented. Organizations accumulated transactions on their side before sending them in batches to partner organizations through the VANs. The primary reason for doing this was to save on costs, as VANs charged higher prices for information transmission during peak hours. The Internet medium, however, eliminates such limitations, and thus enables the use of EDI for time-sensitive transactions. For example, a business could get immediate information on a customer's credit history and provide instant loan-approval services.

- *The establishment of global marketplaces:* Extranets facilitate the establishment of electronic marketplaces where organizations can meet electronically for various business activities such as selecting suppliers, procuring goods, and so on. For example, GE provides numerous extranet services over the Internet that enable an open marketplace for all buyers to search for the required suppliers or submit RFQs and engage in a bidding process.

- *Information sharing:* Extranets facilitate information exchange between various businesses. For example, businesses can publish valuable information on their extranets—such as user manuals and research reports—for access by business partners. Similarly, suppliers can publish their catalogs for immediate review by buyers. These extranets enable external partners to access an organization's systems for activities like checking shipment status, product prices, and other information. The Industry Data Exchange Association (IDEA), for example, provides extranet services for organizations in the electrical industry. Using this extranet, manufacturers can update product and pricing information in an extranet that becomes immediately available to the various distributors that are part of the extranet membership [1].

Organizations that sell e-commerce services to its corporate customers can also provide data related to the buying organization's buying patterns. Such data may be burdensome for the buying organization to record, track, and analyze, and the selling organization can provide special views of that data. The buying organization can access the extranet established by the seller organization and study its personal buying patterns. For example, a bank that has established a corporate relationship with a computer supplier can track the number of computer orders using various criteria (e.g., department codes, dates of order) by accessing the computer supplier's extranet. Similarly, suppliers to a manufacturing organization can get appropriate insights into the inventory levels of the manufacturer, thus enabling the suppliers to deliver appropriate supplies in due time. This can also help the supplier schedule production.

- *Collaboration:* Beyond information sharing, extranets also enable organizations to collaborate on specific business issues. For example, competing networking organizations who need to collaborate in order to work on industry standards can set up extranets to share standards information, track project progress, and track industry events related to those subjects. In this context, extranets serve as private bulletin boards and enable secure collaboration between its members.

- *Global acceptance:* Extranets for business-to-business commerce are continuously on the rise, with numerous standards bodies defining open standards for e-commerce through extranets. CommerceNet, for example, is a nonprofit organization whose membership includes businesses and users that contribute to the resolution of e-commerce- and extranet-related issues.

5.1.1   Classification of Extranets

An extranet implementation can be quite complex depending on the users it serves and the content it provides to those users. Such complexities require varying degrees of procedures and controls. Organizations have come up with intelligent ways of reaping value from extranets for various purposes. Extranets thus fall into two primary categories, depending on the specific implementation. Organizations can either deploy extranets by extending their own intranets or alternately lease and leverage extranet services from specialized extranet service providers. This section describes the two classes of extranets and their uses.

5.1.1.1   Building Extranets by Extending Organization Intranets

Figure 5.1 illustrates the first type of extranet. In this scenario, an organization extends its intranet to its customers and business partners, thus enabling them

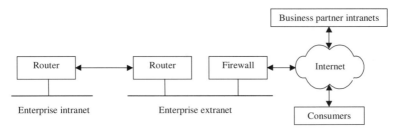

Figure 5.1   Extending an enterprise's intranet to an extranet.

to transact various business services. This implies that customers and business partners connect through the Internet to gain access to an organization's extranet. In this implementation, the organization providing the extranet manages the extranet and is responsible for its implementation and operation.

Business-to-Business Implementation

Organizations can build these extranets to transact with private business partners as well as customers. The term *private* in this case indicates that the organization usually does not open its extranet for public e-commerce activities. Rather, the focus is on proprietary and specific business linkages and transactions. For example, an automobile manufacturer may establish an extranet to enable proprietary data transfers with a supplier that supplies various goods to the manufacturer. As this extranet does not fulfill the needs of other buyers and suppliers, it is proprietary in nature. Another example is that of a health insurance company that may deploy an extranet to allow only its member organization's staff to access various services.

Business-to-Consumer Implementation

Extending organizational Intranets for transacting e-commerce activities with consumers was the first type of extranet to surface on the Internet. Establishing consumer Web sites and selling products and services through those Web sites fits this description of extranets. In this scenario, customers connect to an organization's extranet (Web site) and engage in buying, bidding, and other e-commerce activities.

Business-to-consumer extranets serve the consumer market by providing its users specialized access to select content and services. In this type of an extranet, an enterprise either sells its content or services to customers, or provides transactional services to customers. For example, a vertical industry organization such as *The Wall Street Journal* provides its users access in return for subscription fees. A bank, on the other hand, provides banking and stock investment services only to its customers. In both cases, customers authenticate to the extranet (Web site) using a user ID or password (or some other authentication scheme), to access the extranet's services.

5.1.1.2 Extranet Service Providers

In the second scenario, illustrated in Figure 5.2, the organization leases extranet services from an extranet service provider. In this scenario, the extranet is external to the organization. The extranet resides on the other side of the Internet, implying that the organization has to access the extranet across the Internet. In this implementation, the organization is not responsible for the implementation and operation of the extranet; rather, the extranet service provider facilitates operations of the e-commerce marketplace. An extranet service provider offers its extranet services to a wide range of customers who wish to use such e-commerce services. Extranet service providers promulgate the rules and operating guidelines for electronic transactions and other e-commerce services, and also provide the infrastructure for various services that bind suppliers, enterprises, and customers together cohesively. Such extranets provide both business-to-business and business-to-consumer services.

Business-to-Business Implementation

For business-to-business transactions, these extranets are considered a potential alternative to EDI, as EDI implementation requires the use of value-added networks (VANs) and their associated high costs. VANs are separate organizations that provide businesses with the means to carry out sensitive and voluminous business transactions. A business, for example, can send EDI transactions to another business through the VAN, which is in turn responsible for the distribution of EDI transactions through various electronic mailboxes. The partner organization periodically checks its mailboxes for these transactions and acts accordingly. EDI transactions include purchase orders, invoices, payment

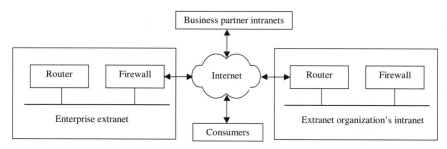

Figure 5.2   Linking to an extranet service provider.

information, and many other kinds defined by the ANSI X12 standards. VANs have certain drawbacks. The cost of using VANs, for example, is quite high. This is because linking various organizations in electronic marketplaces requires large-scale systems and appropriate infrastructure. Therefore, this inhibits small-sized organizations from participating in business-to-business transactions with large organizations. The large enterprise thus has to find other nonstandard alternatives to link to those smaller organizations.

Extranet services provided by GE resolve such problems by bringing buyers and suppliers to the Internet. GE Trading Process Network (TPN), for example, offers Web-based procurement services and is an Internet-based trading network that enables buyers and sellers to do business-to-business electronic commerce, including transactions. Ariba is another example of such an extranet. Ariba, Inc. provides the Ariba Network, which connects various buyers and sellers on the extranet. Sellers can register with Ariba through an online process and thus connect to various buyers over the Internet without incurring high VAN or EDI costs.

Business-to-Consumer Implementation

Business-to-consumer implementation through ESPs enables an organization to provide value-added services to their customers. Transpoint [2] is an example of such an extranet, as it provides end-to-end Internet bill delivery and payment services to a participating organization's consumers. By providing such services, the ESP provides tremendous benefits to consumers, billers, and financial institutions. Customers can sign on to the extranet's (e.g., TransPoint's) Web site, access their bills, and trigger appropriate payment instructions through the Web. Therefore, any customer equipped with the appropriate information appliance and network connectivity can trigger a bill payment (e.g., electric bills), if the electric company leases such extranet services from TransPoint. These extranets also help organizations reduce costs. For example, by offering Internet bill payment and delivery applications, organizations can eliminate the massive amount of work associated with printing, sending, and collecting bills.

5.1.2 Development and Deployment of Extranets

Extranet development and deployment is quite challenging as it entails almost all the issues involved in the development and deployment of enterprise systems. Extranets are about networking, security, legacy and ERP systems, systems management, and a plethora of issues that IT departments contend with on a day-to-day basis. Establishing an extranet (especially an ESP) is similar to starting a new organization and deploying the appropriate infrastructure for its operations. This discussion focuses on the primary issues in the development and deployment of extranets. The specific details of other technology components (e.g., directory services, VPN solutions, etc.) are covered elsewhere in this book.

1. *Define the scope and objectives of the extranet implementation.* An extranet's scope can vary considerably depending upon the business requirements. User base and content offering are some of the factors that drive the scope of extranet deployment. An extranet deployment may start as a small pilot project focusing on a small number of relationships. However, with the success of the pilot, the extranet can soon expand to engulf multiple business relationships. The technological and process infrastructure should therefore be scalable, to handle such expansions. Defining the short-term and long-term objectives of the extranet and associated solutions is therefore quite vital before venturing into an extranet initiative. Variation in scope and objectives could lead to different infrastructure-related decisions such as:

   - Self-development of VPN solutions or outsourcing them from an ISP. Small-scale deployments and organizations usually benefit from self-development.

   - Acquisition of an extranet product, such as Aventail's solution, or development of applications based on Java and other development technologies. Small-scale deployments and organizations usually benefit from internal developments.

   - Choice between a traditional VAN that offers extranet solutions, pursuing the services of a dedicated extranet provider, or deploying an extranet internally. Stability requirements and the nature of mission-critical applications drive this requirement.

   - Choice of directory services to store extranet user information. Volume of users and the nature of transactions drive this requirement. RDBMS, for example, may not suit inquiry applications that are accessed by a high volume of users.

2. *Define the security administration processes.* A vital characteristic of extranets is the formulation of secure access to a set of defined application and network resources. Security is therefore a primary ingredient in the planning of extranet deployments. One of the tenets of security (as will be explained in further detail in Chapter 8) is the definition of

security administration processes and procedures. In an extranet context, this involves activities such as the following:

- Definition of procedures for the identification of security administrators that will manage extranet user administration. This may require the approval of all business partners who are part of the extranet relationship.

- Definition of criteria for extranet user definition. This involves the identification of individuals in various organizations who will authorize the defining of users on extranet systems.

- Definition of user credentials. For example, requisitioners from a buying organization may need appropriate security credentials to place orders on a supplier's systems. To do so requires the issuance of certificates, passwords, and so on. This step defines the authorization of the requisitioners, their purchase limits, and other profile information.

- Identification of specific roles and their respective authorization levels. For example, an extranet system may have the following defined roles in the extranet application: requisitioner, level 1 approver, level 2 approver, security administrator, and system administrator.

- Flexibility in the addition and deletion of entities that will let an enterprise dynamically alter business relationships with customers. For example, if an enterprise wishes to end a relationship with a certain supplier, it should be able to do so with minimal effort.

3. *Identify linkages to backend applications.* Extranet deployment requires the identification of all content that an organization intends to publish on the extranet. This includes content such as:

- Links to backend and ERP systems, if the extranet users require access to transactions or backend data (e.g., inventory databases).

- Link to knowledge repositories for information sharing.

- Groupware functions for collaborating with business partners.

4. *Identify systems management infrastructure.* An extranet that supports mission-critical business transactions requires an extensive systems management infrastructure to ensure the availability of the extranet and to manage various service levels mutually agreed upon by participating organizations. Subsequent sections discuss systems management issues in further detail.

5. *Devise responsibilities for updating extranet content.* Delegating responsibilities for updating extranet content is more vital and challenging than it is for an organization's Web site. This is because delegation of responsibilities among business partners involves multiple entities and depends upon the type of extranet. Furthermore, such considerations require mutual agreements from all parties before extranet deployment. Updating content on an extranet may involve many steps. Such issues include:

- The development of content updating methods. The extranet provider should maintain recent content on its servers by using manual or automated schemes. This requires the extranet provider to establish data links with other systems, or to transfer information through alternate means. For example, an extranet provider that offers an electronic marketplace for buyers and suppliers should either ensure that suppliers continually update their electronic catalogs on the extranet or provide links to supplier systems.

- An extranet provider also needs to maintain updated versions of extranet applications, and needs a way to distribute software on extranets. An extranet provider can use various tools to update applications. For example, Castanet's Marimba enables the automatic distribution and updating of extranet applications.

6. *Assess the need for a Public Key Infrastructure.* Extranet applications involving monetary transactions incorporate certificate-based user authentication and other security controls. Certificate-based security controls ensure control of sensitive information and access to critical business transactions (e.g., authorization of money transfers to partner organizations). Such an extranet deployment therefore requires the establishment of a Public Key Infrastructure (PKI), which entails numerous planning steps of its own. Chapter 8 discusses critical elements of establishing a PKI.

7. *Clarify security policy.* An organization's security policy may require updating and/or reformulating to cater to the new e-commerce order. As extranet deployments bind other organizations in business agreements and thus expose an organization's sensitive data to those partners, the security policy should reflect statements and procedures to address such scenarios.

8. *Instituting an elaborate infrastructure for tracing transactions.* Organizations usually deploy extranets to carry out critical business transactions. Such an infrastructure therefore requires that organizations

appropriately record all business transactions conducted through the extranet and adequately backup such data. Such data is of paramount importance for both security reasons (e.g., tracing security hacks, non-repudiation) and business reasons (e.g., maintaining all business transaction details).

9. *Building portal functionality.* Organizations can customize their extranets to provide portal-like functionality. Similar to portals, extranets can provide users with customized, relevant content thus helping them perform their duties more effectively. An extranet serving manufacturers and suppliers, for example, can provide links to appropriate manufacturing industry news. Similarly, an extranet established for scientists to collaborate on specific technical subject matters can provide technical news on the subject, keeping scientists abreast of appropriate developments and alleviating them from the burden of collecting this information elsewhere.

10. *Formulate legal bindings.* All enterprises should enter into legal agreements regarding various operational issues. For example, all organizations should have clearly defined procedures to deal with fraudulent transactions and services. Legal bindings may therefore include extranet-specific policies that establish various limits (e.g., establishing approval hierarchies for transactions beyond certain limits).

5.2   Applications Access Through Portals

Portals provide a unified view of a wide range of content to its users. Such content can provide users with access to a wide repository of information, applications, and other services. By providing access to relevant content tailored to user preferences, portals become a one-stop service for Internet- and intranet-based information access. The underlying principle is to provide Internet and intranet surfers access to a wealth of information and e-commerce services by providing links to other Web sites and applications.

The primary objective of establishing portals is to reduce the time that users take to retrieve information in order to perform their jobs or fulfill other interests. Portals are front-end interfaces to the knowledge repositories and enterprise services, and resolve issues of knowledge management. Portals aid users in accessing required information and services through effective information categorization. The ability of a portal to categorize information facilitates effective navigation and content discovery, and enables users to retrieve the appropriate knowledge in the minimal amount of time.

Enterprises can leverage the strength of portals in two different ways. First, enterprises can develop their own portals to provide corporate users access to all enterprise applications and Internet- and intranet-based content. Similarly, enterprises can build and deploy portals for their customers and attract a large base of customers by empowering them with useful content links and IT-enabled services. Lycos, Yahoo!, InfoSeek, and Excite are among the prominent portals that users connect to when accessing pertinent information and services on the Internet. Portals thus compare to TV channels to which users can tune to access various forms of content.

Most prominent portal sites started as powerful search engines that enabled users to access the Internet and the Web's wide repository of services. Later, portals incorporated richer functionality and content to provide users with a "starting page" for accessing common services such as news bulletins, shopping sites, and content related to users' lifestyles, in general providing them with an information map to locate relevant content. The success of these public portals encouraged various organizations to buy the portal's "Web page real estate" to advertise their e-commerce services. For example, banks offer up-to-date stock information and relevant research services on portals like Yahoo!, Netscape, and others. This enables these financial institutions to attract customers to their services if users click on their links. Portals are therefore enablers of various e-commerce services, as Internet users that require e-commerce services enter through portals to search for the appropriate businesses and industries.

The following summarizes some of the key characteristics and features of portals:

- *Attract customers:* As mentioned earlier, portals hold the potential to increase customer visits to e-commerce sites. Large numbers of hits on Internet portals can enable enterprises that own those portals to sell portal real estate to other organizations that may wish to provide links to their e-commerce sites. Enterprises, for example, can leverage the large customer base of the Netcenter portal and WebTV networks by collaborating with these companies and displaying their presence on their portals.

- *Intelligent information search:* Sophisticated search engines and services have positioned portals to attract large numbers of visitors. As was explained in Chapter 4, the power of knowledge management lies in the ability of organizations to search and present relevant information in intelligible ways that maximize the navigational ability of users.

- *Intelligent categorization of information:* Portals enable the intelligent categorization of information, thus enabling the portal sites to include a large number of site references that link to e-commerce sites and knowledge repositories. For example, high-level categories such as "business and finance" on portal windows pack a wide repository of information links and content related to that category.

- *Access to groupware functions:* Portal sites can also offer sophisticated groupware functions such as e-mail, Internet telephony, and other groupware services. For example, Internet-based public portal sites enable users to apply for individual e-mail accounts. Portals also incorporate features to help generate communities for exchanging information, conducting e-commerce, and sharing knowledge.

- *Automatic alerting:* Users can use specialized portals to subscribe to services that alert them when portals retrieve certain content. For example, a marketing person may receive special alerts when a competitor announces special products and services on the Internet through various news bulletins. For example, the Content Server from Autonomy, Inc. [3] enables continuous searching and monitoring of the Web for news bulletins and other information and then aggregates the retrieved information and for display on the user's portal. Similarly, Autonomy's ActiveKnowledge product monitors a user's activities and retrieves and displays information related to that user's activity. The user could be composing a report, for example, in which case the software agent will attempt to retrieve knowledge from various sources related to the topic and display the information in a small window on the user's screen.

5.2.1   Classification of Portals

The success of public portals has triggered the duplication of the concept to other areas of e-commerce and has resulted in the formation of various forms of portals. The following sections delineate the various types of portals and the services that they offer.

5.2.1.1   Internet-Based Public Portals

This class of portals provides all Internet users with general categories of information that suit their interests. This includes categories such as weather, news bulletins, sports, entertainment sources, etc. The early comers such as Lycos, and Yahoo! fall into this category of portals. Specifically, these portals provide the following services:

- Information categories that include content on sports, entertainment, and so on. Public portals provide such content by collaborating with relevant organizations, such as weather organizations.

- Provide tailored information for users in different countries to offer content that suits the tastes of different demographics and interests. For example, Yahoo! provides www.yahoo.co.uk for UK-based users, www.yahoo.se for Swedish users, and so on.

- Provide basic groupware services such as calendaring, personalized e-mail, Internet telephony channels, and so on.

- Offer specialized services such as people finder services, Internet auction services, and so on.

- Provide features that enable users to personalize portal interfaces. This enables visitors to receive relevant information in desired formats. Yahoo!, for example, provides myYahoo! service, and Netcenter provides myNetcenter services.

5.2.1.2   Internet-Based Vertical Portals

Vertical portals focus on presenting specialized information related to an industry. For example, www.zdnet.com provides links to various computer- and IT-related news and events. The differentiating indicator is that the portal site tends to focus on attracting users related to the computer industry and offers services and products that suit those user demographics. For example, zdnet provides users with links that offer free software downloads, technology tutorials, and other IT-related services. This also enables zdnet to offer its own products and services on the same portal site, thus maximizing the exposure of its products and services to its visitors. Similarly, www.finance.com, a portal service offered by Citibank, provides users with various references and links to financial services, products, and related information. Finance.com offers services that enables visitors to get answers on various aspects of their financial interests (e.g., home purchasing, insurance, credit, etc.). Similarly, Microsoft's MoneyCentral is a financial Web site (www.moneycentral.msn.com) that offers financial news and links to financial services organizations and their offerings. Various financial organizations (e.g., Citigroup, American Express) collaborate with MoneyCentral to offer their products and services to MoneyCentral's visitors and members.

Various vertical industry organizations have continued to build functionality into their Web sites to portray them as portals. For example, a bank enhances its Web site by providing all financial services and links to those services for its visitors. The bank thus ventures to become the portal of the financial industry. Similarly, Amazon.com has become a book-industry portal by offering value-added services related to books. An enormous percentage of Internet users have linked Amazon.com to the experience of buying books, even though numerous other retail sites offer similar or better services. However, recently Amazon.com has diversified its offerings to become a more generalized shopping site and is venturing to become the online shopping portal for a wide array of products (e.g., toys, music) and services (e.g., auctions).

5.2.1.3 Intranet-Based Corporate Portals

Organizations can also establish portals for their internal use. Portals can give a facelift to an organization's intranet by helping corporate users find appropriate information on the intranet. Portals therefore form a critical element of an organization's knowledge management strategy. The corporate portal should link to internal applications in order to retrieve data from internal data stores (ERP data), data from the Web, and other vertical portals. The portal should also display the retrieved data in a manner that enables corporate users to make knowledgeable decisions. A salesperson, for example, should be able to retrieve information on products in the minimal amount of time, as this will empower the salesperson to make effective sales decisions.

In line with these trends, ERP system providers have started to offer products that facilitate the development of such portal services. These portals provide a browser-based graphical user interface to ERP users. ERP portals provide their users with customized screens that suit their business needs. Popular examples of ERP portals include Peoplesoft's e-business portals, SAP's EnjoySAP, and JD Edwards' ActivEra portal strategy.

An organization can build a corporate portal for its entire staff and thus enable them to access the appropriate information for their specific job functions. However, such generic portals may not prove to be very useful at times. This is because catering to the needs of an organization's entire user base may result in complex hierarchies and categorization of information. This may require the staff member to drill several levels down before they can invoke a particular application that will provide them with access to the data they seek. Organizations can resolve this issue by providing customization options. Various departments can therefore use the same portal and produce customized interfaces for various users. A manufacturing department, for example, can receive a portal that shows its applications in the portal window, whereas a marketing department's users will see a different set of applications suited to their needs.

5.2.1.4 Wireless Portals

The maturity of the wireless industry has triggered the emergence of various wireless portal initiatives by various wireless and software organizations. Wireless portals primarily target mobile users that connect through wireless networks and require access to various services. The popularity of Internet portals has enabled various wireless service providers and vendors to partner with public portal service providers and offer similar services to mobile users that connect through wireless networks. For example, Lucent and Netscape have recently joined forces to introduce a new portal initiative for wireless and mobile users. This initiative, referred to as Zingo, provides wireless mobile users with Netcenter's user interface. This initiative also includes technology from Spyglass, which (as explained in Chapter 3) maps HTML content into content suitable for wireless and other mobile information appliances. Similarly, Microsoft and Nextel Communications have teamed up to offer a comparable wireless portal initiative. The portal will extend Microsoft's MSN initiative to wireless devices and will offer specialized content such as MSNBC to wireless users. Yahoo! offers another wireless portal service. Yahoo! Everywhere and Yahoo! Mobile services allow users to get Yahoo! content over their mobile information appliances.

Although these initiatives are geared toward the general Internet masses, these portal services will enable enterprises to extend enterprise portal services to their mobile users as well. For example, enterprise users will be able to access extranet and other enterprise applications through their mobile portals.

5.2.2 Development of Portals

Developing portals requires an understanding of the user population needs and content requirements and establishing the necessary technical infrastructure to link users to such content. The following describes these steps:

- *Provide relevant content.* An organization offering appropriate content should be sensitive about its customer's demographics when offering content. For example, if an organization decides to offer multimedia content to its home user base of customers—who may not have access to high-speed lines—it will receive little acceptance. Similarly, if an organization targets certain content at corporate users, the presence of firewalls at corporate sites may block certain types of content such as applets or RealAudio files.

- *Provide customization features.* As portals offer a high-level view of a variety of content, they ought to incorporate appropriate customization features. Various factors such as user interactions, preferences, security, and demographics drive portal customization requirements. In order for a portal to offer customization features, it must record user preferences and profiles. One method requires the registration of users, collecting pertinent information during the registration process, and assigning them a user ID and a password. When signing on to the portal, the portal retrieves the profile and displays the appropriate user interface to the customer. For less stringent security requirements, user preferences can be stored on the user's PC.

- *System development efforts.* The breadth and depth of a portal's content determines the effort required for bringing backend services such as access to knowledge repositories and enterprise applications to corporate users. Portals essentially provide the user interface to all the backend services and information. The development of portals may thus require an enterprise to undertake all the steps delineated in Chapter 4 for deploying enterprise applications and knowledge repositories. This includes integrating systems and providing links to various knowledge repositories and applications. This also includes providing a framework for searching Web content and storing the results in an appropriate fashion into a window, as was explained in Chapter 4.

5.3 E-Commerce Systems Operational Management and Control

Systems and network management tools provide the means to manage and control three primary issues related to the operations and maintenance of systems. First, these tools enable an enterprise to operate their systems at desired service levels. For example, systems and network management tools enable NSPs to ensure the availability of their services per established service level agreements (SLAs). Second, systems and network management tools facilitate the management of ancillary issues related to the operations of systems. These issues include software distribution, inventory management, and more. Finally, these tools facilitate the detection of problems and provide various means to rectify them.

Problems with networks and systems can be detrimental to a business. The unavailability of systems can gravely impact customer satisfaction and business performance levels. For example, the unavailability of AOL systems a few years ago impacted customer satisfaction levels as customers could not access AOL systems due to network and systems management problems. The unavailability of a pager's satellite system in 1998 stranded millions of paging service subscribers. Similarly, an approximately 22-hour outage in mid June 1999 caused eBay, Inc. (an online auction house) to take a financial hit in millions of dollars due to the unavailability of its systems [4].

Network management and systems management have distinct operational issues. Vendors, however, have provided centralized tools and frameworks to address those issues, which have in turn blurred the distinction between the two disciplines. This section will therefore use the two terms interchangeably. However, it is vital to understand the differences between the two issues. Network management deals with the overall well-being of the network. Specifically, it addresses issues related to the availability and performance of the network. Network management issues include networking software, hardware, networking protocols, and traffic flows. Numerous factors can deem a network unavailable, including software faults, hardware defects, and misconfiguration of devices. An appropriate network management infrastructure provides a mix of proactive and reactive mechanisms to prevent such outages.

Systems management, on the other hand, deals with ancillary issues related to the successful and cost-effective operations of an IT system. These issues include distribution of software to remote sites, change management and version control, and storage management. Systems management encompasses numerous issues. Numerous factors such as the architecture of installed systems, mission criticality, and information sensitivity drive the investment levels for each. Table 5.2 presents a detailed list of systems management and network management issues.

The primary objectives for deploying a systems and network management infrastructure are as follows:

- Minimize the total cost of ownership (TCO) of operating networks and systems.

- Facilitate the maximum availability of systems by providing a systems and network management framework for detecting problems before they occur and providing rapid resolution of problems soon after they occur.

- Enable an organization to meet quality of service levels.

Network and systems management issues became quite complex with the dawn of client/server computing. The primary reason for this complexity was rooted in the distributed computing model of client/server systems. Issues such as remote management and control of IT assets were at the heart of such issues. With the emergence of the Internet-computing model, these issues have become even more complex. Earlier chapters highlighted the new characteristics of e-commerce networks and applications. This paradigm shift has also placed a burden on network and systems management tools and processes to

Table 5.2

Systems and Network Management–Related Parameters

Network and Systems Management Parameters	Description
Performance monitoring	Deals with collecting performance related parameters for systems and networks
Problem management	Deals with noting, tracking, and resolving problems. Typically used in service centers such as call centers
Policy-based management	Deals with managing networks and systems per defined centralized policies
Security management	Deals with various security issues related to the management of networks and systems, e.g., allowing authorized system agents to configure systems and networks
Software metering	Deals with tracking used software licenses across an enterprise
Web site management	Deals with various systems management issues related to the maintenance and operation of Web sites, e.g., content updates, tracking number of hits.
Content management	Deals with content updating, creating and other operational activities on a web site independent of the design of the Web site
Storage management	Deals with managing the limited storage disk space using different schemes
Asset management and inventory control	Deals with tracking and managing an enterprise's software and hardware assets
Fault management	Deals with instituting reactionary measures of dealing with network and systems generated faults
Configuration management	Deals with tracking configuration parameters for software and hardware
Software distribution	Deals with automatic software distribution to local and remote systems
Change management	Deals with managing software and hardware changes across an enterprise's network and systems
Accounting management	Deals with tracking computer and other systems usage parameters, e.g., network usage time
Capacity planning	Deals with issues related to managing, measuring and forecasting optimal and required network and system capacity and utilization issues

maintain control and availability of e-commerce systems. Specifically, network and systems management tools and processes have to counter the following challenges related to e-commerce systems:

- An enterprise's network spans multiple organizations, further challenging the task of managing systems and networks. An enterprise, for example, may have Web sites located at an NSP. Thus, the need has evolved to manage enterprise applications, networks, and data located on various service providers' premises along with self-hosted enterprise systems.

- Systems management issues have to extend to support legacy systems, client/server systems, and the new e-commerce systems.

- Network management tools provide unified views of network topologies to better enable administrators to resolve problems. The emergence of the enterprise virtual network that spans public networks, extranets, and networks of external organizations makes it difficult to get such unified views. As e-commerce systems include numerous applications and operating systems, enterprises need frameworks that will seamlessly provide network managers with a centralized view of the well-being of their virtual networks.

- With the emergence of wireless computing, the enterprise has the additional burden of supporting remote users that connect through wireless networks. The low bandwidth of wireless networks complicate the extension of system management functions such as software distribution, inventory control, and so on.

- The proliferation of mobile computing demands truly "hands-free" remote management because users do not have the privilege of being in the vicinity of in-house experts who can help resolve issues locally.

- Systems management models that have solely relied on push methods for management tasks such as software distribution may be frustrating for users who want to connect to the network for brief periods to check e-mail or perform other activities. For example, a software distribution task for a word processing application usually takes longer, thus requiring the mobile user to stay connected, thus making the entire process impractical. Mobile connectivity also suffers from stability issues. For example, connections tend to break up often, and the entire activity may not complete in one connection.

- The need for mobility requires Web-based management to facilitate systems management (e.g., view of network topology) from other information appliances rather than monitoring the network from a static site like the network operations center (NOC).

5.3.1 Network and Systems Management Issues

The list of network and systems management issues is quite extensive. Table 5.2 tabulates most of those issues. The following section discusses the prominent issues that relate to e-commerce systems and pertinent solutions.

5.3.1.1 Asset Management

Asset management or inventory control provides centralized control and a unified view of an organization's hardware and software assets. These include maintaining information about the various information appliances, servers, installed software, and the appropriate configuration settings of those devices. Traditional client/server tools focused on collecting inventory information related to PCs and servers only. However, the emergence of various information appliances in the enterprise's systems portfolio also necessitates asset management functions for such appliances. For example, maintaining settings of personal data assistants (PDAs) on centralized servers is crucial to ensure their recovery when a user calls to report software problems with the appliance.

The new line of systems management tools in this arena includes Mobile Automation [5], which extends systems management principles to mobile devices such as Windows CE and Palm devices. Organizations, for example, can distribute software to their mobile appliances, collect software inventory, manage their connectivity to the enterprise networks, and manage configuration of devices (registry, .ini files, etc.). An intelligent agent on the client side interacts with the server for managing such information. Mobile automation facilitates these operations by installing the mobile automation server on the enterprise's e-mail system (e.g., Microsoft Exchange, Novell Groupwise), and communicating those changes to the agent installed at the client by sending e-mail messages. The product also plugs into popular systems management products such as Tivoli and Microsoft's Systems Management Server (SMS).

Callisto Software, Inc. is another vendor that provides mobile systems management. Callisto's product, Orbiter 3.0 [6], offers a suite of systems management functions for PDAs. Orbiter provides software distribution, event management, inventory management, and job scheduling functions to various information appliances.

5.3.1.2   Performance Management

Performance management deals with performance issues of networks and systems. Performance management tools enable monitoring of networking traffic by protocol type, network segment statistics, packet sizes, and so on. Other tools in this arena also monitor performance at the operating system, application, and database levels. Increased transaction volumes, especially in the business-to-consumer arena, have necessitated the use of these tools to maintain system performance at required levels.

As enterprise's networks get flooded with all types of traffic, performance management solutions come in handy for managing network bandwidths. Tools such as Xedia's Access Point QVPN system, for example, allow the management of bandwidth on extranets and intranets and provide solutions for optimizing inbound and outbound network flows, thus optimizing network bandwidths.

Keynote Systems, Inc. is another vendor that provides services to better gauge and improve the performance of Web sites by installing software agents. Agents provide granular information on various aspects of a Web site. Such information includes the effectiveness of cache servers, the amount of time it takes for customers to access content and conduct transactions, performance information on various network bottlenecks, and other parameters. These services enable enterprises to get answers on critical issues such as whether an enterprise should install mirror sites to improve performance [7], isolating network congestion problems on various ISP network backbones, assessing performance issues on Web pages, and so on.

5.3.1.3   Software Distribution

Software distribution deals with automated software distribution to various servers, desktops, and information appliances. Systems can distribute software per management policies specifying time of day, nodes, and so on. The emergence of various tools such as Microsoft's SMS and Tivoli considerably simplified these issues for client/server systems that required frequent software installations and updates. The e-commerce computing paradigm with a multitude of appliances requires extension of the same discipline by keeping the additional complexities of the e-commerce computing paradigm in perspective. Within the context of software distribution, the low bandwidth of wireless networks and discontinuous remote connections are the primary issues.

Tools have surfaced to address these issues. For example, Mobile Automation, Inc. provides products that enable software distribution through an enterprise's e-mail system. The system distributes software components whenever the user connects to the e-mail system to check messages. To counter bandwidth limitations of wireless networks, the products divide the download process into various packets. An intelligent agent on a mobile appliance such as a PDA (e.g., Windows CE or Palm) manages the complete download. Similarly, other products such as Sterling Software's CONNECT:Manage family of products use various compression techniques for downloading software onto mobile appliances. The products also divide the download process into various stages. Tools such as Marimba's Castanet and Novadigm's Radia also provide intelligent software distribution features in that they download only relevant portions of code to the information appliance when users connect online.

5.3.1.4   Change Management

An enterprise's IT infrastructure goes through continual changes triggered by new software releases, system upgrades, physical equipment moves, version control, and other IT infrastructure-related parameters. Many times these changes destabilize systems and even result in unavailability of systems. Change management controls the process of managing changes across IT elements. An appropriate change management infrastructure provides due insurance to revert to older configurations, should new changes to systems prove undesirable. For e-commerce systems, this can be even more cumbersome, as the new release can involve many installations on many servers at different physical locations.

5.3.1.5   Configuration Management

Configuration management deals with tuning various network devices to appropriate settings and parameters. This includes the ability to reset devices and load appropriate configurations on network devices as well as on information appliances such as PCs and PDAs that connect to the network. Since undesirable configurations can result in unavailability and various security issues, an organization should approach configuration management functions with great care. This includes granting only authorized administrators the privileges to such capabilities. Other appropriate measures include maintaining appropriate configurations for various devices so that a copy of the configuration may be uploaded to the appropriate device, if required.

5.3.1.6   Fault Management

Fault management deals with reactionary measures of dealing with network-generated faults. A fault management framework includes the identification and prioritization of faults and the definition of an end-to-end process for fault resolution. Faults surface at various tiers and levels of an e-commerce system. Both network devices (e.g., routers, switches) and application software are prone to faults that can impact an e-commerce system's availability. However, not all faults are critical. An enterprise's various network elements carry the load of both mission-critical and noncritical applications. All such devices can generate faults, and it would be impractical to define equal measures for resolving all such faults. For example, resolving the unavailability of a router that affects a 24×7 shopping Web site is more critical than resolving a router fault that impacts a non-mission-critical LAN. It is therefore vital to define and prioritize faults based upon their impact to business performance and appropriate measures defined for their resolution.

Fault detection capabilities of network management tools have to be assessed in order to ensure that they contribute effectively toward the resolution of appropriate faults. For example, certain tools generate alerts only on the network management console. Other tools extend these alerting capabilities to administrator's pagers and other information appliances, while more sophisticated tools automatically resolve certain faults. But fault resolution is not merely about installing appropriate network management tools that detect faults. An organization should ensure that an end-to-end process exists for faults that have to be resolved manually. This may necessitate the set-up of other ancillary tools. For example, certain organizations (e.g., ISPs) institute appropriate problem management tools such as help desk and trouble ticketing software that ensure the tracking of faults and allocation of appropriate resources for fault resolution.

5.3.1.7 Accounting Management

Accounting management deals with allocating costs of various network resources. This may include measuring Internet usage time or measuring the bandwidth used by certain applications. NSPs, for example, employ these tools to offer various services to their customers. An ISP that guarantees a certain amount of bandwidth to its customers uses accounting management software extensively to track customers' network usage.

5.3.2 Standards and Technologies

Before reviewing various network and systems management technologies, it is helpful to review a basic model for systems and network management. Figure 5.3 illustrates this simplified model. The model includes four elements. The first element is the managed object. This is the object that needs to be managed, such as a network device (e.g., router) or a system (e.g., PC). An object database resides on the object that contains various types of management information. An agent resides on the object that interfaces with the database to

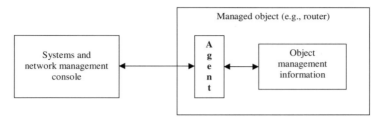

Figure 5.3   Systems and network management model.

retrieve and update management information and relays it to the management station using a specific protocol.

In network management, network devices include routers, switches, and other network devices. SNMP is the standard protocol for managing network devices. The database that resides on the networking devices is called management information base (MIB). An SNMP agent interfaces to the MIB and relays the MIB information to a management station using the SNMP protocol commands.

For systems management, the managed object can be a PC, laptop, or some other similar device. Desktop management interface (DMI) is the standard protocol for managing systems. The database is called the management information format (MIF) and interacts with a DMI agent, which in turn relays the information to a DMI-enabled management application that runs on the management station's console.

A management station enables the monitoring of all network and systems elements. The industry has introduced various mapping technologies that allow management of DMI-enabled applications using SNMP-based frameworks. This is one of the primary drivers for the centralization of network and systems management functions. Furthermore, an integrated console enables enterprises to display network topologies in graphical and intuitive formats, group information for viewing data, and enforce global policies and distributing policies across a network.

Storing information about various devices, defining agents that access that information, and communicating this information to network managers requires standards and technologies that network and systems management vendors can incorporate into their systems to enable their tools to function seamlessly. The Distributed Management Task Force (DMTF) is the organization that defines and publishes such standards. Previously known as the Desktop Management Task Force, this organization is responsible for the formulation of interoperable standards for network and systems management issues. DMTF has expanded its focus from desktop management issues to include network and systems issues related to the enterprise and the Internet in general. DMTF has introduced various standards that include the Common Information Model (CIM) and Web-Based Enterprise Management (WBEM). The following sections review the prominent technologies and standards for managing networks and systems.

5.3.2.1 Simple Network Management Protocol (SNMP)

SNMP is the standard protocol for managing networks and network devices and is widely supported by most network devices. SNMP enables the collection of management information from various networking devices. The most popular version of SNMP is its first version. However, it has several limitations, the most critical of which is security. Version 2 of SNMP never became popular, while SNMP version 3 (SNMPv3) is the new standard introduced by IETF that addresses these deficiencies but has yet to be incorporated into devices and management frameworks.

SNMP establishes a session between the network management station and the managed device (agent) and collects various statistics. Lack of security in the first version of SNMP may not be a major issue for networks restricted to an enterprise's boundaries, but it is a serious issue when network management and monitoring traverses public networks. The networking industry realized that in the continuously expanding network world, security is of paramount importance. The newer version thus allows for authentication as well as privacy. Authentication ensures that the device is receiving an SNMP request from a valid device. Privacy ensures that the nature of the request remains confidential to the manager and managed device. Lack of security can allow external parties to make unauthorized networking configuration changes. Networking vendors incorporate support for both protocols into their latest offerings.

5.3.2.2 Management Information Base (MIB)

A critical element for systems management is to be able to share management data between the managed devices and the systems manager programs. MIB is the information database for the storage of information about managed objects. MIB-II is the latest standard and incorporates various objects that are essential for fault and configuration management. Various MIBs have been created for various devices. For example, Remote Network Monitoring (RMON) is an SNMP MIB for remote management of networks. RMON collects information that traverses through the device (e.g., fault diagnosis, performance tuning) and makes it available to SNMP. SNMP in turn makes it available to the management station.

5.3.2.3   Desktop Management Interface (DMI)

DMI is a DMTF standard that facilitates management of desktop devices. The DMI standard enables systems and network management tools to manage and control user's desktops. The DMI architecture specifies a database for the storage of management information related to the managed device. This database is in management information format (MIF). MIF specifies a structure and format for defining manageable attributes, a set of standard MIF definitions for hardware and software components, and a standard interface in the form of APIs for accessing the MIF database. An MIF is similar to the MIB described in the previous section and it encapsulates descriptions of a managed object such as a PC. MIFs enable inventory management, software distribution, and other system management functions.

DMI is a layered architecture. The management application interfaces with the DMI-enabled device (e.g., printers and computers through the management interface, or MI). The MI is therefore specific only to the management applications. The MI in turn interfaces with the service provider (SP). The SP manages the information in the MIF database and interfaces with the device through the component interface (CI). The CI exposes the device to the layers above, allowing the collection of various device-related data (e.g., printer status). CI thus makes the managed device transparent to the design of management applications. DMI 2.0 is DMTF's latest specification. It is based on CORBA and it enables the transmission of management information using ORBs.

5.3.2.4   Common Information Model (CIM)

CIM standardizes various existing network and systems management data such as SNMP, DMI, and CMIP, and defines a common standard for the representation and exchange of systems management data across devices and systems. The primary reason for the introduction of CIM was to enable the interoperability of various systems management vendors with the enterprise networks and systems. Thus, CIM enables the integration of various systems management tools such as Tivoli, CiscoWorks, and others. CIM builds upon the object-oriented paradigm and enables the representation of managed objects using object-oriented terminology such as instances and classes. CIM defines a model for representing various network elements that include routers, desktops, applications, etc.

CIM Object Manager (CIMOM) handles the interface between management applications and data providers (data from managed objects or devices). Management applications thus access managed objects through the CIMOM. An example of a management application is a framework like Tivoli that displays information related to the managed objects. CIMOM links various managed objects into relationships and associations. CIMOM also has an associated object repository that together with CIM comprises the management framework.

5.3.2.5 Web-Based Enterprise Management (WBEM)

WBEM was founded in 1996 by BMC software, Cisco, Compaq, Intel, and Microsoft. WBEM is a DMTF standard designed to unify various management and Internet standards and technologies. The primary objective of WBEM is to enhance the interoperability of various systems management standards. WBEM facilitates uniform access to management data and devices across the enterprise network. WBEM uses CIM to represent data in XML and uses HTTP as its communication protocol. This enables the Web-based management of all devices and stations. All WBEM-compliant devices or applications meet these specifications.

WBEM enables the integration of various management paradigms. Previously, data for the SNMP-enabled management of network devices was represented in MIBs, whereas management of desktop systems and data was represented in MIFs. WBEM calls for data to be held in CIM and managed through HTTP/TCP/IP. DMTF has standardized on XML for information representation in managed devices and management applications.

5.3.2.6 Java Management API (JMAPI)

JMAPI is a protocol-independent application protocol interface (API) that facilitates the development and implementation of network and systems management applications. JMAPI supports major network and systems management standards (e.g., WBEM, SNMP). For example, developers can use JMAPI to develop SNMP agents. The development of management applications and Java-based agents thus facilitates portability across platforms.

5.3.2.7 Windows Management Instrumentation (WMI)

WMI [8] is Microsoft's initiative for managing Windows environments. WMI extends DMTF's WBEM initiative to the Microsoft Windows environment and allows access to various management data through the WMI interface. As WMI builds upon WBEM, it uses DMTF's CIM for representing managed objects in the Windows environment.

Providers supply data from the managed objects to CIMOM through the WMI interface. Thus, WMI acts as the interface for funneling data between the management applications (i.e., send queries to WMI) and the managed objects (i.e., send event notifications to WMI).

Intel's LANDesk Client Manager is an example of a product that implements the WMI/CMI specification. The product enables administrators to manage desktop and mobile systems. Administrators can interrogate system features such as the operating system version, BIOS information, capacity, and various other types of information.

5.3.3   Formulating an E-Commerce Operational Management Strategy

An enterprise's network and systems management framework should be aligned with its business and IT objectives. The following steps will help enterprises build an effective systems and management framework:

- *Clarify objectives.* The network and systems management discipline includes numerous parameters. The management of all such parameters can be quite expensive and requires numerous tools, and managing all those parameters may not be necessary. (Table 5.2 tabulates all popular parameters that comprise the subject of network and systems management.) An organization should therefore assess the value that it will get by managing the appropriate parameters. For example, an ISP's network and systems management goals and objectives will be quite aggressive, as the delivery of high-quality IT related services is an ISP's primary mission.

- *Define scope.* Identifying the scope of network and systems management tools is quite vital as the span of management includes in-house services and infrastructure, ISPs, NSPs, content delivery organizations, and many others. The definition of scope will ensure the identification of the appropriate parties to fall under the enterprise's umbrella of systems and network management infrastructure.

- *Pursue phased deployment.* The installation of systems and network management tools is more complex than for traditional software and hardware components because these tools connect the entire enterprise's IT operations. It is therefore vital to pursue a parallel and phased implementation approach to minimize the downtime of critical IT elements.

- *Perform requirements analysis.* A formal requirements analysis helps in the identification of appropriate parameters required for network and systems management and associated details. There is a plethora of network and systems management tools on the market that pack sophisticated functions. Matching the tools with an enterprise's requirements is an essential step. Network monitoring, for example, may not be vital for an organization that outsources all its network operations to an external NSP. Perhaps an appropriately formulated SLA would be sufficient. An organization may find that the basic set of systems management functionality bundled with the new operating systems (e.g., Windows 2000) is sufficient for the organization's requirements.

- *Tie in tools with business processes.* The ultimate objective of investing in network and systems management tools is to ensure availability and maintain control of the IT infrastructure of the organization. The complexity involved in the installation of tools can easily distract an enterprise from its primary objectives. Furthermore, tools alone rarely help achieve such objectives. For example, while many tools pack sophisticated functions such as proactive alerts, self-diagnosis, and auto-recovery, it is vital to institute processes, procedures, and work flows that will initiate the appropriate actions upon the triggering of such alerts. For example, a sophisticated systems management tool could alert the operator when a certain disk reaches a certain threshold of its maximum capacity. In response to this alert, the organization should have appropriate reactionary processes (e.g., inform the acquisitions department to place an order for a disk drive).

  The formulation of appropriate processes and procedures is also applicable for scenarios when an enterprise develops applications internally. An internally-developed application rarely incorporates appropriate network and systems management traps into its core. In such instances, if a user experiences problem(s) with an application, the organization should institute appropriate levels of support within the organization to appropriately deal with the situation.

  These processes and procedures are even more vital today considering the large span of e-commerce systems. An organization should clearly collaborate with partner organizations with which it shares systems on the mechanisms of rectifying a situation if a problem occurs.

- *Brainstorm on problem scenarios and devise recovery plans.* Building appropriate resiliency in e-commerce systems is a prerequisite to minimize downtimes. Brainstorming can aid in the formulation of appropriate scenarios should the business suffer specific incidents. Consider a scenario where a bank co-locates its Web site at an ISP, and the ISP detects hacking incidents at its Web site. Specifically, the ISP may notice that external hackers managed to break in and steal critical passwords of certain customers. The ISP should have specific instructions from the bank to deal with such scenarios. Depending on the bank's security policies or the nature of the incident, the ISP may have instructions to either deny services to all users and rectify the situation or maintain services and take alternate actions as stipulated by the bank.

- *Foresee and plan the operational management of new IT initiatives.* A common issue associated with network and systems management is the insufficient consideration given to operational aspects when planning new IT initiatives. An organization's IT network and systems management infrastructure may not be able to cope with the new IT installations. It is therefore very important to tie-in the planning of new IT initiatives that a business is pursuing with the planning of the network and operations center infrastructure. For example, a business may be planning to install a large server farm to support Internet bill payments and processing in the data center, whereas the data center may not be equipped to handle network monitoring of all those elements. Early planning will enable parallel upgrades of the network and systems management infrastructure to handle the new systems and applications.

- *Identify various levels of support.* As the deployment of e-commerce systems may span many organizations, it is vital to plan for various tiers and levels of support to handle systems and network management issues. Traditionally, when network and systems management was centralized within an organization's data center, the data center staff usually took the responsibility for initiating first level of support. This meant that the staff reacted to the problem by performing certain actions to rectify it. If the problem could not be resolved at that level, the data center staff invoked other levels of support. For e-commerce systems, there are two vital issues. First, an organization should ensure that it clarifies the role of all organizations in providing the necessary first, second, and subsequent levels of support. An ISP, for example, may decide to be the first level of support for its network connections but not for other elements collocated at the ISP's site. The second issue is to ensure that all of the e-commerce system's network connections and systems are covered by the various levels of support. The reason is that it is not uncommon to overlook certain elements in establishing such levels of support.

For a self-managed systems and network management infrastructure, an organization should establish support links with appropriate networking and systems vendors. This is because when equipment fails, only equipment vendors may be able to rectify the faults. To ensure minimal downtime, an operations center should call the vendors, who can immediately arrive at the site to rectify the problem if required.

- *Provide intelligent data retrieval and analysis.* Systems and network management tools provide a very rich set of reporting features that include information on various aspects of IT operations. These reports can be very useful in early problem notification and can thus prevent future incidents. However, an enterprise should be careful in formulating reports that are necessary for its use. For example, the pertinent department should decide on daily, weekly, or monthly reports.

5.4 E-Commerce Hosting

Deploying an e-commerce system entails various stages. An e-commerce system is an IT system that follows a typical IT system's development and deployment life cycle. Figure 5.4 illustrates a simplified IT systems development life cycle. Although each of the three stages illustrated in the figure entail detailed processes and procedures (some of which will be discussed in Chapter 9), the three stages depict trends for the outsourcing or insourcing of e-commerce projects. The first stage deals with business planning, which involves assessing an enterprise's business processes and conceptualizing the e-commerce systems required to further the respective business processes. The second stage involves the building or acquisition of systems (applications) that implement the desired logic formulated in the previous stage. Finally, the last stage involves deploying systems in a production environment (e.g., a data center or NOC that carries out and oversees the daily operational aspects of the system).

Enterprises have matured through various phases in the implementation of the IT systems life cycle over the past few years. In the first phase, enterprises typically performed all stages in-house. This involved conducting all activities of the planning and design stages, developing source code and testing and deploying systems in self-deployed and self-operated data centers. However, as various cost models forced enterprises to re-think the ROI reaped from such do-it-yourself strategies, enterprises ventured to outsource various stages of the IT life cycle. The data center was the first element that enterprises outsourced, as its operations involved more mundane tasks than the other stages. Although enterprises struggled to outsource other elements of the IT life cycle as well, they could not successfully decouple themselves from such elements for a number of reasons, including immature development technologies, lack of the integration discipline, and immaturity of the software development processes.

Figure 5.4   Simplified systems development phases.

The emergence of the object-oriented paradigm gave new life and hope to development initiatives and related integration efforts. For the first time, enterprises could build or acquire systems and truly could plug components together to realize integration with little or no effort. This maturity, along with an increase in trends to specialize in various areas and disciplines of IT, triggered the emergence of various service providers that offered various specialized IT services (e.g., network services, application services). These trends continued unabated with the emergence of the Internet, which witnessed the surfacing of various ISPs and other service providers. For example, ISPs began to offer Web-hosting services in addition to network services. ISPs are continually maturing to offer other e-commerce- and Web-related services as well. Later, other service providers surfaced to provide complimentary services in the materialization of full-fledged e-commerce systems.

The contemporary IT life cycle is a mix of these trends, and it depends upon enterprise business plans, maturity of technology infrastructure, and related processes. Figure 5.5 illustrates the various stages and paths that enterprises could potentially pursue to materialize the hosting and operations of e-commerce systems.

The first stage in general has stayed unchanged, as enterprises possess a better understanding of their business processes and requirements. The second stage, however, presents various options to enterprises aiming to deploy appro-

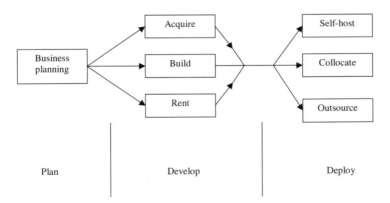

Figure 5.5   Systems development phases for e-commerce applications.

priate e-commerce systems. In this stage, enterprises have the following three options:

1. *Acquire the application.* In this option, enterprises acquire off-the-shelf solutions from external vendors. Most business processes such as human resources, procurement, and accounting are quite common across all enterprises. Enterprises can therefore acquire these applications instead of developing them. Various vendors offer these packages and provide enterprises with the flexibility to make site-specific modifications. ERP applications and many e-commerce applications (e.g., Netscape's BuyerXpert) discussed in Chapter 4 are examples of such systems.

2. *Build the application.* In this option, an enterprise develops the application internally. Application development requires an extensive infrastructure that includes human capital, hardware, development software, support software (testing, configuration management), and other elements. The building of applications also involves various scenarios. In the first scenario, an enterprise self-develops the application. In the second scenario, a service provider provides the necessary infrastructure to enable enterprises to build applications online. The value that these service providers offer to their customers is the ability to establish an online presence rapidly without extensive technical expertise. By providing the necessary technological infrastructure for hardware, software, and other networking services, these providers let small- and medium-sized businesses focus on the business aspects of the application (e.g., building customer relationships). iCat, a division of Intel Corporation, is an example of an organization that enables small- and medium-sized enterprises to build and operate e-commerce sites on the Web. iCat offers software wizards that help build Web sites and host businesses without the need for any extensive technical knowledge. iCat, for example, offers an E-Commerce Provider Program that helps companies build branded online stores without having to develop, design, administer, or host their own solutions [9]. This program also offers merchants the ability to use Visa Purchasing Card payment services.

3. *Rent the application.* Easy access to the Internet and Web has facilitated means of renting applications from other vendors, referred to as application service providers (ASPs). ASPs offer applications ranging from ERP applications to groupware and various e-commerce applications. The thin-client model has greatly facilitated the emergence of this

model, as users can point browsers to specific URLs that contain their customized rental code. Enterprises, for example, can lease an already deployed application for a periodic or fixed fee payable to the vendor. KnowledgePoint, for example, is an organization that lets enterprises access its HR base of applications, which is comprised of critical business functionality commonly used by organizations. Another example is that of NaviSite that in addition to providing regular hosting services also offers application rental services. USinternetworking, Inc. (USi) is another example of an ASP that enables customer organizations to lease enterprise-level applications without the cost of maintaining those applications. USi offers an iMAP (Internet Managed Application Portfolio) service that includes various end-to-end applications in the areas of human resources, relationship management, Internet selling, financial management, ERP, and messaging [10].

The next and last stage of the IT life cycle involves the deployment of application operations. As Figure 5.5 illustrates, organizations have multiple options for this activity as well. These include:

1. *Self-hosting the application.* This requires the enterprise to erect its own data center and NOC for operating the system. An enterprise in this case is usually responsible for deploying the necessary infrastructure, including network and systems management software, formulating appropriate operational processes and procedures, deploying operations staff in the data center, and performing other related activities. In this case, the organization is responsible for formulating and maintaining various operational service levels.

2. *Outsourcing operations.* This option involves outsourcing operations to external entities that provide the necessary technical and process infrastructure for running systems operations. Numerous examples exist that reflect these trends. These include a recent deal by IBM and AT&T, in which AT&T will run IBM's operations for a certain number of years. Other organizations including EDS offer the infrastructure, services, and expertise to enable organizations to operate their operations centers. Enterprises can also outsource other e-commerce services, including PKI services and extranet services.

3. *Collocate operations.* This recent trend involves enterprises hosting their hardware or software at a service provider's premises. It is quite common for example to collocate Web sites at an ISP's or NSP's locations. These collocation services include 24×7 network monitoring, a suit-

able climate for operating equipment, and battery backups. Collocation involves various scenarios as well. In the first scenario, an organization supplies its application software and hardware to the ISP, which merely takes the responsibility for operating the Web site. In the second scenario, organizations rent applications from the service provider that operate at the same service provider's Web sites. Knowledge-Point's example fits this scenario.

Notes and Web Sites

[1]   www.idea-inc.org/index.html

[2]   www.transpoint.com

[3]   www.autonomy.com

[4]   biz.yahoo.com

[5]   www.mobileautomation.com/

[6]   www.callisto.com/fr_brief.html

[7]   www.keynote.com/downloads/demo/demo.html

[8]   www.microsoft.com/ntserver/management/Techdetails/TechSpecs/
       WMIandCIM.asp

[9]   www.icat.com/aboutus/pr/990503visa.htm

[10] www.usinternetworking.com/products/imap/

Additional Web Sites

www.cisco.com

www.ge.com

6

E-Commerce Systems Technology Infrastructure

Earlier chapters discussed vital components of e-commerce systems. This chapter introduces the readers to ancillary technology elements that an organization requires for the effective development and deployment of those systems. The distributed nature of e-commerce systems has further necessitated developments and innovations in technologies that bind the various components of these systems together as a functioning unit. Middleware is one of those components that offer ancillary services such as security, transaction processing, time services, and so on for the building of complex e-commerce systems. The design of e-commerce systems requires many middleware services, each of which fulfills a unique role for the effective functioning of an e-commerce system. This chapter elaborates upon the many services that fall into the realm of middleware. Directory services is another such component that enables systems and users to locate each other on the network. Various standards and products in this arena have challenged organizations to choose the right standard or product for their applications and business processes. This chapter presents an overview of the popular directory services commonly used in the development and deployment of e-commerce systems.

A groupware technology infrastructure is the cornerstone behind the establishment of virtual communities and the enabling of e-commerce applications with features that enhance communication, collaboration, and information sharing. Groupware includes features such as e-mail, information sharing over the Internet and intranets, and workflow. This chapter elaborates upon their role within e-commerce systems.

The multiple components and modules that shape an e-commerce system coupled with the dynamic business requirements necessitates component and object models for analysis and software development. The object-oriented standards and frameworks facilitate such development. This chapter delineates some of those models and development paradigms.

6.1 E-Commerce Systems Middleware

Middleware encompasses various technologies and products that facilitate the availability of backend network resources (e.g., databases) for frontend applications. Middleware components also include software that triggers backend applications to achieve end-to-end automation of business processes (e.g., chaining the order-entry business process with inventory and distribution processes). Middleware technologies facilitate integration with backend enterprise services that range from simple cases of information repositories to complex ERP applications (e.g., SAP R/3, Peoplesoft). An example in this case would be an e-commerce–based ordering system that interfaces with a production scheduling system that, in turn, ties-in to inventory, distribution, and shipping systems.

Multiple definitions exist for middleware. The IT industry has failed to standardize on the specific terms regarding middleware. The only definition that the industry tends to agree upon is that middleware is anything that resides between the client (user) and the server (database and application resources). These components could be data access components, communication protocols, specialized servers, or a mix of all of the above. The on-going evolution of computing architectures is one of the reasons for the ambiguous definition of middleware. Middleware played different roles in the three computing paradigms of legacy, client/server, and Internet-based computing. Transaction processing (TP) monitors were the primary middleware technology for legacy applications. The client/server era introduced multiple versions of middleware that included everything that resided between the two layers of the client and the server. Furthermore, a broader definition surfaced in the Internet era due to the vast distributed nature of Internet applications.

This book categorizes middleware as all the software components and applications that reside between the user (connecting through information appliance applications that solicit input from users) and the backend service repositories. These include access gateways, TP monitors, and other similar software. For example, if a user requests fulfillment of an e-commerce service through an information appliance, all software that resides between an information appliance (e.g., a PC) and the backend applications falls within the realm of middleware.

Middleware from the mainframe era primarily focused on the availability of backend resources. Middleware products such as transaction processing software fit that description of middleware. The business application was a hodge-podge of business logic, network and database interfaces, and other application services. Only transaction processing software resided outside the realm of the business application to provide controlled access to databases.

The dawn of client/server computing triggered the emergence of additional middleware technologies. These include various database and network interfaces to link the various client and server pieces. This is where the definition of middleware started to blur. Distributed client/server computing resulted in additional middleware technologies and services.

E-commerce computing added more technologies and services to the middleware category. The primary reasons for this were the increase in the number of tiers that fall between the information appliances and the backend services and applications. Another reason was that the increase in the number of clients in e-commerce computing and the distribution of backend services across various networks required specialized network and application services. The popularity of building modular programs had decoupled these services from application programs and pushed them to the middleware pack.

With these trends, middleware has come to encompass various products, technologies, and services, some of which include:

- Access gateways;
- Database interfaces;
- Network and communication interfaces;
- Application interfaces to facilitate interoperability between distributed applications;
- Network/application services (e.g., security services, directory services, transaction services);
- Computer telephony integration (CTI) software;
- Middle-tier business logic implemented using traditional software technologies and programming languages or object frameworks such as Enterprise Java Beans (EJBs);
- Application execution services such as those that support large numbers of users, fault tolerance, workload balancing, session and state management, multithreading, accessing multiple resources, and others.

As mentioned earlier, middleware technologies and services play a vital role in e-commerce computing as they integrate a greater number of tiers in order to materialize an e-commerce system. Middleware, for example, provides the following services to e-commerce systems:

- It supports diverse client-side environments such as Java clients, ActiveX clients, and other graphical environments.

- The decoupling of business logic from underlying application services hosted by different tiers requires new middleware technologies and services. Common Object Request Broker Architecture (CORBA) is an example of a framework that defines such services.

- By integrating applications and data distributed throughout the Internet, middleware enables a variety of IT-enabled services and products to customers. Organizations thus can offer better and more robust content by delivering an integrated view of applications.

- It supports various types of information appliances

- It integrates various tiers of e-commerce systems.

- It manages complex transaction flows.

- It can cope with the diversity of platforms (legacy, ERP, and others) and databases.

- It incorporates better failure detection and recovery capabilities.

- It handles an increased load of users.

Thus middleware technologies and products in an e-commerce framework enable the formation of business relationships through the integration of applications and services. The evaluation, selection, and acquisition of middleware is thus quite vital to build proper functioning e-commerce systems. The following describes the popular middleware frameworks.

6.1.1   Middleware Frameworks

Middleware frameworks are standards-based frameworks that enable the development of various middleware services. The two most popular camps defining these standards are Microsoft and the Object Management Group (OMG). The OMG defines various standards related to object and component technologies. Microsoft is pushing its COM architecture, whereas OMG is spearheading the CORBA initiative. This section reviews the two frameworks and their capabilities.

6.1.1.1   Common Object Request Broker Architecture (CORBA) Services

The OMG devised the CORBA specification, which is an object-oriented middleware specification facilitating object communication in a distributed paradigm. As described earlier, the middleware landscape includes numerous other services that include security services, naming services, and time services. CORBA not only defines the client/server middleware services present in the earlier generation of middleware products but it also provides a framework for the development of additional middleware services based on an object-oriented framework. The power of CORBA thus lies in enabling the building of distributed and object-oriented applications and facilitating interoperability between software objects regardless of their implementation details.

The object-oriented approach of the CORBA specification facilitates simpler integration of business applications. Organizations can focus on building objects that represent business functions and then combine these business functions with others through CORBA interfaces. This is unlike procedural middleware in which low-level APIs do not provide a higher-level abstraction and thus make the process of assembling applications more cumbersome.

OMG has defined the Object Management Architecture (OMA). OMA defines four primary middleware interfaces for the development of object-oriented applications: Object Request Broker (ORB), CORBA services, CORBA domains, and CORBA facilities.

ORB is what enables objects to communicate with each other. The ORB specification does not focus on the implementation of objects. Rather, it focuses on the interfaces that enable a client object to communicate with a server object. The interface definition language (IDL) represents the interfaces between objects. The Internet Inter-ORB Protocol (IIOP), on the other hand, is the communication protocol between ORBs. By invoking methods on another ORB through a client/server relationship, ORBs can communicate in a heterogeneous and distributed environment. This facilitates the development of objects regardless of implementation details such as platform, language, and location. When a client needs to communicate with another server object through an ORB, the ORB locates the necessary server object, establishes communication, invokes methods on the object, and returns the results to the client object that requested the services.

CORBA services are the system-level middleware services that are necessary to build enterprise-wide applications and systems. Table 6.1 presents some of the services defined by the OMG [1]. CORBA facilities provide higher-level services usable directly by application objects (e.g., printing). These facilities

Table 6.1
CORBA Services

CORBA Services	Short Description
Life Cycle Services	This service facilitates the life cycle services of objects, namely the creation, copying, deletion, and moving of objects.
Persistence Services	This service provides the interfaces that allow common operations to manage and retain the persistent state of objects.
Naming Services	This service enables the locating of objects, e.g., files, folders. It specifies the binding of names to objects and resolves those names.
Event Service	This service allows client and server objects to communicate. The event defines the supplier and consumer roles. The supplier object produces event-related data, whereas the consumer object processes the event data.
Concurrency Control Service	This service controls concurrent access to objects and supports both transaction and nontransactional modes of operations.
Transaction Service	This service supports transaction services as defined by the ACID properties of a transaction (defined in Section 6.1.2.).
Relationship Service	This service maintains the relationships between various objects, e.g., customer (object) buys (relationship) a car (object).
Externalization Service	This service is used for externalizing and internalizing objects.
Query Service	This service enables applications to query a set of objects and also allows for insertions, deletions, and other types of object manipulation.
Licensing Service	This service allows for the control of intellectual property rights on objects per business requirements.
Properties Service	This service allows for the manipulation of an object's properties.
Time Service	This service provides time-related services such as current time, interval between two events, order of events, etc.
Security Service	This service provides security features such as authentication, authorization, auditing, etc. to objects.
Trader Service	This service allows objects to interface with trader objects. Trader objects advertise an object's services to others (export) and also allow the matching of services on the request of a client object (import), so that the client object can use those services.
Collection Service	This service allows for the grouping of objects and enables a set of common operations on that group of objects.

include specifications for information management, systems management, and task management. OMG is working to define additional facilities related to mobile agents, data interchange, and internationalization. CORBA domains define specifications for specific vertical industries (e.g., finance, telecommunications, and manufacturing). These specifications define OMG-compliant interfaces for these vertical-industry systems.

CORBA does not limit itself to the definition of middleware services. CORBA also includes another component that defines the behavior of applications. This component is called CORBA Application Objects. CORBA Application Objects are not middleware services; rather they are the ultimate consumers of other CORBA services (i.e., application objects that are the users of middleware services). These objects are comparable to the Enterprise Java Beans (EJB) model described later in this chapter. Used in conjunction with EJBs, CORBA defines all object services necessary for the development of distributed object-oriented applications. CORBA defines these objects to enable the definition of business objects that are independent of the application details. These objects can define business constructs such as customer, order, payment, and so on. Various business objects can integrate and interoperate to define higher-level business processes (e.g., a stock trading application over the Internet).

Since the EJB framework defines the design of modular business objects within the Java paradigm, CORBA also defines the CORBA Beans framework to enable the integration of both technologies. The primary difference is that CORBA Beans framework allows the definition of business components in any language, as opposed to EJB, which relies on the Java language. CORBA bean containers can host and run EJB objects. Similarly, EJB containers can host and run CORBA bean objects if they are implemented in Java.

Several vendors have committed to support CORBA and have rolled out products that implement the CORBA specification. One of the first vendors to do so was IONA Technologies, which integrated CORBA support into its Orbix product. Orbix is a comprehensive implementation of CORBA that includes interfaces that facilitate the integration of software using CORBA middleware. IONA offers two Orbix products. The first, Orbix C++, is for C++ programmers, and OrbixWeb is for Java programmers. Orbix has found its way into the development and implementation of various types of vertical solutions such as telecoms, manufacturing, finance, healthcare, and many others [2].

Other popular products based on CORBA include IBM's Component Broker and BEA's WebLogic server.

6.1.1.2   Component Object Model (COM)

COM is Microsoft's object-oriented technology and infrastructure for the building of distributed object-oriented components and applications. COM grew out Microsoft's earlier Object Linking and Embedding (OLE) initiative. OLE provided a framework for building software components (called OLE controls) and the interoperability of those components. OLE's objective was to enable Windows applications to communicate effectively within the Windows environment by offering capabilities such as cut-and-paste operations and drag-and-drop functions.

The COM model provides various products and services for building distributed and object-oriented applications through COM products—unlike the CORBA architecture, which defines specifications for various services (e.g., naming and security services). Various products implement CORBA services based on these specifications. For example, the transaction service is an integral part of the CORBA services architecture/specification, whereas the Microsoft Transaction Service (MTS) product provides the same service in the COM architecture.

COM's design is closely tied to the Windows operating system, which is the primary reason for the popularity of the COM architecture. CORBA's design goal, on the other hand, was to be platform independent. However, Microsoft continually provides interfaces and gateways to open COM to other platforms. For example, COM has already been ported to various platforms such as AIX, Solaris, and OpenVMS.

COM+ is an extension of the COM model that further simplifies the creation of software components. COM+ adds runtime and services layers to the COM architecture. The runtime layer includes features such as dynamic invocation and memory management, and the services layer includes services such as security services, transaction services, and load balancing.

As both object-oriented frameworks (COM and CORBA) are quite popular in the industry, the OMG ratified a set of standards that enables interoperability between the two frameworks. Several tools are also available based on these standards that facilitate the integration of applications based on respective frameworks. OrbixCOMet is an example of an offering from IONA Technologies that is compliant with the OMG's standards and facilitates such integration.

6.1.2   Transaction Processing Middleware

Transaction processing programs facilitate controlled connectivity and service access to a large number of users with limited backend services. These include databases, other applications, and files. Transaction monitors link user applications executed on or through various information appliances to backend resources by providing services such as workload balancing, multithreading, and memory management.

Transaction monitors control the state of various transactions. Transactions are computer operations that change the state of network services such as databases. Their use in mission-critical applications is vital to ensure that changes to databases and other services defined within a transaction are either committed or left untouched. For example, a money transfer transaction involves deducting money from one account and adding it to another account. Transaction monitors ensure that both these operations defined in the transaction complete successfully (i.e., money leaves one account and goes to another account). If the transaction monitor detects a problem with any of these operations, it rolls back all operations and returns the databases and other modified services to their initial state.

TP monitors provide other services in addition to transactional integrity. TP monitors provide workload balancing, persistence, service management, queuing, fallback and recovery, and other services required for executing applications in a multiuser and multiservice environment.

The TP industry defines transactions as having four distinct properties, known as ACID properties. ACID stands for atomicity, consistency, isolation, and durability. Atomicity implies that all operations within a transaction form one unit of work. As explained earlier, the success of a transaction depends upon the success of all operations defined within the transaction. Failure of one requires the rolling back of all modifications. Consistency ensures that a transaction leaves the system in a stable state after its execution. If the transaction does not leave the state in a stable condition, all operations defined within the transaction abort and get rolled back. Isolation ensures that every transaction is unaffected by the actions of other transactions, while durability ensures that all operations of a transaction are permanent in nature.

TP monitors date back to the centralized mainframe applications era and have continuously evolved to meet the demands of new computing environments and architectures. With the emergence of the client/server computing paradigm, TP monitors bring the same transaction discipline to client/server applications. Popular TP monitors include BEA's Tuxedo, Transarc's Encina, and IBM's CICS.

With the emergence of e-commerce computing, TP monitors and associated technologies and standards evolved to handle e-commerce applications. For example, the new generation of TP monitors include support for ORBs. This enables TP monitors to bring ACID and other properties of TP monitors to the Internet and object world. In the object world, the discipline of transactions becomes even more essential as invocation of one transaction may involve multiple databases and other network services spread across the Internet and enterprise intranets. Popular products from vendors include Microsoft's MTS, IBM's Component Broker, and BEA's M3. The new generation of TP monitors now support multiple backend resources, including relational databases, HTML repositories, and proprietary databases such as Lotus Notes.

Various technologies have surfaced to offer transactional support for object-oriented e-commerce applications. For example, as mentioned earlier, CORBA defines Object Transaction Service (OTS) as part of the CORBA services component. OTS enables distributed objects to access various transactional services. Java Transaction Service (JTS) is the Java implementation of OTS. JTS specifies the implementation of a TP monitor that supports the Java Transaction API at a higher level and implements OTS at a lower level [3]. EJB, to be discussed later, includes standards for the development of transactional beans. Microsoft, too, through its COM+ architecture, provides functions for the development of transaction processing applications on Windows platforms. All these object-based transaction processing standards facilitate the development of multitiered and object-based e-commerce applications.

The application development trends have continued to evolve with TP monitors as well. For example, TP monitor vendors have partnered with development tool vendors to facilitate the development of TP monitor applications from GUI tools. Similarly, various TP monitors have evolved to Web-enable their TP monitors, thus making them directly accessible to Internet and e-commerce frontends.

6.1.3   Communication Middleware

Communication middleware enables full-fledged applications to communicate with each other, unlike communication software that merely binds various modules of an application. The following delineates the popular types of communication middleware that facilitates such integration.

6.1.3.1   Message-Oriented Middleware (MOM)

Message-oriented middleware (MOM) is a class of middleware that enables client/server applications to communicate. MOM applications run by using queues. Each application places a message in the queue that the other application retrieves and then acts upon. Popular MOM products include IBM's MQSeries and BEA's MessageQ.

MOM applications differ from TP monitors as client and server applications do not maintain a logical connection. These applications are therefore suited for time-insensitive applications. After the client application places a message in the queue, the server program can retrieve it at its leisure. MOM middleware allows programs to run offline, place messages in a queue, and then connect to the server for transmission of those messages. An example is a banking application that allows customers to open accounts. Since opening accounts requires banks to obtain certain authentication papers from their customers, accounts do not have to be available immediately for access on the Web. For such implementations, the bank can implement a MOM product that queues account-opening requests and later updates the bank's backend systems, thus alleviating the load of the backend systems.

Since MOM architecture suits different situations and applications, the OMG has introduced an object-based CORBA messaging standard using MOM that facilitates the development of object-based e-commerce applications and services.

6.1.3.2   Internet Inter-ORB Protocol (IIOP)

IIOP is a client/server protocol defined by the OMG as part of the CORBA specification for enabling CORBA objects to communicate over a network regardless of their location. The definition of the IIOP specification as part of the CORBA 2.0 specification facilitated the design of CORBA objects that can interoperate over the Internet. The IIOP protocol defines data formatting rules and various message types to enable ORBs defined per CORBA specifications to communicate with each other. The IIOP details are transparent to developers. IIOP works on various data transport protocols such as TCP/IP and SNA. IIOP facilitates communication between client and server object-based components (e.g., applets on the client-side and servlets on the server-side). IIOP thus provides an alternative to HTTP for facilitating application data flows between the client and server components.

6.1.3.3   Distributed Component Object Model (DCOM)

Similar to IIOP, DCOM is another protocol for interapplication communications across a network for communication with COM objects. DCOM is Microsoft's standard, which has in turn been licensed to other vendors. DCOM runs not only on Windows, but is available for UNIX, Macintosh, and other platforms. Developers can deploy software components across platforms distributed over a network and facilitate communication among the components using DCOM. DCOM is language-neutral and therefore a viable technology for building e-commerce systems that usually involve multiple application components developed in diverse languages on multiple platforms. These software components can include both applets and ActiveX components.

DCOM uses a DCE-RPC–based communication protocol to enable this communication and integration.

6.1.3.4   Hypertext Transfer Protocol (HTTP)

HTTP protocol is the Internet's primary communication protocol that simply conveys information from the client (e.g., Web browser) to the server (e.g., Web server). HTTP primarily facilitates communication between an Internet's Web browser and the Web server that hosts requests from the Web browser. HTTP is usually used to establish early connection between the browser and Web server. For example, Internet browsers use HTTP to request an HTML page from a Web server. Subsequently, application components such as Java-Beans, ActiveX controls, and applets—loaded through HTML files—communicate with backend server-side application components using other enhanced communication protocols such as DCOM and IIOP.

6.1.4   Database Middleware

Database middleware mask database-access complexities from applications, and also mask the implementation differences of various databases. For example, although Oracle and Informix are both RDBMS databases, they each provide native access mechanisms. Database middleware standards and products mask those differences from applications that access those databases and thus increase the portability of applications. Java Database Connectivity (JDBC) and Open Database Connectivity (ODBC) are both popular database middleware standards for the development of database access applications.

6.1.4.1   JDBC

JDBC, similar to ODBC, is an open specification that enables Java clients to access backend databases such as DB2, Oracle, and IMS. Using this method, Internet browsers, for example, can execute application components and directly connect to backend databases. Typically, JDBC is used to incorporate calls within backend application interfaces—such as servlets that access databases—and then send results to the client through access gateways such as Web servers. JDBC defines two interfaces. The first defines the application's interface to a JDBC-compliant database, and the other defines the rules for database vendors to make their databases JDBC compliant.

6.1.4.2   ODBC

ODBC is an open database middleware standard developed by Microsoft that enables database access across various vendor databases. This means that any client application can access an ODBC-compliant database as long as the client application uses appropriate ODBC drivers. Similar to JDBC, ODBC defines the application- and database-specific interfaces.

Major database vendors such as Oracle, Sybase, and IBM provide ODBC interfaces for their respective databases in addition to offering their own internal interfaces. For example, Oracle's native interface is called Oracle Call Level Interface (OCI), and Open Client is Sybase's native interface. Both also offer ODBC connectivity. Since ODBC is Microsoft's standard, Microsoft's SQL server uses ODBC as its native interface.

6.1.5   Application Middleware

Application middleware technologies and products enable the triggering of other applications, extend the functionality of applications, and provide various runtime execution services (e.g., memory management and workload balancing). The following delineates the various forms of application middleware used to develop e-commerce applications.

6.1.5.1   Common Gateway Interface (CGI)

CGI is the traditional way to invoke server-side programs, and it is portable. CGI programs are usually Perl scripts that invoke server-side programs developed in any language. CGI programs interface with the Web server using environment variables that the Web server populates by receiving input from the Web browser's URL. The Web server starts the CGI program that typically resides in a cgi-bin directory on the server platform. The CGI program interacts with the backend application or database, retrieves the result, formats it into an HTML or appropriate format and presents it to the Web server. The Web server then funnels the response to the browser for display. For example, a registration form filled out by a user while browsing on the Internet could be submitted to a CGI program on the server through the Web server. The CGI program, in turn, will process the form input by accessing a backend database and returning the results to the user's browser through the Web server.

CGIs are usually suitable for less intensive e-commerce applications. CGI programs are complete programs that invoke a separate process for each browser request and are thus inefficient for large-scale applications. This causes enormous overhead on the servers. CGI programs are not suited to run intense Web applications that service a large number of users and handle large transaction volumes. Nor are CGIs suitable for architecting and chaining complex applications that involve multiple transaction monitors and backend ERP applications.

6.1.5.2   FastCGI

Open Market, Inc. developed FastCGI to resolve some of the performance issues of CGI programs. FastCGI facilitates this by allowing programs to stay active in memory. FastCGI programs maintain persistent connections with applications and wait for server requests thus increasing the performance of applications. FastCGI compares favorably in performance with programs developed by the vendor's proprietary APIs (discussed next). However, FastCGI's primary advantage is its portability between servers, as most servers support CGI. Fast.Serv is a product from Fast Engines, Inc. that implements FastCGI for Netscape and Microsoft Web servers.

6.1.5.3   Vendor Proprietary APIs

Developers can use vendor proprietary APIs as alternatives to CGI scripts for invoking server-side programs. These APIs extend the functionality of various Web servers by allowing developers to write the additional modules needed for the execution of backend applications and programs. Developers can also develop modules that completely replace services that come bundled with various servers (e.g., Web servers). Most of the major vendors that sell Web-related servers provide APIs to extend the functionality of their servers. Popular examples include Netscape's Netscape Server API (NSAPI) and Microsoft's Internet Server API (ISAPI). NSAPI extends the functionality of all Netscape servers. Using NSAPI, developers can write modules or replace existing modules on those servers. Developers can write modules that provide various functions and services including security, logging, and other services.

Microsoft provides the ISAPI to extend the functionality of ISAPI enabled Web servers. For example, developers can use ISAPI to develop modules and applications that can retrieve data from backend databases. ISAPI applications are more efficient than CGI programs as ISAPI programs run in the same address space as the Web server and thus do not incur the overhead of communication between processes and invocation of processes.

6.1.5.4   Application Servers

Application servers embody various applications, network, and database services. Earlier application server implementations focused primarily on transaction-based services. However, the proliferation of Internet- and client/server-based services and technologies triggered the emergence of full-fledged application servers that encapsulate most functions in one middle tier.

Application servers provide capabilities for the development and deployment of server-side applications. These solutions are analogous to user interface development tools or integrated development environments (IDEs) that initially surfaced for the development of user interfaces. Application servers extend the same paradigm to networked and integrated applications. Some of the server-side application development and deployment issues—which would otherwise require extensive manual development and deployment practices—include the following features:

- Web server software functionality.

- Select groupware functions and services (e.g., support for e-mail, calendaring functions, newsgroups, and others).

- Transaction services functions.

- Runtime environment for the execution of various application logic, such as that of EJBs, COM servers, and others.

- Visual, IDE-like environment for the development of server-side applications.

- Interfaces for backend applications and diverse data sources.

- An integrated environment that enables the development, deployment, and integration of server-side applications.

- Ability to automatically partition applications based upon presentation, business, and data access logic.

- Application servers that handle browser requests and other requests from information appliances.

- Ability to facilitate the deployment of scalable applications across multiple servers and systems. Scalable deployment usually requires the moving of various system modules to various servers and the building of appropriate interfaces between various applications and modules of the systems. Application servers provide the basic foundation that allow for a flexible deployment application and system modules across servers and systems, yet provide a common glue that binds these modules transparently.

- Similar to providing an environment for scalability, a common platform base for application servers provides performance features such as automatic load balancing, stability of network and data connections, cooperative processing of multiple application components across platforms, and others.

- Numerous technologies exist for the integration of applications and ultimately the development and deployment of e-commerce systems. These include object technologies such as Java, CORBA, C++, IIOP, JDBC for data access, and so on. Application servers provide templates and frameworks that facilitate a seamless development of applications that mask some of the complexities associated with the development of these applications and systems.

- Application deployment usually requires a security environment to control authentication, authorization, and other issues. Native application development requires the establishment of a security environment to provide such services. However, the application servers provide a security framework and environment for applications developed under the specific application server.

- Ability to integrate with legacy and ERP applications such as Peoplesoft and SAP.

- Support for multiple databases.

6.2   Directory Services: Glue for E-Commerce Systems

Directory services store information about network resources and the users that access those network resources. No sooner did enterprises start linking computers on networks that the need for directory services emerged. Users needed a means to locate other users to transmit information and needed to locate other computers to access the various services resident on those nodes. Enterprises implement directory services to connect various users and applications. Following are some of the typical applications that require directory services:

- An e-mail directory enables users to locate other users for sending e-mail. Today this directory exists in many forms. Enterprises often maintain a unified e-mail directory for proprietary deployed e-mail systems, or in the case of Internet e-mail addresses, users maintain local address books on their information appliances. Similarly, public portals provide a set of services for locating Internet e-mail addresses, which either locate e-mail addresses present on their domains or attempt to provide a listing of e-mail addresses present in the Internet universe.

- A LAN directory stores user information about people on the LANs and facilitates various functions such as connecting to the Internet, sharing printers, facilitating LAN chats, browsing through information published on the LAN and the intranet, and so on. Common implementations include NT domain directory infrastructure, Netware Directory Services (NDS), and so on.

- Developers require access to source code repositories to check out source code. Various source code configuration management tools maintain various information about developers in a directory to enable them access to such intellectual assets.

- To access networks through firewalls, enterprises require users to authenticate to the firewall database that stores information about the users and their respective access rights.

A directory services platform includes various components such as:

- *Lookup database:* This is the database that stores information about network resources and user profiles. A user profile includes a user name, address, department code, telephone number, and so on. Network resources and services information includes printer names, computers, applications, and others.

- *Lookup functionality:* These functions enable the requesting of programs and services to get information about network resources and services.

- *Security services:* These services ensure that all accesses are within the defined security limits as set and defined by the security administrator. A directory service can employ various forms of security controls that can range from simple static password authentication to public-key cryptographic systems. Chapter 8 discusses various forms of authentication schemes.

E-commerce applications further fueled the need for additional directories. E-commerce computing relies on the existence of one large ubiquitous computing network with users and network resources. Carrying out transactions and various services as well as locating and searching for information on this large network demands the existence and establishment of various directory services at various legs of the network. For example, the following describes sample scenarios within e-commerce computing to demonstrate the importance of directory services:

- The establishment of a secure application infrastructure based on public-key cryptography requires the existence of a public-key directory that includes the names and certificates of relevant users. For example, connecting a browser to a server using an SSL session involves authentication of the key loaded onto the browser. This authentication is performed by the certificate authority that maintains certificates for the key loaded on the browser.

- An organization sets up an extranet for external partners and suppliers to transact business-to-business e-commerce services. A directory service can be used to store the appropriate security credentials of all parties that have access to the extranet resources. The directory service handles all login requests and provides users with access to appropriate network functions.

- Users on the enterprise's intranet that require access to various network services (access to applications located on various network servers) and devices. For example, printers can use a directory service that facilitates the means to locate those resources on the network and enables transparent connectivity.

- To access corporate mission-critical applications, users can sign-on to a portal or portal-like directory that provides users with a view of authorized applications.

- Various e-commerce applications may require their respective directory services. For example, an organization that provides Internet bill payment functions may use a directory that enables customers to sign-onto the to Web site to make appropriate payments.

Enterprises usually maintain multiple directory services for different applications, which has resulted in islands of directories. This trend is on the rise, as enterprises aggressively launch various e-commerce services that require the deployment of relevant directory services. Almost all e-commerce transactions and services require the establishment of directory services.

The problem of multiple directory services and the inability to effectively locate network resources through directory services will continue to worsen unless enterprises launch a formal initiative to resolve these issues. The directory service problem has worsened over the years due to the numerous applications and associated computing platforms that have surfaced. It is therefore vital that enterprises formally decide upon the particular strategies for directory services deployment. Some of the issues associated with multiple directory services are as follows:

- Users must sign-on multiple times to use various network services.

- Users cannot logon due to lack of streamlined administrative processes that cover multiple directories. For example, a user may be allowed to use the e-mail services in a directory but is denied access other enterprise applications.

- Users cannot locate appropriate services or applications.

- Users find it difficult to launch enterprise applications because they need to interface with multiple directories that require multiple levels of authentication.

- Organizations have to implement multiple security administration processes to enable synchronization of those directory services spread throughout the enterprise.

6.2.1 Centralization of Directory Services Through Metadirectories

The preceding discussion is indicative of the dilemma that enterprises face in moving to an e-commerce computing paradigm, which requires a more unified way to locate network resources. An obvious option is to centralize directory services. The advantages of deploying a unified directory services infrastructure are many. Based on the issues that enterprises have with the islands of directory services, one can clearly deduce the advantages of such a solution. Centralized directory services provide users with a single sign-on (SSO) solution, because a one-stop authentication service provides users with access to all authorized functions. A centralized directory also simplifies management of various functions (e.g., definition, modification, and deletion of users and network resources).

Whether centralized directory services would result in the appropriate ROI or not depends upon various factors, which include:

- Nature of the enterprise's business processes (more applications usually result in more directory services).

- Forecasted business growth (increasing number of users requires scalable directory services).

- Challenges that enterprises are facing with the current directory service infrastructure.

- Nature of e-commerce applications planned for the future.

- Various platforms installed in the enterprise may drive the choice of a strategic directory services platform, as certain directory service frameworks may be tied to specific platforms.

A popular option for centralizing directory services is to implement meta-directory services. Metadirectories logically or physically unify various directory services and their contents that include people, roles, their credentials, department codes, and so on. Metadirectories, besides providing a unified information model for the representation of namespaces, also reduce maintenance costs. For example, the addition or deletion of an enterprise employee could require the updating of multiple directory servers (e.g., the human resources directory, e-mail directory, and enterprise application directory servers). Metadirectories achieve this integration by maintaining relationships between various directories. An enterprise can therefore unify its directory services that may include NDS, NT directory services, IBM Resource Access Control Facility (RACF), Lotus Domino, Netscape's directory services, and many others.

An enterprise can implement the concept of metadirectories in various ways. One method involves acquiring (or building) a product that automatically synchronizes directory information with multiple directory services. For example, this product would automatically update the information of a new user in multiple directory services, alleviating the requirement to create the user in multiple directory services. Another method involves a procedural approach in which directory service entries in one server replicate to multiple directory services as a batch process, thus making basic user information available in multiple directory services.

Isocor, a leading provider of metadirectory solutions, offers a product called MetaConnect. This product provides metadirectory functionality by reducing the maintenance associated with multiple directory services. The product is based on the Lightweight Directory Access Protocol (LDAP), and operates with most LDAP servers, thus centralizing most directory-services-related functions (e.g., user addition, deletion, profile management). Meta-Connect's architecture enables linking to disparate directory servers and centralizing information into one model (e.g., map the "user address" and "address" fields in multiple directories to one logical attribute). This provides a unified view of various directory services. MetaConnect unifies directories by storing all user information from multiple directory servers into one meta-directory (e.g., e-mail address in one directory and telephone numbers in another directory). These features along with others alleviate various management and administrative burdens associated with the maintenance of multiple directory services.

6.2.2   Directory Services Requirements for E-Commerce Systems

As mentioned earlier, the primary functions of a directory service are to enable the definition and lookup of users and network resources. E-commerce computing requires various features from a directory service that provide e-commerce services to its users, including the following:

- *Scalability:* Certain e-commerce implementations, especially in business-to-consumer domains, require the definition of a large number of users in order to give them access to appropriate e-commerce services. For example, an organization providing online bill-payment services will need to define all of its users' profiles in a centralized directory service. This directory should be scalable to meet the exponential increase in customers. Scalability should therefore be a factor in the choice and deployment of appropriate directory services.

- *LDAP support:* As will be explained later, LDAP has become a directory access standard in the world of directory services. The directory service should therefore support the LDAP standard to allow interoperability between various directory systems on the Internet and intranet.

- *Architecture robustness:* The directory service should offer a robust architecture to enable a richer definition of users and network resources. For example, the definition of a printer should allow for its various properties, physical location, and so on, such that a user looking for a color printer should be able to locate the appropriate one. The directory service should also represent relationships between various entities to facilitate lookup of entities and their respective properties. Furthermore, a directory service's architecture should be flexible enough to represent organizational structures. For example, the directory service should require minimal actions to redefine a user in another department with a different set of credentials. NT 4.0's directory model has proven to be complex enough to cater to such features in large environments.

- *Performance:* The directory service should perform adequately under a large number of network connections. Performance becomes an even bigger issue if the server performs complex operations before fulfilling a user's request. For example, a server could suffer performance degradation if it performs complex security functions (e.g., SSL) for every transaction. To handle such scenarios, certain directory vendors provide interoperability with hardware accelerator boards on the servers that boost the performance of such directory servers.

- *Fault tolerance:* Directory services constitute a vital link for accessing critical e-commerce services. The directory service therefore should have built-in fault tolerance or the organization should build alternate measures of fault tolerance to avoid business unavailability problems for its Internet-based customers.

- *Management issues:* Management and administration are vital issues in deploying directory services. User and object additions, deletions, and modifications of profiles can add up to an enormous number of tasks. The administration of a directory service solution for a given application or business process should not incur high costs. For example, NT directory service builds on a domain and trust relationship model. This model works quite well for small numbers of users and domains, but the administrative burden to manage an enterprise with a large number of users and domains becomes extremely burdensome.

6.2.3   Popular Directory Services

The past few years have witnessed the emergence of numerous directory services standards. This section reviews some of the popular standards and products related to directory services.

6.2.3.1   X.500

X.500 was one of the early initiatives that surfaced to standardize various directory services. The X.500 standard is the culmination of the efforts of ITU and ISO in the early 1980s. The X.500 directory service standard surfaced to address the complex operations inherent in the design of a directory service. However, due to the complexity and bulkiness of the protocol, most vendors did not fully implement the X.500 protocol, resulting in incompatibilities among various implementations and a fragmented market of directory services. Vendors that followed the standard in various forms include Microsoft (Exchange), Lotus (Notes), and Novell (Netware Directory Service).

6.2.3.2   Lightweight Directory Access Protocol (LDAP)

Netscape along with some university researchers defined the LDAP standard in mid-1990s. LDAP is a protocol for accessing X.500 directory services and runs on TCP/IP. RFC 1777 defines the initial specification for LDAP, while RFC

2251 through RFC 2256 define specifications for LDAP v.3. The primary reason for the definition of LDAP was to define a lightweight version of the X.500 standard, as the Directory Access Protocol (DAP) implementation of the X.500 standard, due to its complexity and bulkiness was challenging to use and implement. For example, LDAP queries generate much less traffic than the X.500 DAP. The DAP component of the X.500 defines the access mechanism between the client and the server. A server that supports LDAP is the Directory System Agent (DSA). The term *DSA* is also based on the X.500 terminology. The DSA submits the user's query to multiple directories and retrieves the response from the LDAP-compliant directories. The LDAP standard thus facilitates the integration of disparate directory services such as NDS, NT, and so on, and is a lightweight implementation of DAP that enables lightweight clients to access X.500-based directory services over the Internet.

LDAP's directory schema uses a tree-like structure to represent its entries. This tree is also called the directory information tree (DIT). The topmost entry is the "root" that includes subhierarchies of countries, organizations, people, and so on. The entries in an LDAP server can vary depending upon the specific implementation. For example, an extranet application can implement an LDAP server to authenticate its users. The LDAP supports basic commands that can be executed against an LDAP-compliant server. These commands enable search of the directory for objects, maintenance of objects in the directory (add, modify, delete, etc.), establishing sessions with an LDAP server for communication, and many more. These commands can be used between client applications and LDAP servers as well as between various LDAP servers that collectively implement a larger directory of information.

LDAP version 3 adds new features to the original LDAP protocol. These include better security features and a robust command set. The security features include the addition of a Simple Authentication and Security Layer (SASL) for additional authentication. The enhanced command set enable LDAP clients to discover the directory schema of the directory database, unlike the earlier version of LDAP in which clients had to have the knowledge of the directory schema before accessing the directory entries.

LDAP has become a very popular standard, and most directory vendors support it. These include Novell's Netware Directory Server (NDS), Microsoft's Active Directory Server (ADS), and Sun-Netscape's iPlanet Directory Server.

6.2.3.3   NDS 8 Directory Services—Full Service Directory (FSD)

Novell has been an active technology supplier of enterprise directory services for a number of years. The vendor enjoys a large customer base when compared to other players such as Microsoft. Novell's popular directory services architecture is called Netware Directory Services (NDS), and it forms the backbone of numerous large enterprises. NDS emerged as a popular directory service primarily due to the large installed base of Netware networking software.

Novell recently introduced NDS 8 and refers to it as a full-service directory (FSD). Novell's goal in introducing this directory service was to incorporate all the desired functions of an ideal directory service. Although this approach seems quite promising, the biggest challenge in introducing directory services in an enterprise is its interoperability with other directory services.

Novell clearly defines four issues as being vital in a directory service and defines appropriate functions within those areas [4]. Those areas and appropriate functions are as follows:

- *Discovery:* Discovery is the process of lookup of network resources. A user or application should have the ability to appropriately browse and search the various contents of a directory service. NDS 8 provides appropriate features for the discovery functions and employs the LDAP standard for this function.

- *Security:* NDS 8 defines robust levels of security. A directory service should provide appropriate security services for the information stored in the directory service and for the storage of security information to enable users to engage in various security functions such as secure e-commerce transactions.

- *Storage:* A directory's database structure is equally important as it not only stores information about users and network resources but provides the appropriate structure for the storage of this information.

- *Relationship:* An essential feature of any directory service is its ability to store relationship between various entities. NDS 8 provides a robust infrastructure for the definition of such relationships.

NDS 8 provides the following distinctive features:

- The ability to define a large number of objects (e.g., users and other network resources). NDS claims this number to be close to a billion objects. This provides organizations with the benefit of deploying NDS in the business-to-consumer paradigm, which may require the definition of a vast number of users that access an organization's services.

- From a performance viewpoint, NDS 8's architecture measures adequately for this number of objects.

- NDS 8 builds upon the latest version of LDAP (version 3).

- NDS 8 provides robust tools to import a large number of directory entries into its core.

- NDS has the ability to run on the NT system and override NT's directory service by providing NDS services. This is especially useful for environments that have mixed NDS and NT environments and plan to migrate toward a full NDS implementation.

- NDS provides administration and management tools that meet the challenges associated with the definition of a large number of network objects.

- NDS supports the JNDI initiative.

- NDS provides transparent replication services that allow users to access profile replicas from various locations. This enables an enterprise to provide adequate support for users who access network resources through various mobile information appliances from various locations.

6.2.3.4 Microsoft's Active Directory Services (ADS)

Microsoft's latest initiative in the area of directory services is the Active Directory initiative. Active Directory is an integral part of the Windows 2000 initiative and will replace the earlier NT platform directory services and structure. Microsoft introduced the Active Directory initiative to resolve various issues inherent in the NT directory services and to cater to the new directory services required in the e-commerce computing paradigm. Specifically, Active Directory provides the following primary features [5]:

- Facilitates the establishment of a virtual network by providing a unified view of the network.

- Compared to NT directory services, ADS requires a reduced number of trust relationships for managing domains.

- Provides a mechanism to enable the establishment of a unified messaging structure. This enables the definition of attributes related to various messaging applications (e.g., e-mail, telephone, fax, and others).

- Provides a mechanism for managing other directory services without regard to their platform details.

- Provides an API for accessing other directories.

- Provides support for open standards.

- Builds upon the LDAP protocol for accessing directory services.

- Provides the Active Directory Services Interface (ADSI) specification to enable external applications to access the Active Directory service.

- Active Directory, like most other directory services, does not conform to the X.500 standard but does incorporate certain features.

- Active Directory incorporates Microsoft's Intellimirror management technology, which provides administrators with various system management features. These include automatic software distribution and maintenance, centralized desktop configuration management, and remote operating system installation.

- Active Directory is backward-compatible with NT directory services.

- Provides various security features that include kerberos authentication, x.509 certificates, and LDAP over SSL.

The debut of Windows 2000 presents users with an opportunity to standardize their LAN-based directories, especially if they are based on NDS and previous versions of NT directory services. Microsoft and other vendors are offering numerous tools that enable organizations to achieve synchronization between various directory services and ADS as well as to migrate from diverse directory services to the ADS. For example, Microsoft Directory Synchronization Services (MSDSS) synchronizes data between NDS and ADS. However, since directory services are a key infrastructure element of an organization's network strategy, organizations should wait for the maturity of ADS and test directory migration strategies before rushing into a mass migration of users and network resources and standardize on ADS. This prudent course of action for migration should be followed not merely for non-Microsoft–based directory services such as NDS but also applied to migration from NT directory services to ADS. The reason is that although ADS is backward-compatible with NT directory services, numerous new features and new security modes in ADS necessitate detailed planning.

6.3   Internet Domain Name Service (DNS)

DNS is the Internet's directory service. DNS facilitates the unique identification of an organization or entity on the Internet. DNS maps the domain name of an organization (e.g., www.shopname.com) to its IP address (e.g., 123.456.789.012). Various DNS servers distributed across the Internet facilitate the mapping of the friendly domain names that users enter into their browsers to connect to an IP address. An organization may install local servers that connect to the Internet name servers to resolve a domain name. On locating the IP address, the Internet application then uses that IP address to route the information to the appropriate hosts.

DNS assignment tasks include:

- *DNS name registration:* This refers to obtaining "friendly" names for e-commerce sites (e.g., yahoo.com). Network Solutions, Inc. (NSI) has primarily acted as the default global registrar for the .com, .edu, and .net domains. Enterprises obtain domain names from NSI. Recently, however, other registrars (e.g., register.com) have begun to provide comparable services.

- *IP address block allocation:* Internet connectivity requires every host to have a unique IP address on the Internet. This activity deals with maintaining the entire IP address space of the Internet and with allocating of unique IP addresses to Internet hosts.

- *Protocol parameters:* This activity deals with the assignment of miscellaneous protocol parameters and values required for the seamless operation of the Internet.

DNS maintenance comprises multiple activities. Although the primary control of DNS has been under U.S. government's supervision, various organizations have also provided these services. Recently, the U.S. government has triggered initiatives that promise to push DNS and its activities into the public sector. This section provides the background and current state of affairs of the DNS control.

6.3.1 DNS Structure

DNS is a hierarchical database that consists of various domain names (e.g., yahoo.com, artechhouse.com), their respective IP addresses, and other relevant information. DNS uses a tree-like structure for organizing data fields. This is similar to the UNIX or DOS file system. For example, in the DOS file system's directory listing C:\ABC\DEF, "C" is the top-level name (drive name), and ABC is the directory name under that drive. DEF then falls under the ABC directory. The DNS follows a similar convention but represents the hierarchy in the reverse order. For example, for the domain name yahoo.com, "com" is the top-level domain (TLD), that includes yahoo as the subdomain or second-level domain (SLD). TLDs are of two types: generic TLDs and country-code TLDs (ccTLDs). Table 6.2 describes the generic TLDs and their respective meanings. The country code domain names were created for use by entities

Table 6.2
Generic TLDs

TLD	Description
com	Commercial entities (e.g., www.amazon.com)
org	Nonprofit organizations (e.g., www.icann.org)
net	Networking organizations (e.g., www.gte.net)
edu	Educational institutions (e.g., www.mit.edu)
gov	Governmental institutions (e.g., www.irs.gov)
mil	Military institutions (e.g., www.darpa.mil)
int	International treaty institutions (e.g., www.reliefweb.int)

within each country. Examples of country code domains are .uk (United Kingdom), .jp (Japan), .fr (France), and .dk (Denmark).

The Internet includes millions of hosts, and access to any of these hosts requires DNS entries to remain up-to-date. Thirteen DNS root servers distributed around the world provide transparent access to any host on the Internet. Local DNS servers subscribe to these servers and obtain the relevant information to connect to Internet hosts. The U.S. government controls these root servers. The thirteen root servers are A.root-servers.net, B.root-servers.net, and continue through M.root-servers.net. These servers are distributed throughout the United States and Europe, although most of them are located in the United States.

The DNS database (also referred to as the registry) consists of resource records (RR) to handle the mapping of IP addresses and host names and other information. Each RR within the DNS registry includes the following elements:

- *Type:* Indicates the type of resource. Numerous values can appear in this field. For example, if the RR contains the value MX (Mail eXchanger), it would indicate the name of the mail servers that receive Internet mail on the host's behalf. Similarly, if the type is A, it designates the IP address (or addresses) of that host name. Other types include HINFO (Host information), CNAME (Canonical name or nicknames or aliases), and TXT (Text strings), which can include any type of information.

- *Class:* The protocol family of this record. For all Internet records, the indicator *IN* is used.

- *TTL:* Time to live indicators hold time intervals that various name servers use to cache data and to poll other servers to get updates
- *Data:* This represents the data in the field and depends upon the Type and Class fields, as different values of Type and Class will have data pertaining to those fields. For example, for the MX type, the data would indicate mail server names, and for the A type, the IP address of the host would be included.

6.3.2 History of DNS Operations and Organization Control

Chapter 3 explained the networking structure of the Internet, which highlighted the Internet's structure as a conglomerate of independent networks that interconnect through entities called NAPs and ISPs. The existence and overseeing of NAPs is essential to ensure the seamless functioning of the Internet. Initially, the U.S. government had full control of the NAP architecture. However, most functions have shifted toward industry in general and are controlled by ISPs. Similarly, DNS was initially under the full jurisdiction of the U.S. government. However, the government has recently launched initiatives to privatize DNS operations to enable competition and promote innovation in the DNS system.

Before delving any further into the current state of DNS-related activities, it is vital to review the background information and history of DNS. ARPANET developed the DNS system in 1985 to cope with the exponential growth of Internet users and network activity. The formal specification for the DNS (RFC 1034) was published by Jon Postel at the University of Southern California's (USC) Information Sciences Institute (ISI) in 1987. This specification provided detailed standards for the technical design of the domain system and other related issues. Postel also maintained the addresses of the Internet working in conjunction with SRI International, which was the primary contractor to the U.S. government for this activity. With increased activities, these functions resulted in the formation of the Internet Assignment Number Authority (IANA).

The U.S. government later delegated some of the DNS-related services to other organizations. For example, NSI (formerly known as InterNIC) started to handle the domain name registration process. In 1993, the U.S. government officially began funding NSI. Later, in 1995, NSI started charging for the DNS services and became the sole entity to profit from DNS services.

Until very recently, NSI was responsible for most tasks associated with DNS administration, including registering domain names, updating the registry (DNS database), performing registrar services (domain name assignment), and controlling the root name servers. NSI performed these activities under a cooperative agreement with the U.S. government. NSI has been the primary entity to provide domain name registration services for the .com, .net, and .org TLDs and ccTLDs. NSI maintains the master file of domain names and distributes the domain data to the thirteen root servers around the world.

NSI also controlled the task of allocating IP addresses. The U.S. government funded the operation of NSI for this endeavor. IANA was the primary governmental body that authorized the NSI to allocate IP address spaces. Later, yielding to pressure from the Internet community that its members should control the address space allocations, NSI helped form separate bodies by getting appropriate approvals from the National Science Foundation (NSF). These bodies were under IANA's control, and ISPs contacted these bodies for IP allocation services. IANA allocated larger IP address blocks to regional Internet registries (RIRs), which in turn allocated IP addresses to large ISPs. ISPs suballocate those addresses to smaller ISPs, which in turn suballocate specific addresses to end user organizations [6].

The following bodies control IP address allocation:

- American Registry for Internet Numbers (ARIN), in the United States, South America, and other regions (www.arin.net);
- Reseaux IP Europeans (RIPE), in Europe (www.ripe.net);
- Asia Pacific Network Information Center (APNIC), in the Asia Pacific region (www.apnic.net).

NSI still supports the ARIN initiative. However, once ARIN becomes fully independent, NSI will relinquish all control. This means that all organizations that want an IP address on the Internet (within the IPv4 address space) will have to contact ARIN in the United States. However, in reality only larger ISPs directly request these services from ARIN. These larger ISPs then allocate IP addresses (within the authorized address space) to smaller ISPs, which then allocate address spaces (all within the limits of the larger address space allocated by ARIN) to other smaller businesses and customers.

6.3.3 Future Directions for DNS Operations

In 1996, the ISOC (Internet Society), IANA, and other governmental and public sector entities formed the International Ad Hoc Committee (IAHC), to launch a process for the self-governance of the Internet and competition after the NSI monopoly ended. The IAHC later evolved into the Policy Oversight

Table 6.3
New Generic TLDs

TLD	Description
.firm	Businesses or firms
.shop	Businesses offering goods to purchase
.web	Entities emphasizing activities related to the Web
.arts	Entities emphasizing cultural and entertainment activities
.rec	Entities emphasizing recreation/entertainment activities
.info	Entities providing information services
.nom	Entities wishing individual or personal nomenclature, i.e., a personal *nom de plume*

Committee (POC), with a Policy Advisory Board (PAB) to lead the evolution to self-governance.

The process resulted in a generic Top Level Domain (gTLD) Memorandum of Understanding (MoU). More than 200 organizations throughout the world signed the MoU. The MoU lead to the creation of the Internet Council of Registrars (CORE), a nonprofit organization for administering seven new TLDs identified during the MoU process. Table 6.3 describes the additional TLDs.

The U.S. Department of Commerce, through a white paper published in June 1998, requested the establishment of a nonprofit organization to perform administrative activities related to domain names and management of the Internet's address space. Later, in November 1998, the Department of Commerce recognized the Internet Corporation for Assigned Names and Numbers (ICANN) as that organization by signing a Memorandum of Understanding [7]. ICANN has taken over the U.S. government's role and function (earlier performed by IANA) for such functions. IANA has therefore moved under the ICANN umbrella.

ICANN defines itself as the new nonprofit corporation formed to take over responsibility for IP address space allocation, protocol parameter assignment, domain name system management, and root server system management functions now performed under U.S. Government contract by IANA and other entities [8].

The U.S. government white paper distinguishes the term *registry* from *registrar*. In essence, the registry is the database that contains DNS entries, whereas the registrar accesses the registry for various activities (e.g., defining a new second level domain (SLD) name under the .com, .net, and .org TLDs). ICANN's primary objectives are the formulation of policies and procedures for the establishment of registries and registrars to manage and control the root name servers. ICANN is responsible for the accreditation of the registrars that register the Internet domain names. Since registering domain names is a critical activity for e-commerce sites, ICANN is formulating various guidelines for the qualification process. ICANN has also proposed itself to be operationally involved in the process for valid registrars. If the systems providing those services fail or if the DNS registrar fails, ICANN will provide appropriate resiliency and minimize the impact on e-commerce sites and businesses.

One of the objectives of ICANN is to open competition for the registering of domain names. ICANN has focused on taking a pragmatic approach to the problem, as decentralizing the process is prone to surface many issues. For example, domain name holders require the assurance that their domain names will resolve to the appropriate IP address anywhere on the Internet and will be stable. Traditionally, since only one entity (NSI) managed the process, there were lesser issues.

As mentioned earlier, NSI has always been the default registrar. ICANN is currently pursuing initiatives to open the domain name registration process to other registrars as well. ICANN is responsible for formulating guidelines for the accreditation of various registrars and to accredit registrars based upon those guidelines. To ensure a smooth transition, ICANN initially selected five registrars that will operate and sell domain name services alongside NSI. In parallel, the ICANN has also begun to accredit additional registrars that will take over operations once the transition phase is complete. NSI's responsibility in this transitional opening of the DNS to competition is to relinquish control of its registry by creating a Shared Registration System (SRS) that will enable new registrars to access the DNS registry in the same manner as the NSI itself. As of this writing, Register.com (www.register.com) has become the first registrar to offer registration services per the new guidelines defined by ICANN. Other registrars to take part during the transition phase include America Online, CORE (Internet Council of Registrars), France Telecom/Oléane, and Melbourne IT [9].

Other registrars will also provide generic TLDs to define domain names other than the .com, .net, and .org TLDs. For example, CORE, another registrar selected by ICANN, will provide the names .firm, .shop, .web, .arts, .rec, .info, and .nom, which will be available through independent CORE registrars on five continents [10].

6.3.3.1 ICANN Structure

To address the three issues of domain name registration, address space allocation, and protocol parameters, ICANN has formed three supporting organizations to help address policy matters and other issues. The three organizations to address those issues are the Domain Name Supporting Organization (DNSO), Address Supporting Organization (ASO), and Protocol Supporting Organization (PSO). DNSO advises ICANN on issues related to the domain names used for referencing Internet sites and deals with the appropriate policy issues. ASO ensures the uniqueness of the assigned addresses. Internet IP addresses are similar to telephone numbers and provide connectivity to the Internet. The current Internet address assignment is controlled by the three regional internet registries (RIRs) mentioned earlier (ARIN, RIPE, and APNIC). ASO will perform most of the functions of the RIRs and may formulate additional policies per ICANN directives. Under the new model, ICANN will allocate the address space and delegate the responsibility to further subdivide the address space to RIRs [11]. PSO deals with general issues related to communication over the Internet.

As of this writing, most of the aforementioned activities are still in transition. NSI still manages most of the DNS related activities and is in the process of relinquishing control, thus enabling other organizations to actively participate in the IP allocations and domain name addressing functions. Deregulation of these activities and opening the process to competition will simplify the various steps involved in registering e-commerce sites and will provide organizations with more robust services from the new registrars.

6.4 Enabling Groupware for E-Commerce and the Internet

The term *groupware* encompasses many things. Groupware can be considered a group of technologies that enables various members of the value chain of an organization to communicate, share information, and perform activities per defined workflows. Communication in this context can take many forms, including the sending of e-mail and interactive conferencing through various information appliances. Information sharing refers to accessing archives of historical discussions and e-mail, corporate policies, and so on, as well as sharing information through whiteboards and other collaboration tools. Workflow refers to defining work steps through the implementation of various business rules (e.g., filling purchase requests, submitting them for approvals to other parties, routing them to the next step, etc.).

Historically, groupware tools and technologies have been considered part of the communications category. However, recent advances in groupware tools incorporate functions from other categories as well. Still, no groupware vendor packs all functions that fit the above-mentioned three categories into one package. Various solutions from numerous vendors have surfaced that seem to offer different pieces of the groupware puzzle.

Groupware tools and infrastructure facilitate the automation and streamlining of an organization's business processes not implemented in an organization's legacy and ERP systems. An organization can use groupware tools to streamline and automate intrabusiness processes that may not be ideal candidates for implementation through larger-scale application frameworks (e.g., ERP systems and applications). For example, an organization can implement a groupware-based procurement application that fulfills an organization's basic procurement needs. The groupware application can provide the basic workflow (incorporating rules for approvals) and basic communication tools (e-mail for sending requests, sending alerts on approvals) for implementing the application. Conversely, an organization may implement a large-scale EDI/e-commerce application to automate and streamline the organization's procurement process, which involves millions of transactions. Groupware applications thus complement enterprise applications implemented on ERP and other legacy platforms. Enterprise applications automate business processes for large-scale and high-volume data business applications. Groupware products, on the other hand, provide workflow functionality and automate business processes that rely on functions such as electronic communication, meetings, discussions, document forwarding, and so on.

Groupware has undergone three phases of evolution. In the first phase, the term *groupware* did not exist. However, functions that make up groupware today existed as separate applications. For example, e-mail, calendaring software, and other similar applications existed as separate applications. The second phase gave birth to the term groupware, and it encompassed all applications that enabled teams to communicate, work, and share information. Groupware packages provided organizations with the opportunity to structure activities that relied upon the usage of those tools. For example, e-mail has become one of the most common communication tools today, enabling users to discuss various project-related and non-project-related matters. Groupware enables the structuring of these activities by facilitating the means to aggregate e-mail related to one department into one repository and thus enable all authorized users within that department to view department-specific discussions. Groupware is thus a vital element of an organization's knowledge management strategy, as it facilitates access to the intellectual capital and knowledge repositories of the enterprise.

Groupware in this phase enjoyed its own niche of business processes. Recent months, however, have witnessed the blurring of this niche as groupware applications merge with enterprise applications to provide an end-to-end automation of business processes. It is common to see groupware applications that tap into enterprise databases, just as the new enterprise e-commerce applications incorporate groupware functionality. The two camps seem to have joined hands to serve the customer collaboratively.

To illustrate the first case, consider a department within an organization that ventures to maintain its own budgetary process and implements a simple application based on groupware products from Novell or IBM/Lotus. The application incorporates basic rules for approvals, reviews, and so on, and serves the need of the department. The department can also implement hooks that enable the application to tap into the enterprise's ERP database to extract key budgetary values and numbers, and then uses those numbers in its own application. This is a typical example of a groupware application that relies on the ERP application database to obtain certain information.

On the other hand, consider a shopping Web site that offers customers the opportunity to collaborate and communicate with sales agents before buying a product. The Web site offers customers the opportunity to send e-mail to agents or collaborate through an electronic whiteboard to share information. In this case, the e-commerce application relies on groupware functions to offer a complete value-added solution to customers.

Groupware strengthens an enterprise's internal and external linkages by leveraging the Internet and other electronic channels. This facilitates ceaseless information flow through the enterprise and thus boosts the effectiveness and efficiency of business processes. To summarize, groupware products and services:

- Enable the setup of electronic communication channels between an organization's users and business partners.

- Establish a technology infrastructure that maximizes information sharing within the enterprise as well as across an enterprise's external linkages.

- Enable the streamlining of business process workflows not feasible or practical through other applications and systems (e.g., legacy and ERP systems).

- Provide an infrastructure for the management and sharing of unstructured information such as e-mail, discussions, and so on.

- Provide an infrastructure for the management and sharing of structured information such as documents.

- Provide the foundation for a knowledge management infrastructure, as covered in Chapter 4. Groupware solutions are easy to set up by departments whereas the knowledge management infrastructure discussed earlier requires extensive planning and involves global information sharing. These solutions provide enterprises with an entry into the knowledge management paradigm.

Groupware has the following three primary functions:

1. Group communications, enabling communication between teams and individuals;
2. Electronic information sharing, enabling groups to share information through electronic means;
3. Workflow, enabling the definition of work processes in an orderly manner.

The next section will elaborate upon each of those functions separately.

6.4.1   Group Communications

This section discusses the various communication methods enterprises can deploy to enable robust communication among parties internal and external to the organization. All communication methods discussed in this section are not necessarily part of a groupware product. The section discusses details regarding e-mail, voice communication, and new trends of unified messaging that enable a one-stop mechanism for storing and retrieving various types of messages.

6.4.1.1   Voice Communication

Voice communication continues to be the most popular and ubiquitous form of communication. Businesses rely heavily upon this form of communication. The PSTN infrastructure, value-added services offered by telecommunication organizations, and internal PBX-based infrastructure provides enterprises with enhanced communication services. These services include group conferencing, call waiting, and multiway calling.

With the popularity and increased proliferation of data networks, organizations are exploring opportunities to migrate PSTN-based voice communication services to data networks. IP telephony or Voice over IP (VoIP), as explained in earlier chapters, provides organizations with such opportunities. The primary applications of IP telephony, however, have involved the installation of telephony gateways that map between data and circuit-switched voice. True migration to IP telephony would rely solely upon data networks to fulfill an enterprise's voice communication needs. This would alleviate enterprises from the cost and administration headaches associated with maintaining two disparate networks. Although, as explained in Chapter 3, networking technologies have matured considerably to enable such migration, the industry is still wresting with multiple issues to make IP-based voice communication mainstream. These issues include:

- Matching IP voice quality with circuit switched voice quality;
- Offering PSTN-like services (e.g., group conferencing, call waiting);
- Finalizing the cost models of telecommunication organizations for charging IP voice services;
- Infrastructure overhaul costs;
- Ubiquity of information appliances that would be as easy to use as the telephone.

The use of VoIP has generally been the greatest among freelance users on the Internet. Using tools such as Microsoft's NetMeeting, users communicate with friends, family members, and other users on the Internet. Recognizing this empowerment has triggered organizations to offer IP telephony–based services on their e-commerce sites, from customer service support to entertainment and other forms of content.

For the intraenterprise domain, VoIP implementations have started to surface in an effort to deliver a converged data and voice network to users in order to reduce an enterprise's costs associated with the maintenance of separate voice and data networks, but it is too soon to predict whether VoIP will replace the current methods of circuit-switched voice communication available through the telephone network. Thus far, few enterprises have ventured to use VoIP to enable their corporate users to communicate over the Internet or an intranet.

6.4.1.2   E-Mail

E-mail is the one of the most common and popular forms of computer communication, for business and nonbusiness users alike. The process of e-mail involves users formulating an e-mail message using a client e-mail application. The message then travels to a mail server that transmits the message through the Internet (or intranet) to the destined e-mail server. The recipient periodically signs-on to the mail server and retrieves the message sent by the sender. Various groupware products such as Novell's GroupWise, Microsoft Exchange, and Lotus's Domino deliver robust e-mail communication engines that support various e-mail formats. Some popular e-mail client applications are Netscape's Messenger, Microsoft Outlook, and Qualcomm's Eudora.

Figure 6.1 illustrates a simplified e-mail process flow. E-mail, much like other applications, has evolved through multiple protocols and standards. The following describes the more popular e-mail communication protocols and standards.

6.4.1.3   Simple Mail Transfer Protocol (SMTP)

SMTP is a TCP/IP-based mail transfer protocol over the Internet between one mail server and another. SMTP deals solely with the sending and receiving of e-mail messages. Once the mail reaches the mail server, POP3 or IMAP protocols (described later) deliver the message to the user's client application. The Internet standard RFC 821 specifies the SMTP protocol.

A UNIX application known as sendmail is the most popular SMTP implementation. However, recently many mail servers have surfaced that implement the SMTP protocol. Popular mail servers include Microsoft Exchange, Netscape Messaging Server, and HP's OpenMail.

6.4.1.4   Multipurpose Internet Mail Extensions (MIME)

The MIME protocol is an Internet standard that deals with the sending and receiving of mail in multiple data formats (besides plain ASCII text). The various formats include video, audio, graphics, and other formats. MIME works by

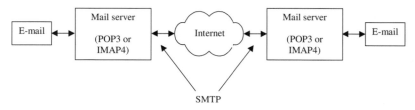

Figure 6.1   Internet e-mail architecture.

including special headers in the e-mail message. These headers indicate the type of message included in the data packet and enable client-side mail applications to load an appropriate application (player) to display or output the content of the e-mail message. Most client-side mail applications include support for popular data formats (e.g., GIF, JPEG, etc.) and automatically display the result of the message. For other data formats, the mail application loads another player application (similar to browser plug-ins) to display the contents of the message. The Internet standard RFCs 2045, 2046, 1047, and 2048 specify various details of the MIME protocol.

6.4.1.5 Post Office Protocol 3 (POP3)

POP3 is used by e-mail clients to retrieve e-mail messages from an e-mail server. POP3 is a client/server protocol that enables e-mail clients to retrieve and download e-mail messages from the server to the client. Therefore, POP3 defines the interface between the client application and the mail server and includes specifications for manipulating mailboxes on the user's client mail application. POP3 uses store-and-forward techniques that work by storing the mail on the server until the user signs on to the mail server, when the mail is forwarded to the client application. The Internet standards RFCs 1725, 1734, and 1082 specify various details of the POP3 protocol.

6.4.1.6 Interactive Mail Access Protocol (IMAP)

IMAP is an alternate mail protocol for accessing e-mail on the e-mail server. However, unlike POP3, IMAP provides the additional benefit of leaving e-mail messages on the e-mail server, rather than forcing the download to the client application. IMAP also extends features of the POP3 protocol to manipulate e-mail messages and mailboxes (folders, etc.) on the e-mail server. Using IMAP, users can view the sender's information and subject heading of the e-mail message and have the option to download the message. The Internet standard RFC 1730 specify various details of IMAP.

6.4.1.7 Unified Messaging

Enterprises usually implement multiple forms of message communication. These include e-mail, voice, paging, fax, and other types of messages. All forms of communication differ in the manner in which the messages are composed, delivered, stored, and retrieved. For example, users send e-mail messages using an e-mail client, whereas pager messages are composed either through the telephone interface or through a two-way pager device. This raises multiple issues. For example, to send messages, users need access to multiple information appliances to compose these messages (e.g., to send an e-mail message, users require access to an e-mail client that connects to a mail server). Similarly, to send a voice message, the user calls a telephone number and leaves a message on the recipient's voice mailbox. Furthermore, to receive messages, users have to access those messages through various information appliances. Supporting these forms of messaging and communication protocols needs maintenance of different technology platforms and infrastructure. For the enterprise, this increases the total cost of ownership resulting from the cost of platforms as well as administration activities.

The unified messaging paradigm provides a universal platform for the storage of all types of messages in one user mailbox while providing a standard interface for the user to retrieve the messages in any desired format. For example, Bob can send an e-mail message to Susan that will be stored in the universal messaging repository. In the unified messaging model, Susan can retrieve the e-mail message through the e-mail application. Alternately, Susan can use her telephone and in conjunction with the unified messaging server's text-to-speech technology, hear the message through the telephone. For example, MailCall E-Mail Reader is an application by Phonesoft that allows the use of telephone to hear e-mail messages stored in a Lotus Notes database. Similarly, Unified MailCall is a complete voice mail system that allows the storage of voice mail messages in a Lotus Notes mailbox and allows user access through the telephone as well as the desktop. Lucent Technology's Octel Unified Messenger is another unified messaging solution that allows users to consolidate fax, e-mail, and voice messages in one server.

Various technology standards enable the building of unified messaging solutions. These technologies enable the integration of non-compatible technologies and data formats. For example, the Telephony Application Programming Interface (TAPI) enables the conversion between circuit-switched voice and IP data packets. Similarly, the Speech Application Programming Interface (SAPI) provides technologies for speech recognition and speech-to-text technologies.

6.4.2   Group Information Sharing and Collaboration

Internet collaboration refers to the live interaction between two or more users through their information appliances. Collaboration enables users to share information through chats (sending text messages), sharing documents, whiteboarding, and other means. Collaboration also enables users to work together using various scheduling and calendaring tools. The following delineates the popular collaboration and information sharing tools:

- *Group scheduling:* Group scheduling enables users to access a common repository to schedule meetings. Using group-scheduling tools, enterprise staff on the same enterprise groupware infrastructure can use their client applications to plan work activities, schedule meetings, and invite other users on the messaging network to those meetings. All invited users can check their names for the specific meetings through their groupware client applications (e.g., Microsoft Outlook). Various messaging servers (e.g., Microsoft's Exchange server or Netscape's Calendar server) provide the underlying infrastructure to connect all users and maintain their schedule information in one common repository. With the increase in mobility, users can use various information appliances to schedule such work activities. For example, an executive can sign on to the enterprise's central groupware server using his PDA (e.g., a Windows CE computer) and download schedules for meetings that his or her assistant has uploaded onto the server.

- *Information sharing:* Using a groupware client application, users can establish public folders and publish information to those folders, enabling users to share information. Users can also set up folders for which access is restricted to a certain group of people.

- *Discussion groups:* Groupware products also allow the setup of discussion groups that allow messages from various parties to appear on the discussion group window. Users can access these discussion groups using news-reader applications such as Netscape's Messenger or other groupware client applications such as Microsoft Outlook.

- *Maintaining a common contact database:* Groupware products enable users of a certain group to maintain a common contact list applicable and useful to users within that group. For example, the marketing department can maintain a list of customers. Setting up this list can enable all mobile sales staff to look up information through their mobile information appliances without sending e-mail messages to their department assistants or calling peers, thus tremendously saving time and increasing productivity.

- *Web collaboration:* Web collaboration refers to two parties sharing information over the Web. For example, Bob could send Web pages or Web links to Susan over the Web through a Web collaboration server that connects all collaborative parties over an intranet or the Internet. These solutions are very useful in business-to-consumer solutions in which call-center agents can up-sell and cross-sell solutions to their customers. Agents can also discuss the contents of various documents with customers over the Web by looking at the same information (e.g., a utility bill) as the customer. Organizations can also use Web collaboration to display products to their consumers as consumers shop on their Web sites. For example, Lands End, Inc., a direct marketer for clothing and other products, enables consumers to interact with sales agents who can push specific products to consumers' Web browsers as consumers chat with the agents through the Lands End Web site. For intraenterprise solutions, businesses can also use these solutions for product demonstrations and training sessions. Webline Communications, Inc. offers a suite of products that enable collaboration over the Internet and intranets.

- *Whiteboarding:* This feature enables networked users to share information on a common window that appears on all user's screens. All information posted to the screen either by typing on the window or cutting and pasting from other applications is visible to all users who are sharing the whiteboarding window. This feature enables the user to enter into discussions as if they were sitting in a common room. NetMeeting provides this feature and enables users to collaborate in this fashion over the Internet and intranets.

- *Application sharing:* This feature allows various users to share the execution of a program. Users can view the output of an application as it executes on another user's computer. In this scenario, the user who shares the application can control the application (e.g., terminate the application).

6.4.3   Workflow

Workflow provides intelligent sequencing and structure to activities that are not implemented through an enterprise's application systems. Workflow activities usually incorporate either nontransactional computer activities or functions that rely upon pure human intervention and cannot be computerized easily (e.g., reviewing the contents of a document, approving documents, verifying approver's signatures). In providing this structure, workflow tools leverage e-mail and other communication and information sharing tools. For example, by using a groupware product a department may define a specific workflow that would enable buyers (requisitioners) to enter an order through e-mail and route it to the supervisor. The supervisor, upon receiving the e-mail, would approve, disapprove, or request additional information from the requisitioner. The next workflow step (depending on the defined rules) may page the requisitioner if the supervisor requires additional information. A workflow product therefore provides the structure to the ad hoc activities already being followed in the organization.

Workflow matches people with various activities and information. Workflow steps carry information and activities from one user to another and empower users to work effectively within each process step. As explained earlier, an enterprise's core business processes are usually implemented through its backend applications. For example, various business processes related to the supply chain are facilitated through business applications that reside on an organization's ERP or legacy systems. However, various work steps such as getting approvals and interacting with customers require employees to devise their own business processes and workflow. Groupware solutions enable users to structure those activities. A supply-chain system, for example, may generate a report through a batch system on orders that a department needs to process on a specific day. Instead of an employee signing on to the system to generate that report, an enterprise can devise workflow that would fax or e-mail the report to the employee. The workflow product would then track that activity and generate reminders for the employee if the orders were not processed in due time. After certain reminders, the groupware product would generate an alert for the supervisor on the status of those orders.

Workflow and business process design require graphical tools that can illustrate the entire process flow and facilitate additions, modifications, or deletions of various process steps. Development tools enable designers to design forms and define the various data elements on those forms. The development tools allow designers and administrators to define various users of the business process as well. And designers can define various rules for the workflow application (e.g., route the form to the supervisor for approval if the form data includes a purchase request of more than a specific dollar amount).

The new groupware products that provide these functions offer Web-enabled functionality, thus enabling enterprises to include external business partners and customers as part of the desired workflow. A central administrator can thus control the flow of the application across the Internet and include external parties' part of the work steps. Netscape Application Server: Process Automation Edition is an example of one such workflow engine. Lotus Domino Workflow is another example of a product that allows groups of users to define workflow and process for various activities.

The Workflow Management Coalition is a nonprofit organization of workflow vendors, users, analysts, and university/research groups whose mission is to formulate standards for various workflow products. The Workflow Management Coalition (WfMC) defines *workflow* as a series of steps during which documents, information, or tasks are passed from one participant to another in a way that is governed by rules or procedures [12]. The organization cites improved efficiency, better process control, and business process improvement as being some of the primary benefits of workflow.

Enterprises can devise workflows of varying complexities. A workflow may incorporate simple steps that, for example, involve sending e-mail, and do not include any business rules. On the other hand, a workflow may interface with backend applications and require extensive application development. The following scenarios illustrate the complexity involved in devising the two types of workflow.

The following procurement process is quite simple and leverages the enterprise's e-mail infrastructure:

- A user sends e-mail to a supplier requesting a quote for a product.

- The supplier responds via e-mail (or telephone) with the quotation.

- The buyer sends e-mail to the supervisor requesting authorization.

- The supervisor sends e-mail indicating the approval.

- The buyer sends the request to the supplier, who then processes the order and e-mails the invoice to the department secretary.

- The secretary processes the invoice and routes it to the accounts department, which then issues a check to the supplier.

The technology infrastructure required for this scenario is minimal as the department can rely upon the traditional communication channels of e-mail and telephone to process transactions. The workflow also applies to situations that involve a limited number of transactions. However, as the number of transactions increase, the supplier will not be able to cope with the volume. Similarly, the accounts department will require a more streamlined system to issue and track payments.

To cope with increased transactions and transactions involving high dollar amounts, the enterprise can deploy a system that relies upon an e-commerce application, a set of groupware products, and a set of interfaces to backend legacy systems. For example, consider the following workflow:

- A user (requisitioner) signs-on to an intranet-based procurement application.

- The requisitioner browses the online catalog, selects the desired product, and enters the order on a form provided by the application.

- The system automatically validates the form by checking the data fields and routes the application to the requisitioner's supervisor.

- The supervisor receives the order through the e-mail system in his or her e-mail inbox.

- The supervisor approves or disapproves the order by typing the word *approve* or *disapprove* on the subject line.

- For approved orders, the e-mail system automatically routes the message to the e-commerce application's backend, which then generates an EDI order and routes it to the supplier's mailbox through the Internet.

- The supplier processes the order and routes a payment to the buyer's accounts systems, which, after proper authentication and authorization, issues the payment.

- The e-commerce application also generates an internal transaction and routes it to the enterprise's legacy system, which then updates the department's budgetary files.

- For disapproved orders, the e-mail system—through groupware functionality—triggers a pager message to the requisitioner indicating the reason for disapproval.

- At any time during this process, the requisitioner can sign on to the groupware database and check the progress of the workflow or status of approval. The requisitioner can sign on to the groupware database through various information appliances such as wireless telephones and PDAs (e.g., Lotus's Wireless Domino Access allows access to Notes databases from cell phones and PDAs).

Implementing the above process requires the following:

- Definition of the workflow through a groupware product.

- Definition of the roles of people (requisitioner, approvers, and security administrators who can define the various roles in the product).

- An e-mail system that interfaces with the groupware's workflow product and the e-commerce application.

- An e-mail system that incorporates functionality and rules to trigger pager messages to appropriate users.

- Data connectors that interface with the enterprise's legacy systems (e.g., Lotus provides Domino Enterprise Connection Services).

- Implementation of an e-commerce procurement application (e.g., Netscape's BuyerXpert).

- Interface of the e-commerce procurement application with the VAN that processes EDI transactions.

- Online validation of the purchase orders at the supplier's site.

- Issuance of the invoice to the buying organization (potentially through an EDI system or e-commerce application).

- Linking of the seller's application to the backend shipping application.

This list by no means provides all of the steps required to automate the procurement application. Rather, it is meant to illustrate the manner in which groupware applications can interplay with various enterprise applications in order to realize an end-to-end business process.

In conclusion, it is vital to understand that not all groupware products provide this degree of functionality. Some tools are better at providing communication and collaboration features, whereas others provide robust workflow functionality. Enterprises may have to implement various tools from various vendors to build a solution that fulfills the requirements of their business processes.

6.5   E-Commerce Application Development Standards

This section highlights the various industry frameworks and tools used for the development of backend e-commerce applications. For internal deployment of e-commerce applications, enterprises usually have three choices. First, an organization can use various off-the-shelf application suites to integrate and deploy e-commerce applications such as storefronts and simple information access services. Organizations can also acquire suitable vertical application suites related to banking, health care, sales force automation, and so on. And finally, organizations can use various development toolkits by industry players such as Microsoft, Netscape, and HP to develop specialized modules, interfaces, and services.

E-commerce application suites and backend ERP applications are sometimes not sufficient to deploy full-fledged e-commerce applications and systems. In such cases, organizations must develop the necessary business functionality and integrate with appropriate prepackaged e-commerce applications. Numerous technologies and programming frameworks have surfaced to enable the development of backend business logic and components. These technologies suit different business requirements. This section presents some of the technologies necessary to build backend business logic for e-commerce applications.

6.5.1   Java Servlets

Java servlets are programs developed in the Java programming language based on Sun's servlet API specification. Java servlets run on the server and extend the functionality of the Web server, just as applets extend the functionality of the browsers by providing an executable code for the client. Servlets usually host requests from clients and service those requests by either connecting to a database directly through database interfaces like JDBC or invoking other applications that service all or select portions of the user requests.

The industry considers the servlet model to be a prime candidate for replacing CGI-based scripts and programs. (CGI-based scripts are the primary means of invoking backend applications and access databases.) This is because servlets provide a more efficient paradigm for invoking backend applications and connecting to backend databases. For example, unlike CGI scripts, servlets do not require a separate process for each browser request and hence provide an efficient mechanism for invoking server-side programs, especially for handling higher-volume traffic. Java servlets also maintain consistent connections with backend databases and other programs. Servlets also have the ability to interface directly with enterprise databases, which enables enterprises to write simple and medium-sized interactive e-commerce applications. Due to their roots in the Java programming language, servlets are portable across various platforms and interoperate with a multitude of Web servers, as their APIs enjoy wide support.

Servlets run under a servlet engine that is usually included in an application server. Various vendors provide application servers that support execution of servlets. These application servers are either full implementation of Web servers that support servlet development and execution or provide plug-ins and add-ons to existing Web servers to support servlets. Full Web server implementations include Lotus Domino Go Webserver, Java Web server, and Konsoft Enterprise Server. The IBM WebSphere Application server provides an add-on for execution of servlets.

6.5.2   Enterprise Java Beans (EJB)

EJB is a Sun specification used for building portable application business logic. EJBs are similar to JavaBeans, which are reusable client-side components used to develop information-appliance GUIs and other client-side application logic. EJB's extend the same model to the server side and provide a framework for the development of reusable server-side application components.

While servlets are ideal for developing simple e-commerce applications, the development of complex business logic requires a computing paradigm such as the one offered by EJBs. The EJB specification enables the development of object-oriented application modules and masks numerous development complexities from the developers, including transactional complexities. For example, EJB includes the specification of "entity beans" that enable developers to focus on developing business components and logic without regard to the underlying logic of accessing databases and transaction services. By coupling entity beans with databases, developers can write applications against entity beans. This makes EJBs portable across platforms and developers do not have to write SQL to access databases directly.

EJBs require "containers" that provide services for their execution and management in a runtime environment. Application servers usually provide these containers, which are similar to servlet engines, to support the execution of EJBs. Any application server that implements the EJB specification can support EJBs. This facilitates the porting of EJBs from one application server to another without any modifications. This is unlike the current implementation of enterprise transaction-based applications, which are locked into one vendor's implementation so that porting application logic from one server to another requires extensive modifications. For example, an application component that is developed to run in a CICS environment requires extensive modification before it can run in BEA Tuxedo's application environment. Legacy application environments such as CICS and Tuxedo do not interoperate with EJBs since they do not implement the EJB specification. However, their vendors have recently released new versions of their products that support such interoperability.

Three versions of the EJB specification have surfaced as of this writing: 1.0, 1.1, and 2.0. Application servers have implemented specifications through version 1.1 with plans to implement version 2.0. The 2.0 specification packs more functions for robustness, performance, and scalability.

Various development tools from vendors enable the development of EJBs. These tools are either standalone tools or are IDE as part of application servers. Examples of application development tools are IBM's Visual Age, Symantec's Visual Café, Inprise's JBuilder, and BEA's WebLogic.

6.5.3   Active Server Pages (ASP)

ASPs are HTML files that contain references to various scripts and programs that execute on the server when the user requests the ASP page. Upon executing an ASP page, the Web server produces an HTML file and sends it to the

Web browser. The server-side programs are usually VBScript and JScript programs that handle user input from a browser and can perform various operations such as accessing backend databases, retrieving results, and presenting the output to the user's browser as an HTML page on the appropriate information appliance. ASP files have the .asp file extension.

6.5.4   Java Server Pages (JSP)

JSPs are similar to ASPs. JSPs control the appearance of HTML pages on the user browser. JSPs have embedded Java code (Java servlets) that run on the server when accessed. Within the HTML file, appropriate HTML tags contain bracketed Java code. This approach differs from the one where backend Java programs runs to generate HTML code and sends it to browsers on information appliances.

Notes and Web Sites

[1]   www.omg.org/cgi-bin/doc?formal/98-12-09

[2]   www.iona.com

[3]   www.javasoft.com

[4]   www.novell.com/corp/strategy/fsd/whatis.html

[5]   Microsoft Exchange Server, Microsoft's vision of unified messaging white paper.

[6]   Management of Internet Names and Addresses, United States Department of Commerce, Docket number 980212036-8146-02.

[7]   Status Report to the Department of Commerce, June 15, 1999.

[8]   www.icann.org

[9]   www.icann.org/registrars/icann-pr21apr99.htm

[10]   www.corenic.org

[11]   Internet Governance, ICANN and ASI information (www.arin.net).

[12]   www.wfmc.org

7

The E-Commerce Payment Infrastructure

Payments are an integral component of any commerce activity. Consumers pay businesses for products or services. Similarly, businesses pay consumers and other businesses for various activities. Both consumers and businesses use a variety of payment instruments and mechanisms to pay for commerce transactions. Popular payment instruments include cash, checks, credit cards, and electronic transfers. Each payment instrument is suitable for different conditions. For example, consumers usually use cash for relatively small-amount transactions. Checks are used for payments of any amount. For very large transfers, however, individuals or businesses use electronic wire transfers, a service available through local financial institutions.

Both financial institutions and the government regulating bodies (e.g., the Federal Reserve System in the United States, Bank of England in the United Kingdom) play a central and vital role in processing various payment mechanisms. Government institutions regulate issuance of currencies and control their flow. Financial institutions provide end-user services to merchants and individual consumers that include depositing, withdrawal, and other movement of payments. With the economic growth and associated rapid increase in commerce transactions and related payments, both governments and financial institutions established an electronic infrastructure to process various types of payments and hence facilitate the free flow of commerce. In fact, governments and financial institutions have relied upon an electronic payment infrastructure for the past three decades. Credit cards, for example, surfaced in the 1960s. Similarly, various clearinghouse-based payments that credit and debit the accounts of various parties involved in transactions rely on an electronic infrastructure that has been in existence since the 1970s.

However, the initial legs of the payment process in the business-to-consumer paradigm did not begin to become electronic until a few years ago. It is still quite common, for example, for customers to manually issue checks for payments or submit checks manually to their financial institutions for processing. Similarly, carrying cash to pay for purchases continues to be one of the dominant methods for paying for products and services. Point of sale (POS) terminals at merchant locations and automated teller machines (ATMs) were among the first mechanisms to submit customer transactions electronically. The POS terminals accept plastic cards (debit or credit cards) as input from the customers. Financial institutions issue these cards and enable customers to conduct various financial services, such as withdraw cash from their accounts (debit cards), or extend credit to customers by providing them the flexibility to pay later (credit cards).

The dawn of business-to-consumer e-commerce–enabled financial institutions and merchants pushed the existing forms of payment instruments such as checks and credit cards onto the e-commerce bandwagon. Not optimized for e-commerce, use of these payment instruments posed numerous challenges (e.g., security risks associated with submitting payments over the insecure Internet channel). Security technologies such as SSL provided interim solutions but were not foolproof.

The next wave of technologies considerably address those issues. Introduction and use of SET for credit card transactions fully secures all legs of the credit card transactions. Similarly, the introduction of eChecks by the U.S. government provides an electronic means for customers to submit checks and have them processed electronically by merchants and banks.

For business-to-business transactions, large organizations have been at the forefront in the processing of large payments by relying on the EDI infrastructure. Financial EDI, as it is usually called, enables organizations to exchange large sums and volumes of payments between participating organizations through appropriate financial institutions. With the emergence of various electronic payment models, small businesses can participate in an electronic exchange of payments as well, using the Internet and the Automated Clearing House (ACH) payment networks.

This chapter discusses the various payment instruments making headway in the various e-commerce initiatives. The last section focuses on various payment models and the associated technology infrastructure that facilitates payments over the Internet.

7.1 Federal Reserve System

The Federal Reserve System acts as the central bank of the United States. The Federal Reserve sets and controls the monetary policies of the country within the framework of policies set forth by the government. The Federal Reserve also distributes currency (notes and coin) and removes unfit currency from circulation.

The Federal Reserve System plays a central role in controlling the payment infrastructure of the country and controls the various financial institutions. The system consists of twelve regional Federal Reserve Banks that among other functions operate the nationwide payment system. Each regional bank acts as a depository for the financial institutions under its district. The Federal Reserve monitors financial institutions' compliance with applicable laws and regulations. The Federal Reserve also acts as an intermediary between banks in settling accounts between payers and payees. One reason for the Federal Reserve to play this role is the the fact that it maintains accounts for financial institutions. Settling accounts therefore involves the Federal Reserve debiting one financial institution's account(s) and crediting the other institution's account(s).

The Federal Reserve system provides two primary means of processing electronic funds transfers (EFTs): Fedwire and Automated Clearing House (ACH). Banks use the Fedwire system to transfer large amounts of money themselves or on their customer's behalf. Fedwire transfers are processed immediately and the Federal Reserve debits and credits accounts of the banks involved in the transfer appropriately. The ACH system is used to transfer relatively smaller amounts. ACH payments take longer to process and settle. The next section discusses the ACH system in greater detail. The Federal Reserve provides ACH services to most financial institutions.

7.1.1 Automated Clearing House (ACH)

ACH is a private nationwide network for the collection and settlement of funds. The ACH network provides the electronic payment infrastructure for the transferring of payments. It is operated by the primary ACH operators, which include Arizona ACH Clearing House, Federal Reserve, New York Automated Clearing House, and VISA USA. Both the government and commercial sectors are active users of the ACH network services. The Federal Reserve System developed the ACH network in cooperation with the private sector in the early 1970s. ACH payments usually include the transfer of high-volume and low-dollar amounts between parties. Unlike Fedwire transfers, which are immediate, ACH transactions take approximately two days to settle.

ACH provides the electronic infrastructure for various types of debit and credit transactions. The ACH network allows crediting and debiting of customer accounts through EFT. The ACH network acts upon ACH instructions that contain the payer's name, account number, financial institution routing number, amount, and transaction date. Credit transactions include payroll payments and payments to other parties, whereas debit payments include payment for mortgages and insurance premiums. In ACH, one bank initiates the transaction while the other receives the transaction. ACH processing provides an advantageous alternative to payment instruments such as checks. As will be explained in later sections, it is expensive for checks to process due to various factors that include check printing and reconciliation. Other issues such as lost or stolen checks add to the expense.

The ACH network also plays a vital role in financial EDI. Financial EDI refers to EDI transactions that include the transmission of payments or other financial information. Many banks offer financial EDI services to their corporate customers. This means that banks can receive payment and remittance information from their corporate customers and pass on the remittance information to the other participating corporate customers. The bank processes the payment information through the ACH network.

Most banks and other institutions operate Checkfree's ACH system. Checkfree's Paperless Entry Processing (PEP+) is the online mainframe system that receives and originates ACH instructions for the ACH network. This system provides functions such as stop payments, automated settlement and scheduling, billing statistics, and other functions.

The National ACH Association (NACHA) promulgates ACH rules [1]. All organizations interfacing with the ACH system must abide by the ACH rules to use its services. NACHA represents thousands of financial institutions through its thirty-five regional ACH associations. NACHA also develops operating rules for the ACH Network and for emerging electronic payment solutions in the areas of Internet commerce, bill payment and presentment, financial electronic data interchange, cross-border transactions, electronic checks, and electronic benefits transfer. NACHA has formed six councils to address the mentioned solutions. These councils are the Bankers EDI Council, Electronic Check Council, Bill Payment Council, Cross-Border Council, Electronic Benefits Transfer Council and the Internet Council.

ACH processing is executed by the following four primary parties:

- *Originator:* This refers to the person or entity that initiates the ACH transaction through the ACH network.

- *Originator's depository financial institution (ODFI):* This refers to the financial institution that originates ACH entries into the ACH network on the originator's behalf.

- *Receiver:* This refers to the entity that either pays or receives the payment as a result of the ACH transaction.

- *Receiver's depository financial institution (RDFI):* This refers to the institution that receives the ACH entries from the ODFI.

As an example, in the ACH direct debit process, the originator submits written authorization to the receiver and submits an appropriate transaction through the originator's financial institution (ODFI) over the ACH network. The ACH network deducts the appropriate amount from the receiver's bank (RDFI) and deposits the money into the originator's account. The Direct Deposit process is just the reverse of the debit process.

7.1.2   Fedwire

Fedwire is an online communication network operated by the Federal Reserve System that facilitates money transfers between financial institutions. Due to the high cost and security of Fedwire messages, it is primarily used for transferring large amounts of money. Financial institutions use Fedwire for its customers and for settling its accounts with other banks. All Fedwire payments are final.

Financial institutions use various methods to access the Fedwire system. The Federal Reserve System, for example, offers institutions the Fedline system, which is the Federal Reserve's proprietary software for initiating Fedwire and other forms of payments. Alternately, financial institutions can connect their mainframe systems directly to the Federal Reserve System for large-volume transfers.

7.2   Credit Card Payments

Credit card payments are one of the most popular methods of payments. By receiving a credit card, the customer establishes an account with the financial institution and can make purchases. Credit card payments involve payments to other parties against that account. Once the card issuer's bank authorizes a payment for a customer, the merchants receive their payment in the bank account established at the merchant's bank, also referred to as the acquirer. The customer clears the balance in that account by paying the issuing institution through other means (e.g., checks or cash). For credit accounts, the customer has the option to pay a monthly fixed amount, whereas the institution charges the customer interest on the unpaid balance.

VISA and MasterCard are the primary credit card issuing organizations. VISA is a card issuing institution that includes approximately 21,000 member financial institutions. VISA member institutions can issue VISA cards (with the VISA logo) to their customers. MasterCard is a similar organization and has a larger membership of financial institutions. Other credit card organizations include Novus/Discover, Carte Blanche, American Express, and Diners Club.

7.2.1 Credit Card Process Flow

Several parties are involved in the credit card payment process. The parties include the customer that owns the credit card, the customer's card-issuing bank, the merchant, the merchant's bank (acquirer) that processes credit card payments for the merchant, and the transaction processor through which these payments are processed. The merchant bank is also referred to as the acquirer, as it acquires the purchase transaction from its enrolled merchants. The customer applies for a credit card at a bank that is a member of a credit card organization (e.g., VISA and MasterCard). The issuing of a card provides the customer with the flexibility to use a credit card as a payment instrument at any merchant's location that honors credit cards of that particular brand (e.g., VISA). The merchant, on the other hand, opens a business checking account at an acquirer institution that offers merchant credit card services to its customers. This allows the acquirer to deposit customer's payments to the merchant's account related to credit card sales. The acquirer charges the enrolled merchants a fee for this service. The acquirer also leases (or assists in leasing) the merchant the necessary terminal and network connectivity that the merchant uses to accept customer credit cards.

Figure 7.1 illustrates the existing credit card process flow. As illustrated, the merchant can accept a credit card in multiple forms. The traditional modes involve acceptance at POS locations and through telephone and mail orders. For POS payments, the customer physically presents the card to the merchant who then swipes the customer's card through a card reader. Sometimes the merchant also takes an imprint of the credit card. The POS terminal hooks to the

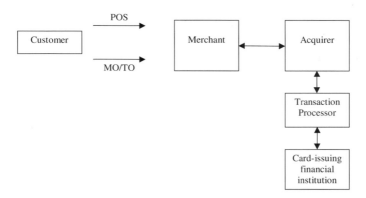

Figure 7.1 Traditional credit card process flow.

acquirer through dial-up or other network connectivity means. Upon receiving the merchant transaction, the acquirer forwards the transaction to the transaction processor for authorization and subsequently for settlement.

For mail order/telephone order (MOTO) transactions, the customer is not physically present at the merchant's location. Rather, the customer communicates the necessary card information to the customer through the mail or telephone. This raises several security and authentication issues. To counter such issues, merchants require customers to furnish additional information such as a street address and other information. In some cases, merchants require that merchandise be shipped to the customer's billing address.

After submitting the credit card for payment, the acquirer forwards the transaction to a transaction processor for authorization. Transaction processors are essentially the interface between the acquirers and card issuing banks. Acquirers usually seek the services of these transaction processors to authorize payments submitted by the customers to merchants that have opened accounts at the acquirer's institution. Transaction processors (also referred to as payment gateways) process credit card transactions on the acquirer's behalf. These processors also perform various merchant-processing functions that include billing and reporting and settlement services. Most acquirers operate these gateways locally. For example, Paymentech and Wells Fargo are financial institutions that operate these gateways internally. Other institutions outsource these services to third-party processors. Third-party transaction processors include First Data Corporation; Vital, which operates the VisaNet payment processing network (VISA and Total System Services, Inc., launched VisaNet to provide merchant processing services); First Data Merchant Services (FDMS); GPS; Amex; and First USA.

A credit card sale consists of three phases. *Authorization* refers to getting an approval from the card issuing financial institution about the validity of the card and the permission to charge the appropriate amount to the customer's card. The merchant can receive three possible authorization codes through the transaction processor. The three codes indicate authorization of the transaction, denial of the transaction, or a request for additional information before responding to the request. For requesting authorizations, merchants can request online authorizations and charge the customer's card simultaneously, as is in the case of buying a product. Alternately, the merchants can request a pre-authorization and charge the customer's card later (e.g., reservations for hotel accommodations and car rentals).

The second phase of the credit card transaction involves *capturing* the transaction. Capturing a transaction means that the merchant charges the credit card when it ships goods (or delivers services). Merchants are required by the laws of the financial institutions to charge the customer's card (capture the transaction) only after delivering goods or services. The third phase involves *settlement*, in which the acquirer receives money from the issuing institution and deposits the money into the merchant's accounts.

Transaction processors work in three modes referred to as terminal capture, host capture, and universal capture. In the terminal capture mode, merchants accumulate various credit card transactions in a batch file and then upload them to the transaction processors periodically. Online connections to transaction processors occur only for immediate sales authorizations (e.g., customer making a purchase). For other transactions (e.g., preauthorized transactions such as those used for hotel reservations and returns), merchants use the terminal capture mode to capture transactions in a batch file that later uploads to the transaction processor for settlement. Terminal capture mode differs from host capture mode, in which the transaction processor (instead of the merchant's terminal) captures all transactions as they occur online. In universal capture mode, the transaction is captured at both the host and at the merchant. On submitting transactions for capture, the acquirer deposits money into the merchant's account.

7.2.2 Secure Socket Layer (SSL)–Based Internet Payments

Commerce on the Internet necessitated associated payment mechanisms. The obvious choice for merchants was to enable customers to pay for products and services using credit cards. Due to the inherent security issues with the Internet, Netscape developed the SSL protocol. SSL has become the de facto standard to guarantee privacy and integrity for Internet sessions. Merchants use SSL to accept credit cards from customers when accepting payments over the Internet. When accepting credit card payments from the customer over the Internet, the merchant site displays an SSL-protected form to the customer. This form prompts the customer to enter pertinent information that includes name, address, card number, expiration date, and other order-related data. Merchants usually require customers to enter their address information in order to prevent fraudulent transactions. Merchants use the Address Verification Service (AVS) as an authentication method to validate the identity of the customer. The AVS method is available only in the United States and uses only the numbers in the address (e.g., street numbers and zip codes) to report the status to the merchant. The AVS supplies a Yes/No response to the merchant, who then decides to accept or deny the transaction. While not a foolproof method of authentication, it provides merchants with a deterrent for preventing fraudulent transactions.

The encrypted SSL session provides confidentiality and integrity for the credit card data sent to the merchant. The merchant server decrypts the SSL session and retrieves the payment and order information. The merchant server then interfaces with a payment server to route the payment information to the transaction processors that then return the authorization status to the merchant. The merchant server then proceeds accordingly with the processing of the order.

The disadvantages of using SSL for accepting credit card transactions are as follows:

- The merchant has access to customer's payment information. The merchant must therefore institute appropriate security controls to ensure protection of customer information (card number, etc.) from internal fraud scenarios.

- The merchant does not have the due assurance that a valid customer is using a given card. An imposter can use another person's card information and address information to pose as a valid customer. This increases the risk to the merchant.

The SET protocol surfaced to counter these issues and is discussed next.

7.2.3 Secure Electronic Transactions (SET)–Based Internet Payments

VISA, MasterCard, and other technology companies developed the SET specification to address secure credit card payment over the Internet. SET builds upon the use of public-key cryptography to validate the authenticity of the various parties involved in a credit card transaction. According to the published SET specifications, SET's primary objectives were the following:

- Provide merchants with the assurance that the customer using a customer credit card is a legitimate user of the credit card;
- Provide customers with the assurance that the merchant can accept branded credit cards and is associated with an acquirer institution;
- Provide integrity and confidentiality of the payment transaction throughout the authorization life cycle;
- Decouple the security of the protocol from the transport security mechanisms;
- Support interoperability between products and services (e.g., transaction processors, merchant's and acquirer's servers, and so on).

VISA and MasterCard formed SETco to oversee various aspects of SET implementation. SETco provides SET certification to various vendors that offer SET products and provides this certification through Tenth Mountain Systems, Inc. Vendors that fulfill all SET compliance requirements receive the approval to display the SET Mark alongside their product offerings. Various products such as certificate authorities, electronic wallets, payment gateways, and merchant software that employ the use of SET can receive appropriate certification from SETco regarding their products. The SET specification also allows for numerous extensions to the protocol without compromising the security aspects of the protocol. These extensions include the addressing of country-specific issues in the protocol, identification, and authentication considerations (e.g., addition of personal identification numbers in the payment instructions, additional credit card verification information, and others) [2].

SET employs public-key technology as the primary method of establishing trust between parties. Public-key cryptography uses a key pair to encrypt and decrypt information. Public-key algorithms are asymmetrical and based on the concept of key pairs, a public and private key. The two keys work in conjunction with each other: one key encrypts and only the other key can decrypt the data. The advantage is that the private key does not need to be revealed, and the public key can be made publicly available. The technology builds upon the premise that data encrypted with one key can only be decrypted with the other key. This facilitates secure transactions in public networks. Chapter 8 discusses public-key cryptography in greater detail.

All parties involved in a SET transaction have SET certificates that match a public key to the certificate owner. The issuing certificate authority within the SET hierarchy digitally signs SET certificates. All parties make their SET certificates publicly available, which enables parties to send information to one another by encrypting (digitally signing) the information using the other party's public key. This allows only the intended recipient to decrypt the information. By using SET, customers can send card data directly to the financial organizations that authorize the payment, and the card data is thus not exposed to the merchant. This is unlike an SSL transaction in which the merchant has clear access to the customer's credit card information.

A SET certificate consists of some of the following information:

- Certificate holder's public key (holder includes the cardholder, merchant, payment gateway, acquiring financial institution, or acquiring payment card brand);
- Certificate holder's unique identification;
- Issuing certificate authority's digital signature;
- Hashed payment card account number and/or a unique serial number;
- Signing algorithm;
- Validity period (typically the certificate issuance and expiration dates).

Figure 7.2 depicts the primary players of an Internet-based payment transaction, whereas Figure 7.3 illustrates the difference between SSL- and

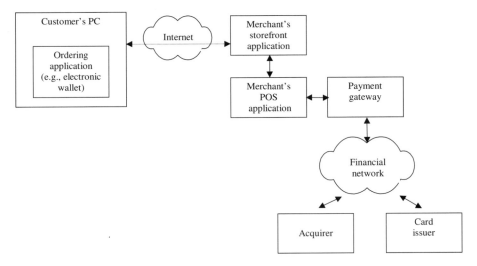

Figure 7.2   Internet-based credit card process flow.

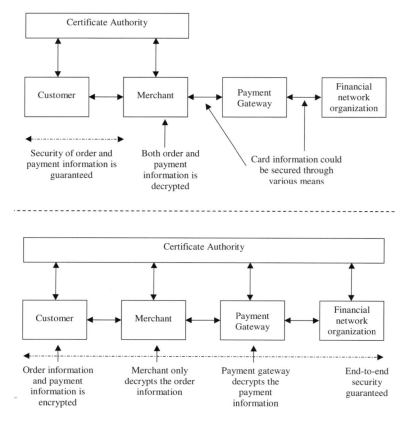

Figure 7.3   Security in a SSL transaction (top), and security in a SET transaction (bottom).

SET-based transactions. As Figure 7.3 illustrates, SSL secures the data path between the browser and the server that accepts the SSL connection by completely encrypting all information that is transmitted over the SSL session. SET, on the other hand, concerns itself only with the payment portion of the session and provides due assurance to all parties of the transaction regarding the authenticity of the customer and the merchant.

An end-to-end SET transaction requires all entities involved in a SET transaction to use SETco-certified products. The products include:

- *Wallet:* Customer's wallet application that submits credit card for purchasing should comply with SET specifications. Microsoft Wallet 3.0, Verifone's vWallet, CyberCash's CashRegister 4.0, and Gemplus's

Gemplus Wallet 1.0 are examples of customer wallet applications that are SET certified.

- *POS application:* The merchant requires a SET-compliant POS application to process customer requests by connecting to the customer's software application. Various vendors have developed such POS applications. These include GlobeSet's POS application and IBM's Payment Server applications. The POS solutions also interface with the merchant's commerce application (e.g., Microsoft's SiteServer). Most POS solutions provide out-of-the-box interoperability with a specific commerce server and provide software development kits to interface to customized commerce sites.

- *Payment gateway:* A payment gateway performs the various functions of a transaction processor. These include interfacing with the acquirer's systems and with the issuing institution's systems. Payment gateways for SET-based transactions provide functionality that includes payment authorization, payment capture, and protocol translation that enables the payment gateway to interface with the financial networks and various acquirer institutions. Certain payment gateways provide the flexibility to support both real-time and batch transactions. The payment gateways support various communication mechanisms (e.g., HTTP and X.25), which allows them to interface with backend financial networks. Payment gateways are usually high-end machines that support robust transaction processing. They handle multiple transactions from one acquirer and support multiple acquirer configurations. Various institutions can deploy payment gateway servers. These include the acquiring financial institutions or third-party transaction processors. Various vendors provide SETco certified payment gateway solutions. These include IBM's Payment Gateway, Globeset's Gateway, Trintech's S/PAY 2.5, VeriFone's vGATE 4.1, Hitachi's Payment Gateway 01-00, and others.

- *Certificate authority:* Certificate authorities (CAs) that issue SET should also be SET certified. Products include Entrust Technologies' Entrust/CommerceCA product and IBM's Payment Registry.

- *Development toolkits:* Various vendors provide libraries and development toolkits to facilitate the development of SET-based applications. For true interoperability, these libraries should be SET certified as well. Trintech's S/PAY is an example of such a toolkit.

The following describes the SET process flow illustrated in Figure 7.3:

1. The cardholder obtains a SET certificate from a designated CA, such as VeriSign. The cardholder provides pertinent information to the CA that includes name, issuing financial institution, and so on.

2. The CA obtains the necessary authentication and validation of the cardholder and the issuing financial institution before issuing the SET certificate.

3. Similarly, the merchant has to request a merchant certificate from the CA by identifying itself and providing pertinent information about the acquirer to the CA. The merchant requires this information before it can process SET payment instructions from the customer and interface with the payment gateway.

4. Once registered, customer and merchant can enter a SET-enabled transaction.

5. The customer visits the merchant's Web site, decides to purchase goods or services after browsing through products, and selects the appropriate option to buy the product(s) or services.

6. The customer software sends the purchase request to the merchant.

7. The merchant responds with appropriate authentication information. The merchant also authenticates itself to the payment gateway that the merchant will use to forward the customer's transaction.

8. The customer software validates the identity of the merchant using its certificate.

9. The customer software bundles the order information along with the payment information and transmits it to the merchant. The customer software encrypts both order and payment information before transmitting it to the merchant. However, the merchant only decrypts the order information and forwards the encrypted payment information to the payment gateway, which decrypts the payment information.

10. The merchant enters a payment authorization mode and communicates with the payment gateway. The merchant uses SET certificates to identify itself to the gateway while validating the integrity and authenticity of the gateway.

11. The gateway receives the payment request from the merchant and sends it to the customer's card issuing financial institution to obtain the appropriate authorization.

12. The gateway receives the authorization response from the institution and forwards it to the merchant, who in turn forwards the response to the customer.

13. After transaction authorization, the merchant processes the customer order.

14. The next step for the merchant is to initiate the purchase capture process. This process is identical to the payment authorization process, except the card issuing financial institution forwards payment to the merchant's account.

7.3 Electronic Cash

Cash is the most popular payment instrument in existence today. Unlike other payment instruments, cash requires no authorization and ITS value is immediately transferable. With the proliferation of e-commerce, various electronic cash schemes have surfaced to allow customers to store and transfer cash electronically. The primary motivation of offering these cash models has been the following:

- The need to pay for small-value purchases for which other payment mechanisms are impractical due to high associated processing costs;
- The need for anonymity when purchasing over the Internet and the Web;
- Replacing real cash with cash-based applications stored on smart cards.

The state of the electronic cash industry has been quite turbulent in the past few years. Various electronic cash payment models have surfaced, but very few have sustained stability and continued progress. Ecash, Millicent, and Net-Cash are a few examples of electronic cash schemes that have surfaced but as of this writing have not succeeded in attracting significant interest in the Internet commerce marketplace. Ecash, developed by DigiCash, Inc. proposed an electronic cash model to allow anonymous purchases over the Internet. However, due to lack of popularity, DigiCash has completely ceased operations. Millicent was developed by Digital Equipment Corporation and debuted as an electronic cash pilot project to offer a purchasing model for very small amounts (as small as 1/10th of a cent up to $5.00). The pilot ended in 1998 and has not progressed since then. The Information Sciences Institute proposed another electronic cash model called NetCash that consisted of currency servers that mint electronic coins, but it did not provide full anonymity. It, too, has failed to attract enough financial institutions and other parties to become a popular electronic cash scheme.

The preceding discussion shows that electronic cash schemes over the Internet have had little success. However, electronic cash schemes based on smart card technologies have had better luck. Mondex and VISA cash are two schemes that have progressed continually since their inception in the 1990s. These electronic cash schemes are also referred to as stored value cards as they store the value of cash on the cards. Special terminals transfer cash from the customer's cards to the seller's cards. The primary advantage of these cash schemes is that they usually work independently of the various financial networks. Electronic cash value can thus be transferred from one card to the other without interfacing with any financial institution or payment networks, and some electronic cash schemes have no associated transaction costs. Similar to cash, however, electronic cash used as smart cards cannot be recovered if lost, unlike credit cards and debit cards for which customers can notify their financial institutions and get replacements.

These electronic cash schemes require the use of sophisticated security technologies to prevent fraud and theft. Electronic cash schemes usually build upon smart card technologies that allow users to store value on smart cards and enable them to transfer value from one card to another. Software loaded on smart cards facilitates these functions through special readers or terminals that transfer value, enable the reading of card balances, and perform other functions. A smart card is a small plastic card approximately the size of a credit card. It differs from a credit card because of the integrated circuit (IC) chip that renders the card "smart." The smart card employs sophisticated security functions through hardware and software that prevent forgery and misuse. By using smart cards for electronic cash, consumers will be able to replace the various tokens (currency notes and other) that they use for daily use. These include subway and bus tokens, driver licenses, and so on. One smart card will enable users to carry cash and other applications.

The following provides a detailed discussion of Mondex and VISA Cash, which are smart card–based electronic cash schemes that have enjoyed greater success than their counterparts.

7.3.1 Mondex

Mondex, a subsidiary of MasterCard International offers an electronic cash scheme that was introduced in the early 1990s. Mondex's primary advantages are that it offers immediate transfer of value to the payee and works equally well on popular electronic channels such as the telephone and the Internet. Mondex also allows the storage and transferring of multicurrency cash onto its card.

Mondex's electronic cash scheme employs a smart card that has a built-in chip to store a value. Users protect the smart cards with a personal password to prevent other parties from using it. The smart card uses operating systems such as MULTOS, a popular operating system spearheaded by Maosco, Ltd., a worldwide consortium whose membership includes American Express, Discover Financial Services, Mondex International, and Motorola. The Maosco consortium is promoting the widespread use of the MULTOS system for use on smart cards.

Various logical slots on the chip enable the storage of multiple currencies. The chip also contains software programs that facilitate various functions such as transferring cash from one card to another or transferring cash from the card to another system. For example, Mondex International has acquired the services of Schlumberger to build the necessary applications for the MULTOS system to be used in Mondex smart cards. Two cardholders for instance can insert two cards into an electronic wallet and transfer value from one card to the other.

Mondex works in the following fashion:

1. Users open an account with a financial institution.
2. The financial institution offers the user a Mondex card.
3. Users approach a Mondex terminal or service point, enter their cards with the appropriate personal code, and load cash value into their cards from their bank accounts.
4. Users can load multiple currencies onto the Mondex card (e.g., the card holder can load US $100, 50 British pounds, etc.).
5. Users can insert the card into a balance reader to read the balance on the card.
6. Users can buy goods and services at merchant locations and machines such as vending machines, and telephones that accept the Mondex card.

Mondex owns the intellectual rights of the Mondex cash scheme and licenses these rights as well as the use of the brand name Mondex to other organizations. The Mondex cash scheme involves multiple participants that include the following:

- *Franchises:* Franchises manage the Mondex cash scheme in a given area (usually a country). Purchasing of franchise rights empowers franchise institutions to build the Mondex cash application (electronic purse) for smart cards and devise an end-to-end payment solution.

- *Originators:* Originators issue and control the Mondex currency in a given area. Originators manage all functions related to the disbursement of value to members and ensure that members comply with various rules and regulations and guard against counterfeit value. Originators monitor value movement between members and themselves and provide functions to support multiple currencies.

- *Members:* Members are licensed by the franchises. They get the Mondex cash scheme from the originators and issue Mondex cards to cardholders and merchants.

- *Manufacturers:* Manufacturers manufacture all equipment that is part of the Mondex cash scheme. This includes cards, POS terminals for merchants, balance readers, ATMs, vending machines, and so on. Specially designed phones allow reading of the Mondex card and can transfer it over phone lines to a similar phone that has a card reader, thus enabling the transference of cash over the telephone.

- *Merchants:* Merchants sign agreements with members that allow them to accept Mondex cards from cardholders. Merchants use Mondex-compatible equipment to process Mondex cash. This includes POS terminals, telephones, Internet server applications, and so on.

- *Cardholders:* Cardholders use the Mondex application stored on a smart card to pay for goods and services.

Many countries are adopting the Mondex cash scheme by licensing the technology from Mondex and introducing the Mondex scheme to consumers in their countries. Early pilots for Mondex have centered on close communities such as university students, specific company employees, and so on. Mondex has also partnered with telephone manufactures such as Alcatel to bring ATM services to user's mobile telephones. For example, as of this writing Mondex and Alcatel are working on a product that will allow users to insert their Mondex card into a GSM mobile phone, connect to their financial institution, and download cash onto their cards. Similarly, Intertrader Ltd. offers a CashBox product, which uses Mondex technology and allows cardholders to pay for various services over the Internet. The scheme works by cardholders inserting their Mondex card into a card reader (e.g., Gemplus card reader) that reads the value and transfers it to a CashBox installed at the merchant location. The approach facilitates the direct transfer of cash value to the merchant without any transaction fees or the involvement of any third parties.

7.3.2   VISA Cash

VISA Cash is another electronic cash technology. It is offered by VISA International and was launched during the 1996 summer Olympics. VISA Cash also uses smart cards to store cash value in a microchip.

VISA Cash offers two methods of payment. The first method involves the use of disposable cards that come preloaded with cash. Users can obtain these cards from card dispensing machines (with the VISA Cash logo). Users can use these cards at merchants that accept such cards. Once cardholders use all the cash on the cards, users can dispose of the cards. During use, the VISA Cash POS terminal at the merchant's location displays the remaining cash value on the card. The terminal prompts the merchant to enter the payment amount on the terminal. Upon verification, the terminal deducts the amount from the customer's card. The POS terminal tracks all transactions and enables the merchant to deposit the VISA cash amount into customers' bank accounts. Certain financial institutions also offer "VISA Viewers," which allow cardholders to insert their cards into those devices to view the remaining balances.

The other method involves the use of reloadable cards that allow users to add value to the cards at special ATMs. Users transfer cash value to the VISA cash cards by paying from their bank accounts or credit cards.

7.4   Electronic Check Processing and ATM-Based Banking

Checks continue to be popular for making both personal and business payments. The term *check* is defined under the Uniform Commercial Code as a payment instrument that is "drawn on a bank and is payable on demand." Traditional checks include the account holder's name at the issuing institution, the issuing bank's name, the bank's location, and fields for entering information such as payee, date, and so on. A check includes a series of numbers printed at the bottom of the check, also referred to as the magnetic ink character recognition (MICR) line. Electronic equipment uses the information in the MICR line to read, sort, and perform other automatic check-handling functions. The routing number is a nine-digit number enclosed between the symbols |:. The first four digits represent the Federal Reserve district where the financial institution is located, and the next four digits represent the code assigned to the financial institution. This code is the transit or routing number of the financial institution and indicates the issuing institution on which the check is drawn. The last digit is derived from an algorithmic calculation.

7.4.1 Check Processing Flow

Figure 7.4 illustrates the traditional check processing flow. As illustrated, a payer issues a check to the merchant, who then submits the checks to its depository bank. The depository bank may send the check directly to the issuing institution or forward it to a clearinghouse or the Federal Reserve office for clearing. The Federal Reserve maintains its own check-clearing offices. Checks submitted to the Federal Reserve prompt the Federal Reserve to debit the payer's accounts and credit payee's accounts in the depository institution. Clearinghouses work by enrolling various financial institutions for its check clearing services. As checks flow in, clearinghouses track amounts owed by various participating financial institutions. Periodically, the clearinghouses either forward this information to the Federal Reserve, which debits and credits accounts of financial institutions, or informs the financial institutions, which use Fedwire to transfer funds.

Check processing is quite expensive due to paper-handling costs. This includes the cost of supplies (e.g., paper, envelopes), postage, and printing as well as the various check-handling costs that involve receiving, sorting, encoding, routing them to financial institutions. Manual activities involved in the handling of paper checks also introduce the possibility of errors.

Recently, certain banks have automated some of the check processing through the ACH network. The process in that case works as follows and is illustrated in Figure 7.5:

1. Payer issues a check to the merchant.
2. The merchant submits its checks to its depository bank (ODFI).
3. The bank encodes the check and retrieves pertinent information to form an ACH transaction.
4. The bank submits the ACH transaction through an ACH operator.
5. The ACH operator settles the transaction by debiting the payer's account at the paying bank and depositing the money into the depository bank.

Other initiatives have surfaced to accept checks through other electronic channels (e.g., telephone, fax, e-mail, and the Internet). The concept behind accepting checks through these channels also rests upon using the ACH network. As most of the information required for processing an ACH transaction is inscribed on the check, the merchant submits the transaction through the

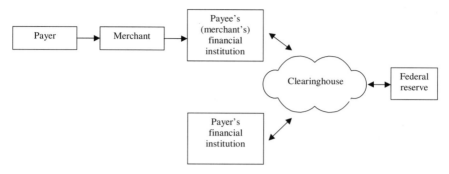

Figure 7.4   Check clearing through the clearinghouse.

ACH network. Accepting checks through the Internet involves presenting an SSL-protected Web page to the user that prompts for information that is inscribed on the check. The merchant uses that information to form an ACH-based transaction and submits it to the financial institution for processing. Accepting checks electronically at the POS involves scanning them. The scanner records vital information such as MICR numbers and the account holder's name. The merchant then voids the issued paper check, prints a receipt, and asks the customer to sign the receipt. This is required to process the check transaction through the ACH network. The merchant then submits the ACH transaction through its bank using an ACH operator.

For electronic check processing through other channels, the process is similar. The primary steps required to process checks include collecting pertinent information, and submitting confirmations to customers.

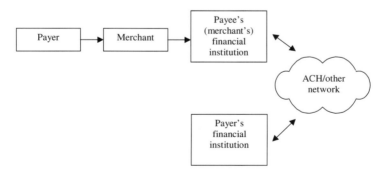

Figure 7.5   Check clearing through the ACH network.

7.4.2   FSTC Electronic Check

Financial Services Technology Consortium (FSTC) is a nonprofit organization of banks, financial service providers, universities, and various research organizations. FSTC sponsors research projects in the area of electronic payments and other aspects of the financial services industry. FSTC developed the electronic check and tested it in June 1998, when the U.S. government used the Internet to pay GTE for rendering various services. The eCheck initiative is a cooperative effort of many financial institutions and other organizations. These include Agorics Incorporated, Bank of America, Certicom, Defense Finance and Accounting Service, Federal Reserve Financial Services, Fleet Bank, GTE, IBM, IntraNet, RDM Corporation, SafeNet, Sun Microsystems, and the United States Treasury [3].

The concept of eCheck builds upon current check processing practices and thus facilitate easier implementation of the eCheck concept. It allows all types of payers to pay all types of payees. For example, individual users can use eChecks to pay businesses or other individual payees, much like traditional checks. eCheck reduces various paper-handling expenses by using electronic information and the Internet to transmit eChecks from one party to another. The use of electronic information institutes electronic controls that reduce errors and automates the process.

The eCheck project incorporates strong security technologies to ensure safe delivery of money to the payees. Security technologies include the use of encryption and authentication using digital signatures and smart cards. The eCheck therefore requires the issuance of certificates to all parties that take part in the eCheck process. This includes users issuing and receiving eChecks and financial institutions that accept and clear eChecks for the customer.

The primary components of the eCheck initiative are:

- *Electronic checkbook:* An electronic checkbook is a smart card that stores the private key of the payer. A PIN protects the use of the smart card. The stored private key is used to sign the eChecks issued by the payer.
- *Certificate authority:* A CA issues the certificates used for digitally signing eChecks, authenticates various parties, and ensures confidentiality. The model uses the standard X.509 certificates. Certificates are required for payers, payees, and all involved financial institutions.
- *eCheck:* eCheck information resides in a file and uses the same information as traditional checks. The design of eChecks allows the delivery of

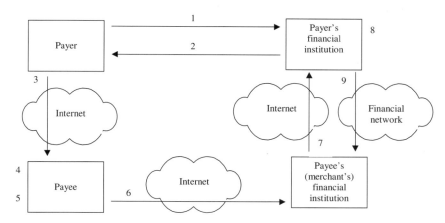

Figure 7.6 FSTC's eCheck process flow.

eChecks between various participants using e-mail or through Web forms. Information in eCheck is stored based on the Financial Services Markup Language (FSML) specification, which is based on the SGML specification. FSTC developed FSML for eChecks but it is extensible enough to support other financial documents as well.

Figure 7.6 illustrates a simplified eCheck flow. The following outlines the various steps involved in the process:

1. Payer applies for an eCheck account at the issuing bank.
2. The issuing bank issues a certificate to the payer for signing checks and an electronic checkbook.
3. To write an eCheck, the payer enters the electronic checkbook's PIN, issues an eCheck to the payee, and transmits the eCheck through either Internet e-mail or the Web.
4. The payee authenticates the eCheck by verifying the payer's digital signature.
5. The payee uses his electronic checkbook to endorse it with his digital signature.
6. The payee uses the deposit slip from the electronic checkbook to transmit the eCheck to the payee's financial institution.
7. The payee's financial institution forwards the eCheck to the issuing financial institution.

8. The issuing financial institution performs various checks such as verifying that the transmission of the eCheck is not a duplicate and that
   the payer's certificate is valid.

9. After verifying the account details, balances, and so on, the issuing
   institution deposits the money into the payee's account.

FSTC is working on numerous other projects with its consortium members to globalize the use of eChecks. FSTC also provides a roadmap to various
organizations that want to participate in the current market trial of eCheck.
Information on this trial is available at www.fstc.org.

7.4.3   ATMs

ATMs are a popular way to withdraw cash from bank accounts and perform
various other banking services (e.g., bill payment, balance inquiry). The ubiquity of ATM terminals makes it convenient for consumers to perform various
financial services. The original motivation behind installing ATMs was to facilitate user access to banking functions and reduce the high costs associated with
erecting brick-and-mortar branches and staffing them with human tellers.
However, organizations have recently begun testing ATMs for marketing purposes as well. Driven by the maturation of network technologies, organizations
are testing a new generation of ATMs that will deliver captivating advertisements on the ATM screens while users' transactions are processed through the
ATM network. FSTC is also exploring initiatives that will bring ATM services
onto the Internet by enabling banks to offer Web-based personalized interfaces
to customers. However, as such initiatives require the addressing of various
interoperability issues, they are still in their infancy.

Figure 7.7 illustrates the typical structure of an ATM network that consists of four primary elements. The ATM terminal that customers use for various banking services is the first component. This ATM terminal may belong to
the customer's financial institution or to another one. Second, the ATM processor is a computer network that processes ATM transactions for the bank.
EDS is an example of an organization that operates an ATM processor. TYME
is another EFT processor that processes ATM transactions. The ATM terminal
connects to the ATM processor through leased communication lines. Recent
trends to decrease costs associated with ATM use are driving financial institutions to explore other options such as dial-up. The high cost of leased lines is
due to the complexity of laying leased lines in certain locations and high
monthly costs. Using dial-up, the ATM dials the number of an ATM processor's local POP location. The connection stays live only during the transaction.

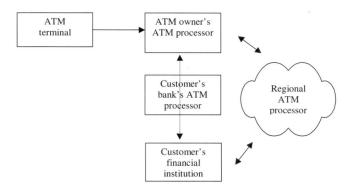

Figure 7.7 ATM transaction process flow.

The third element is the regional computing network that connects various processors and banks, thus allowing customers to use one bank's ATM card on another ATM. Plus and Cirrus are the nation's two largest ATM networks and are operated by VISA and MasterCard, respectively. Both the regional networks and ATM processors act as switches that route and settle transactions between financial institutions. These networks facilitate connectivity between financial institutions, merchants, POS terminals, and ATM cardholders.

The fourth component is the financial institution that holds the account against which the customer requests various ATB-based services.

The following depicts the typical process of an ATM transaction:

1. A customer uses a bank's ATM machine, inserts the debit card, and enters the appropriate PIN number.

2. Information from the ATM is transmitted to the ATM's bank's processor, which forwards the ATM transaction to the ATM processor of the card issuing bank. The ATM processors may communicate directly or through the regional processors (e.g., Cirrus, or Plus networks).

3. The issuing bank's ATM processor communicates with the issuing bank to check the validity of the transaction (e.g., balance in account, limits for withdrawal, etc.).

4. The response travels through the mesh of networks to the appropriate ATM terminal.

5. For cash withdrawal authorizations, the ATM dispenses cash to the user. For others, the ATM responds to the user with an appropriate message.

7.5   Payment Models

Earlier sections focused on the some of the prevailing payment instruments. This section focuses on the various payment models that enable customers and businesses to participate in the payment process. For example, electronic bill payment and presentment (EBPP) enables billers to leverage the power of the Internet to reduce their operating expenses while extending convenience and better customer service to their customers. Similarly, electronic wallets facilitate the process of paying online. The following sections discuss these models in further detail.

7.5.1   Electronic Wallets

Business-to-consumer e-commerce transactions usually require users to pay through the Internet. Customers enter orders on the Internet by filling in their name, address, credit or debit card number, expiration date, and other information necessary to complete the transaction. The merchant uses this information to charge the customer for purchases as well as to ship the goods (if applicable) to the customer's address. As customers expand their Internet shopping horizons to include multiple e-commerce sites, repeated entry of the same information (name, address, and so on.) becomes burdensome. Electronic wallets provide a mechanism to store such information securely and enable customers to invoke or activate the pertinent electronic wallets when entering e-commerce transactions. Electronic wallets automatically fill in the required fields, thus alleviating the burden for the customer. In some cases, electronic wallet applications communicate directly with merchant's servers to transmit wallet information.

An electronic wallet's primary feature is its ability to store primary user profile information, accommodate multiple payment mechanisms, track customers' transactions, and interoperate with popular Web browsers. Advanced electronic wallets also support secure forms of credit card processing (e.g., SET implementations).

Multiple implementations of electronic wallets have surfaced recently. These implementations differ based upon their functionality, portability, convenience, and other features. The following delineates popular implementations of electronic wallet applications.

7.5.1.1   Site-Specific Wallets

In this implementation, each e-commerce site stores user information in its database and reuses this information upon each user's return to the site. For example, a customer may visit a site and order a merchandise by offering the e-commerce site the necessary information to complete the order. The merchant may store this information by keeping a cookie on the user's information appliance with a special user number. The merchant my index this information by the user's e-mail address or user ID. On the user's next visit, when the customer orders again, the site pulls the user profile information (name, address, card number, etc.) based on the e-mail address, user ID, or cookie information. This frees the user from having to enter the information on the forms again. Usually, sites also require users to authenticate to the site before using the wallet information by prompting for a password or other authenticator(s). Amazon.com stores user information (e.g., shipping address, credit card information, and so on) and facilitates subsequent ordering of books by minimizing the time and effort for users.

The security of site-specific wallets depends upon the security of the e-commerce site. The wallet merely populates the form with appropriate user information. For example, if the e-commerce site invokes an SSL session to enter user information, the wallet will merely populate the fields and leave the SSL security to the e-commerce site application. Similarly, if the site does not offer any security, the wallet will not provide additional security. The e-commerce site, however, is responsible for storing the wallet information (user profile information) securely in its databases. Mismanagement or misuse of that information, especially for e-commerce organizations, can seriously impact an organization's reputation and drive customers away. For example, the Federal Trade Commission (FTC) pressed charges against GeoCities, Inc. (now part of Yahoo!, Inc.) for improperly handling customer information and selling critical profile information to other parties.

7.5.1.2 Personal Wallets

Personal wallets store user profile information locally on the user's information appliance. During the initial setup of the electronic wallet application, the user enters all pertinent user information in the wallet and protects the application with a self-selected password. The wallet then stores all profile information in an encrypted file on the user's information appliance and unlocks (decrypts) the wallet information when the user enters the correct password. In this scenario, security of the wallet information is contingent upon the strength of the password that the user chooses to store on the information appliance. The strength of the cryptographic algorithm used to encrypt the wallet information is of vital importance as well, especially if multiple parties use the information appliance. While ordering, the customer can drag and drop the wallet component to the application form to automatically fill in all the required fields of the form. Launchpad Technology's eWallet and Verifone's vWallet are examples. With vWallet, users can create multiple wallets, thus enabling multiple users of a PC to have individual wallets. vWallet also captures users' transactions and enables them to be loaded into PC-based financial packages.

7.5.1.3   Remote Wallets

In this implementation, a remote organization (other than the customer or the merchant) secures user profile information. The customer invokes the wallet information by clicking on the appropriate logo, which then prompts the user to enter the password for their wallet. Subsequent to successful authentication, the merchant application automatically retrieves user information from the wallet and uses it to complete the customer order. This implementation requires customers to register the remote wallet application by entering their information and protecting it with a password. The user can then use this wallet at sites that support the wallet implementation. CyberCash is an example of an organization that offers its customers the InstaBuy wallet application. Various e-commerce merchants support the InstaBuy wallet. Another popular wallet is Citibank's CitiWallet. CitiWallet supports numerous merchants and provides features such as the storage of credit cards and the tracking transactions performed through all those credit cards. With one click, users can use their wallets to fill in all the information on the merchant form and pay for purchases. The disadvantage of this wallet over personal wallets is that customers have to unlock their wallets every time they visit a different site, unlike the personal wallets that remain unlocked on the user's PC for the duration of the user's session.

7.5.2   Electronic Bill Presentment and Payment (EBPP)

Electronic Bill Payment and Presentment leverages the power of the Internet to present bills to customers. Customers can access their bills through various electronic channels (e.g., Internet e-mail) and authorize bill payment to their billers through traditional electronic means (e.g., ACH payments). Certain EBPP service providers also allow customers to download all bill information (e.g., payments, etc.) to various financial packages (e.g., Microsoft Money, Quicken). For the biller, EBPP dramatically reduces the various paper handling costs and provides cross-selling and one-to-one marketing opportunities.

The current methods of sending bills to customers and collecting payment from customers for various products and services is quite burdensome for all concerned. Customers spend an enormous amount of time paying bills

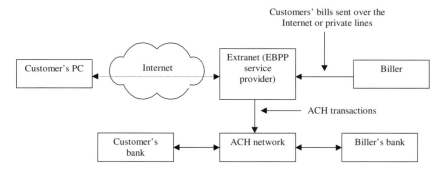

Figure 7.8 EBPP process flow.

through regular means that usually involves checking bills, writing checks, and mailing them to the billers. Billers incur a huge cost in printing bills, mailing them, receiving checks from customers, submitting checks for processing, and other associated paper handling costs. Financial institutions incur high costs in processing those checks.

Figure 7.8 illustrates a typical EBPP model. In this model, an external entity such as an extranet serves as the EBPP service provider and consolidates various billers' bills for customers. The EBPP provider registers billers and customers for the usage of its services. Customers receive account numbers for the usage of its services and provide the EBPP payment information (e.g., bank account number, routing number). The EBPP service provider receives bills in an electronic format from billers and converts them into electronic bills that can be displayed through a user's browser. Some EBPP service providers display the entire bill contents (including fine print). The EBPP service provider consolidates all bills from all of the customer's billers registered to the system and thus provide customers with an overview of all their bill details, amounts due, and payment schedules. The EBPP service providers in this model are also referred to as statement consolidators. Customers can view their bills by logging -onto the Web site and can initiate payments to the respective billers. The EBPP service provider submits the payment through financial networks (e.g., ACH network) to debit the customer's accounts and credits the billers' accounts. TransPoint provides similar bill payment services to its customers.

In another EBPP model the biller offers services at its own Web site. The industry does not expect this model to be popular among customers, as customers prefer to have a consolidated view of their bills rather than having to individually log-on to separate Web sites for the payment of bills. Various vendors offer services and products for this model of EBPP. For example, Cyber-

Cash offers the PayNow service, which enables billers to work in conjunction with CyberCash to present bills and collect payment from customers. The manner in which this solution works is as follows:

1. A biller hosts bill presentment application at its Web server.
2. Customers login to the Web site to view bills.
3. Customers opt to pay the bill through CyberCash's PayNow service.
4. The payment transaction details get routed to CyberCash through the biller's server.
5. CyberCash decrypts the secure information on its CashRegister server and prepares an ACH transaction file.
6. The ACH transaction file is forwarded to the biller's ODFI for crediting the biller's accounts.

Billers thus have two ways to present EBPP services to their customers. The choice of a particular approach depends upon various factors, some of which are as follows:

- By providing their own EBPP-hosting services, billers can use the real estate of their Web pages to cross-sell and target one-to-one marketing solutions. Hosting through the EBPP service providers does not offer them this opportunity, as the service provider uses their Web real estate at their discretion.

- The biller should assess the costs of internally hosting EBPP-related technology operations. These costs include the indirect costs of continued operations and the maintenance of various technology-related services such as data center-related costs.

- Service providers usually charge billers transaction fees for hosting their bills. Self-hosting does not involve such costs.

- Self-hosting enables billers to exploit new technology initiatives and avenues to further improve their EBPP capabilities. Billers, for example, can extend their EBPP capabilities to include mobile information appliances as well. By outsourcing EBPP services, billers become dependent upon the EBPP service provider's technology base and plans unless they negotiate special terms before signing an agreement.

Checkfree is an example of a statement consolidator. Checkfree provides various EBPP services for financial institutions, merchants, and customers by providing its E-Bill services, which deliver electronic bills to consumers and collect payments for financial institutions. Checkfree processes payments and deposits bills into billers accounts through various networks that include ACH and CheckFree's proprietary ECX Payment Network [4]. Checkfree works closely with financial institutions and billers to integrate billers' systems with Checkfree's to offer EBPP services to the billers' customers.

TransPoint is another EBPP service provider. TransPoint is a partnership of Microsoft and First Data Corporation. TransPoint enables billers to provide electronic bills (e-bills) to their customers over the Internet. To do so, Trans-Point offers its billers Biller Integration Software (BIS), which enables billers to integrate their billing systems with TransPoint's e-bills. TransPoint provides links to its customers through certain online banking Web sites and through portals such as MSN.

Implementation of an EBPP system requires the following:

- *Robust directory support:* The directory supports profiles of customers that log on to the system to check their bills. Depending upon the EBPP model (service provider or self-hosting EBPP solution), the directory should be appropriately scalable to support the volume of Internet-based customers. Most EBPP products offer built-in directory support. For example, iPlanet Netscape's BillerXpert bundles Net-scape's Directory Server product as part of the package.

- *Support for e-mail:* E-mail is an integral component of an EBPP solution. Service providers can use e-mail to either notify customers of waiting bills in their EBPP accounts or in some cases send the entire bill to the customers.

- *Support for security technologies:* SSL is the common security technology supported by most platforms. Most EBPP products are also SET compatible. However, for using features such as e-mail, the product should support e-mail security technologies (e.g., S/MIME to send confidential information to customers). Chapter 8 provides additional details on S/MIME and S/MIME products.

- *Data extraction:* Billers store bill-related customer data on their systems in proprietary formats. An EBPP solution requires that billers devise data extraction mechanisms to pump this data into EBPP servers.

- *Support for standard formats:* The EBPP solution should interoperate with various systems for bill extraction. Most EBPP products, for example, interface with legacy systems, extract financial statements, and translate and present the bills to customers in an HTML format.

- *Bill presentation:* The EBPP solution should offer customers flexibility in viewing their bills. The user interface should display user bills in summarized and expanded formats and present a consolidated view of bills from other billers, thus making it easier for users to view and pay their bills. Users should be able to customize their views in viewing and paying bills. The application should also provide customers with access to archived bills.

- *Support for multiple payment types:* The EBPP solution should support multiple payment types (e.g., checks, credit cards).

- *Building of customer care:* EBPP service providers implement customer-care solutions to handle customer inquiries. Using various groupware tools such as Web collaboration and whiteboarding, customers and service representatives can simultaneously view bills and thus help users through their issues and questions. Service representatives can also use this opportunity to strengthen their relationship with customers by cross-selling and up-selling products to users.

- *Network connectivity:* The EBPP service provider has to establish appropriate network connectivity to the billers for receiving bills and submitting payments to financial networks (e.g., ACH network) for payment processing.

- *Business practices:* A service organization that implements EBPP services requires an extensive set of practices to ensure compliance with government regulations, protection of customer information, and other related issues. Formulating good business practices is therefore vital before putting systems into live implementation. NACHA's council for electronic billing and payment has formulated draft business practices that it recommends for various parties involved in the EBPP processes. These parties include the EBPP service providers (also referred to as Biller Service Providers or BSPs), customers, and billers and it covers practices for the enrollment/de-enrollment of customers, bill presentment, payment, and remittance. Issuing of standard business practices also provides interoperability in the world of EBPP, where numerous providers offer such services to their consumers.

For self-hosting EBPP solutions, vendors provide products to help organizations build EBPP solutions. Popular software products include Oracle's iBill&Pay and iPlanet Netscape's BillerXpert. These tools offer varying EBPP features. Organizations can use these tools to directly bill customers or financial institutions can use these tools to consolidate statements from various billers, thus providing a one-stop solution for customers. Oracle's iBill&Pay is a product that supports both these features (i.e., direct billing and bill consolidation). Other features these products support include multiple forms of payment, bill personalization and notification, targeted marketing, and integration with legacy financial systems.

Notes and Web Sites

[1] www.nacha.org
[2] www.setco.org
[3] www.fstc.org
[4] www.checkfree.com

Additional Web Sites and References

www.bog.frb.fed.us

www.mondex.com

O'Mahony, D., M. Pierce, H. Tewari. *Electronic Payment Systems*, Boston: Artech House, 1997.

8

E-Commerce Systems Security

Security has always been a forefront issue in the implementation of IT solutions. However, its complexity and necessity have both risen in the past few years. The information glut and ineffective means to process and disseminate this information has increased information security–related threats. Internet and distributed computing technologies have facilitated information processing across wider geographies and have contributed to this phenomenon as well. Furthermore, increased IT awareness and technological sophistication of users from all walks of life have resulted in the emergence of a wide range of intelligent hackers and cybercriminals. Organizations are wary of such threats and are willing to invest to fortify their defenses against such attacks.

Countering information security–related threats is most effective when addressed from an end-to-end perspective that covers the organization as a whole. This does not invalidate the requirement to individually address information security threats related to every e-commerce system. Rather, a holistic approach fortifies the technology and the business process infrastructure in which the e-commerce system serves its customers. E-commerce systems present more opportunities to cause security violations. However, most of these breaches directly relate to the weak procedural controls inherent in the operations of e-commerce and other IT systems. In essence, the inherent nature of e-commerce systems necessitates higher levels of both technological and procedural security controls.

The challenges inherent in securing e-commerce systems are threefold. First, an organization must be cognizant of the nature of security issues and related security controls. The technological complexity of e-commerce systems and multiple tiers associated with the implementation of these systems involve numerous solutions and technologies at different tiers. A user's request to a specific e-commerce service through an information appliance may have to go through many networks and systems before it can reach the server that would fulfill the customer's request. Understanding all such issues and the technologies that address them are of paramount importance.

Second, an organization has to determine appropriate costs of securing its e-commerce systems. Securing IT systems involves financial and architectural costs. The organization has to comprehend such direct and indirect costs before investing in the security controls and operational environment of the enterprise. Finally, an organization has to adopt the right balance of security controls (both technological and procedural) to secure its systems. Going under or over the appropriate limits can have a direct bearing on the enterprise's bottom line.

An organization faces numerous challenges in implementing appropriate information security controls to secure its systems and information. A successful organization understands relevant challenges and manages them accordingly. Some of the challenges are as follows:

- *Security is not built into IT products:* System-level hardware and software have traditionally lacked security features and have thus forced customers to either buy security packages or develop solutions to retrofit the missing security functionality into the system. Only very recently has the trend shifted to incorporate security features into computing resources.

- *Security is always an afterthought:* In the design and implementation of information systems, information security is usually considered toward the end of development life cycle or just prior to implementation. This lack of planning forces organizations to either compromise on security issues or implement quick and ineffective controls, thus exposing the systems to major internal and external risks.

- *Lack of management commitment:* Attaining higher levels of maturity for information security has a cost, which at times is quite high. Thus it is natural for management to compromise on vital security issues. Increasing management's awareness of such security risks, especially related to e-commerce systems, is vital.

- *Immaturity of software processes:* Numerous surveys indicate that more than 70% of an organization's software processes are not mature enough to deliver reliable and dependable software solutions to their respective businesses. This immaturity has a bearing on information security as well. Initiatives such as those of Software Engineering Institute's (SEI) Capability Maturity Model (CMM), ISO, and IEEE have established software practices that ensure the roll-out of dependable information systems, thus allaying major concerns of security.

8.1 Security Services and Technologies

IT systems have traditionally incorporated a defined set of controls to ensure their security. E-commerce systems are no different in that regard and require the same set of security controls. However, the technological complexities associated with those controls have increased and thus have to be addressed accordingly. This section briefly reviews those security controls before delving into specific security issues and solutions that are covered in subsequent sections.

The primary security controls are:

- Confidentiality
- Access controls (authentication and authorization)
- Integrity
- Availability
- Nonrepudiation.

8.1.1 Confidentiality

Confidentiality refers to the protection of information from unauthorized disclosure to a person or computing entity. Enforcing security controls to ensure confidentiality protects the data residing on shared media and/or in transit from eavesdropping. Cryptography enables confidentiality and protects data from prying eyes. Several encryption algorithms of varying levels of strength have surfaced in the past few years to provide confidentiality for information. Symmetric-key cryptography and public-key cryptography are two popular forms of cryptography. Symmetric-key cryptography uses a key or set of keys to encrypt data and uses the same key(s) to decrypt that data. The practical use of this form of cryptography thus requires sophisticated key management techniques. Public-key cryptography, on the other hand, involves two keys; one public and one private. Data encrypted with one key can only be decrypted with the other key. The public key is distributed to the public, while the user or system holds the private key. Various cryptographic algorithms have surfaced in the industry that serve different purposes. These include DES, RSA public key, International Data Encryption Algorithm (IDEA), CAST, and Elliptic Curve algorithms. The following elaborates upon some of the popular algorithms used in the e-commerce domains.

8.1.1.1   Data Encryption Standard (DES)

DES is a symmetric-key encryption algorithm that IBM introduced in 1975 with cooperation from various government agencies. The DES algorithm is available in public domain and has withstood the scrutiny of many experts since its inception. DES is the most widely deployed commercial algorithm, and numerous organizations and systems employ DES for encrypting critical messages and data. DES is a block cipher, which means that the algorithm encrypts data in 64-bit blocks and uses a 64-bit key. In reality, only 56 bits are used to encrypt/decrypt the data, whereas the remaining 8 bits operate as parity bits. The use of 56 bits provides a large key space (2^{56} potential possibilities for keys). This makes breaking the code difficult with a brute force attack.

The past few years, however, have witnessed many cases where people and organizations have broken DES by breaking the key used for encryption. Many organizations are therefore shifting away from DES toward the use of the Triple DES algorithm, and the U.S. government no longer recognizes DES as the standard. Triple DES employs three keys for the encryption of data, thus expanding the key size. Various modes of Triple DES exist. The first involves encrypting the data three times with three separate keys. The second mode involves encrypting the data with the first key, decrypting the data with the second, and then encrypting it again with the third key. The third mode is similar to the previous two methods except that the same key is used in the first and third operations.

The government has recognized that DES cannot serve the interests of the IT industry for the next century. For this purpose, the government has proposed the use of Triple DES as an interim standard. Meanwhile, the government has announced the Advanced Encryption Standard (AES) initiative as a replacement for DES. As of this writing, the government is still exploring various algorithms that could become an AES standard. These algorithms include MARS, developed by IBM; RC6 developed by RSA Labs; Rjindael developed by Joan Daemen and Vincent Rijmen; Serpent developed by Ross Anderson, Eli Biham, and Lars Knudsen; and Twofish developed by Bruce Schneier, John Kelsey, Doug Whiting, David Wagner, Chris Hall, and Niels Ferguson [1].

8.1.1.2   RSA Public Key Standard

The RSA public key standard is an asymmetric-key system that uses public and private keys to encrypt and decrypt data. The RSA system is based on modular arithmetic operations and builds upon the premise that it is difficult to factor-in large numbers that are products of two prime numbers. Using this as a mathematical foundation, the system derives the public and private keys.

The RSA public key system is primarily used for encryption and digital signatures. For example, when Bob wants to send an encrypted message to Alice, Bob obtains Alice's public key, encrypts the data using that key, and sends the encrypted data to Alice. Since only Alice holds the private key (corresponding to Alice's public key) that can decrypt the data, the data remains confidential in transit.

Digital signatures authenticate the sender of the message. For example, to identify himself, Bob sends a message encrypted with his private key to Alice. When Alice receives the message, she decrypts the message using Bob's public key. A successful decryption confirms Bob as the sender as the message was encrypted using Bob's private key that only Bob has in his possession.

The RSA public-key system and DES (or other symmetric-key systems) are usually used together. This is because RSA is relatively slow for encrypting large blocks of data, for which DES is very appropriate. Systems thus use RSA to exchange DES keys among each other and then use the DES algorithm to encrypt blocks of data. This protocol thus authenticates the two parties and enables a safe exchange of keys.

Many other encryption algorithms exist that provide different features. For example, RC2 is a block-cipher that uses a variable key size, and RC4 is a stream cipher (operating on a stream of bits rather than blocks of bits) that uses a variable key size. RC5 uses a variable key to operate on variable blocks of data and employs variable operations. These three algorithms are widely used in commercial systems, and different security products employ these algorithms. A new cryptographic system called Elliptic Curves has surfaced that builds upon a different mathematical foundation. These algorithms use a shorter key and exhibit better performance for certain operations. For example, when compared to RSA, in some cases these algorithms have shown better performance for decryption and signing operations, as opposed to encryption and signature verification operations, in which RSA has performed better. However, as of this writing, the algorithm has seen limited commercial penetration.

8.1.1.3 Message Digest Algorithms

A message digest is a fixed-size string that is derived from a variable sized input. One-way message digest algorithms are used to check the integrity of the message. One-way implies that given the message digest (hash value), the original message cannot be re-created. Many systems therefore use message digest algorithms to check the integrity of the message. The manner in which this works is as follows. When Bob intends to send a message to Alice, he calculates the message digest of the message, encrypts it with his private key, and attaches the encrypted message digest to the original message. Upon receiving the message,

Alice decrypts the message digest by using Bob's public key and then extracts the original message digest. Next, Alice calculates the message's message digest and compares it with the extracted value. Successful comparison not only assures Alice of the integrity of the message but also authenticates that Bob sent the message to Alice. Thus, these algorithms are also used for digital signatures.

Many message digest calculation algorithms exist. Among them, SHA-1 (government standard), MD2, MD4, and MD5 (developed by RSA labs) are among the most popular ones.

8.1.2   Access Control

Access control authenticates the identity of the entity (human or computer) trying to access a computing resource, and controls the use of the computing resource per predetermined levels of entitlement. The topic of access control thus includes issues related to authentication and authorization.

8.1.2.1   Authentication

Authentication refers to validating the identity of a subject requiring access to a system or network. The subject could be a person, a network application, or a device (e.g., router) that seeks connectivity to another device to use its services. In a distributed computing environment authentication plays a larger role due to the multiple tiers associated in fulfilling user requests. The various computing nodes on the network have to authenticate the subject transparently before granting access.

Various mechanisms exist to authenticate users, the most popular of which are as follows:

- *Static password-based authentication:* This is the most basic and popular form of authentication used in various systems. A user uses an ID and password to authenticate to the system. The system usually stores the user password in an encrypted format. The entered password is encrypted and compared against the stored password. A successful comparison grants the user access to the system.

  The strength of this form of authentication depends upon the encryption algorithm that encrypts the password in storage. Weak encryption poses the risk that an unauthorized user will steal passwords, break the passwords, and sign on to the system. Security of the transmission medium on which a password travels is also important. If the password travels in clear text before it reaches its destination for encryption, it poses a serious risk, as an eavesdropper can steal the password in transit and then masquerade as that user to transact various services.

Enterprises employ this form of authentication for both customers and internal users. Customers, for example, can sign on to various Internet-based applications to access different types of services. Enterprises can provide varying levels of authentication to users for their client platforms. For example, certain Internet-based applications allow users to store user IDs and passwords in a file (also referred to as cookie) on a user's PC. This allows the users to transparently sign on to the system without requiring to reenter the ID and password. For certain applications, especially financial applications, this is a weak form of authentication. A different user who has physical access to the same appliance may invoke the application and sign on without reentering the appropriate authentication information and enter unauthorized transactions.

- *Token-based authentication:* In this form of authentication, the user requires a hardware token (e.g., ATM card) and an associated password (e.g., ATM PIN). This form of authentication is better than a static password scheme as the user requires both the token and the associated password to pass system security.

- *Dynamic password authentication:* Dynamic passwords surfaced to counter the risk of stealing passwords from the network. This scheme forces the user to use a different password for every authentication. Users carry dynamic password tokens that are preprogrammed for every user. The users can unlock the password token by entering a password or PIN and retrieve a password for a session. The authentication server identifies the user and generates a password using the similar algorithm as the password token. Upon receiving the dynamic password from the user, the server compares the two passwords and then grants authorized users access to the system. Any eavesdropper stealing passwords will not gain from having the password as subsequent sessions will use different passwords.

- *Biometrics-based authentication:* Biometrics authentication is based upon the physical characteristics of a person. For example, a system can compare a user's fingerprints against a stored copy of the fingerprints in the system. Biometrics-based authentication thus eliminates the risk of stealing identities. An example of biometrics authentication is a product by Sensar Secure that provides iris recognition software for use in ATMs. Users approach an ATM and the system reads the users' iris image and compares it against a stored image in its database. Another product by Identix TouchNet uses fingerprints to replace passwords when authenticating users at a desktop terminal. Similarly, Visionics Corporation offers FaceIt Face Recognition Technology that uses face images to authenticate users.

- *Public and private key-based authentication:* As mentioned earlier, public key provides an excellent mechanism for authentication. A piece of data encrypted with a user's private key can only be decrypted with a user's public key. This form of authentication is used extensively in applications such as e-mail, where users can send an authenticated message by encrypting the information using their private key. The recipient can then use the user's public key to decrypt that information. This provides the recipient with the assurance of the sender's authenticity. A minimum of 768 bits is usually considered safer for security operations, as opposed to the earlier implementations of 512 bits. The more common implementations today use the 1024 key lengths for security.

8.1.2.2   Authorization

Authorization refers to the access rights of a subject that has access to the system. Authorization in most cases is required subsequent to a valid authentication on the system. Access control lists (ACLs) and rules play a vital role in establishing the authorization levels of the subject that has authenticated to the system. ACLs contain a list of authorized computing resources that the subject is authorized to access and specifies the access criteria for authenticated entities. For example, the NT operating system's security is based on ACLs. Network devices such as routers also use ACLs to direct traffic. For example, a router determines traffic flow by restricting or allowing traffic based on configured ACLs. Based on the defined ACLs, a particular host may have access to a particular subnet, while the router may restrict its access to any other subnet. A router with no ACLs will allow all authenticated entities to all parts of the network. ACLs allow routers, firewalls, and other network security devices to control users' or other computers' access to the network based on their permissions.

Authorization to computing resources is also granted to users based on their security levels and/or groups to which they belong, and on the access rights assigned to the particular group. Directory services play a vital role in the design of an authorization system. Directories store information about the various entities in the system (e.g., programs and/or files that the users have access to). They also store information about entities on the network such as databases and printers to which the users are granted permission of access.

8.1.3 Integrity

Integrity controls protect the data and/or computing resource from any intentional or unintentional tampering. Integrity ensures the accuracy and completeness of information. Integrity violations include the tampering of information before it reaches its intended recipient, destroying information and software, making uncontrolled changes to IT systems in an operations environment, and so on. Security controls such as malicious software control (e.g., antivirus software, mobile software signing), encryption, and hashing technologies and proper access controls guard against integrity attacks.

8.1.4 Availability

Availability refers to the continuity of IT processing and the availability of information. Availability breaches affect an enterprise's business operations and may cause an enterprise to experience financial or customer service impact. Adequate configuration of the systems and controlled processes and procedures guard against denial of service attacks, thus preserving system availability.

8.1.5 Nonrepudiation

Nonrepudiation controls ensure that users cannot deny actions they undertook. For example, a user may enter a financial transaction (e.g., a stock order) and later deny entering that transaction. Nonrepudiation controls (e.g., digital signatures), especially those enforced through public-key encryption, prevent such incidents. Other organizations require their users to read and agree to business agreements that hold users accountable to their authentication tokens (e.g., passwords). Transactions entered through those passwords are thus a user's sole responsibility. Some states have passed legislation that legalize the use of digital signatures. For example, the state of California recently passed the Uniform Electronic Transactions Act. This makes electronic signatures equivalent to nonelectronic signatures and can thus be used to enforce nonrepudiation security controls [2].

8.2 Network Security

An e-commerce system by its nature spans multiple platforms and users distributed throughout a network. These users and systems have to securely connect and communicate with each other in order to carry out e-commerce services. A breach of security can result in forged transactions or a breach of confidentiality. Network security refers to the securing of all traffic at the network level. There are multiple issues in network security. First, the two communicating platforms should authenticate each other. This is necessary so that the receiving platform can trust the messages being sent from the sender platform. This also prevents other computers or users from sending messages to the recipient and pretending to be the sender. The second issue involves the preservation of data confidentiality over the end points of the network.

Traditional methods of securing networks involved installing encryptors at the ends of the network that enabled end-to-end encryption of the network link. However, with public networks such as the Internet, such link encryption is not feasible. The same is also true for some private networks and WANs, which are not encrypted for various business reasons. The emergence of virtual private networks (VPNs) has enabled administrators to selectively secure applications without the need to secure the actual physical network links. VPNs refer to secure data tunnels between two points over public and private data networks. The use of VPNs requires special software protocols that ensure data confidentiality and data integrity over the insecure network between two ends of a network.

VPNs enable enterprises to deploy mission-critical applications for customers, corporate users, and partners over the public Internet. VPNs enable enterprises to leverage the reach of the Internet to extend their connections to customers as well as corporate partners by providing secure channels of communication and data transmission. With NSPs providing the core fabric for future communications, implementation of VPNs is becoming a necessity. This requires enterprises to implement VPNs that enable network users to communicate securely over leased Internet connections.

An enterprise requires VPNs for some of the following scenarios:

- *VPNs for secure communication within an enterprise WAN:* VPNs allow enterprises to reduce their operating expenses by enabling them to build Internet-based WANs. Many NSPs offer such secure VAN connections. For example, UUNET (an MCI WorldCom Company), offers the Uusecure VPN service, which is a fully managed VPN service that uses IPSec for data tunneling. Enterprises can lease this VPN service and include this part of their enterprise WAN.

- *VPNs for remote connectivity:* VPNs enable mobile workers to engage in all types of business activity regardless of their location. The use of tunneling protocols such as PPTP enables secure remote connections.

- *Extending VPNs to extranets:* As enterprises extend their networks externally to include organization partners, regulating their access becomes a priority. As discussed in Chapter 5, VPNs are one of the primary components involved in the building of extranets.

Before delving into the strategies for securing various network topologies, it is vital to understand the security technologies that enable the securing of those network connections. The next section discusses several popular security technologies.

8.2.1 Network Security Technologies

As mentioned earlier, authentication, confidentiality, and data integrity are the primary security controls required for securing network connections. IP Security, Point-to-Point Tunneling Protocol, RADIUS, TACACS+, and firewalls are the primary technologies that enable the securing of the network connections. The following elaborates upon these technologies in further detail.

8.2.1.1 IP Security (IPSec)

IP Security (IPSec) is one of the most common protocols for VPN deployments. IPSec is IETF's security standard for IP-based networking, and it provides authentication, integrity, and confidentiality checks for IP packets. Unlike SSL, which provides security for two applications, IPSec works at the network level. IPSec works between two hosts that have established a security association between them. This means that the two communicating hosts preestablish the session keys and encryption mechanisms. This differs from an SSL session in which although the applications on two hosts communicate privately, there is no trust relationship between the two hosts that have those applications. IPSec can thus provide security between routers, firewalls, applications, or any other IP-based computing platform.

IPSec includes two headers to provide these security controls. IETF's RFC 1826 defines the specification for the first header, which is also referred to as the authentication header (AH). The AH contains security information that provides integrity and authentication of IP packets. The AH resides between the IP header and the IP data contents. Security information in the AH includes a cryptographic checksum and a security parameter index (SPI). The cryptographic checksum is derived based upon the entire packet contents including the original IP header. Systems that implement IPSec use message digest algorithms (e.g., MD5) to calculate the checksum. This helps in check-ing the integrity of the addressing information at the destination. SPI, on the other hand, indicates the use of specific cryptographic keys and algorithms used for the encryption of data within IP packets. The use of SPI makes IPSec independent of the algorithms or key management protocols used.

The Encapsulation Security Protocol (ESP) is the other header, and it provides data confidentiality of the data contents of the IP packets. RFC 1827 defines the specifications for the ESP. ESP focuses on secure end-to-end trans-mission of data. Encrypted data is placed in the data portion of the IP header. SPI indicates the various algorithms used for encryption and the specific key management protocols used.

ESP works in two modes: tunnel mode and transport mode. In tunnel mode, the entire IP packet (IP header, other headers, and data within the IP packet) is encrypted and wrapped around an unencrypted IP header that is used to route the IP packet to its destination. In tunneling mode, the IP head-ers are protected, thus enabling the two hosts that have established security associations to communicate securely as the origin and destination addresses are verified.

At the start of an IPSec session, the communicating parties negotiate the security terms (i.e., algorithms and respective encryption keys). Various key management solutions exist for use with the IPsec protocol. Internet Key Exchange (IKE) combines the original Internet Security Association and Key Management Protocol (ISAKMP) and Oakley protocols. ISAKMP defines the protocol for key and algorithm exchanges and negotiation. Oakley establishes session keys. SKIP and Photuris are other mechanisms for key management used with IPSec installations. The S/WAN initiative is focusing on establishing an interoperable standard for IPSec that will enable vendors to build interoper-able network devices.

8.2.1.2   Point-to-Point Tunneling Protocol (PPTP)

Microsoft developed the PPTP standard in cooperation with other vendors for use of TCP/IP over serial communication lines and to secure Point-to-Point Protocol (PPP)-based communication. PPP provides a mechanism of encapsu-lating and sending data and enables computers connected over a serial interface to communicate with each other in a full duplex mode. PPP works by first establishing a data connection between computers and subsequently transmit-ting IP data over the network link. PPP supports multiple protocols (e.g., IP and IPX).

PPTP tunnels various networking protocols (e.g., IP, IPX, and NetBEUI) in IP packets. PPTP is a network protocol that enables secure communication

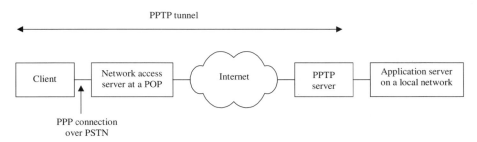

Figure 8.1 PPTP operation.

over the Internet. The major advantage of PPTP is that it enables an enterprise's staff to connect remotely (e.g., through the Internet) to enterprise networks without deploying leased lines. Organizations can also use PPTP to establish secure communication over local area networks. As PPTP provides a secure way to carry PPP traffic, it is very well suited for VPN deployments. PPTP works in client/server mode. PPTP encrypts data and sends it to the other device by encapsulating encrypted data in an IP header. PPTP supports both Password Authentication (PAP) and Challenge Handshake Authentication Protocols (CHAP). PAP is a weaker authentication scheme that simply involves a connecting device to send a user ID and a password in the clear to the connecting station. CHAP offers better authentication.

The architecture of PPTP consists of a PPTP client (referred to as the PPTP Access Concentrator, or PAC) that connects to a PPTP server (referred to as the PPTP Network Server, or PNS). If the PPTP client and server do not connect to the same LAN, then an access server (e.g., RAS) authenticates the client. Normally, organizations use PPTP to enable remote users to access private servers through an encrypted PPTP tunnel. Figure 8.1 illustrates this architecture. A client that requires access to the application server first initiates a PPP connection to an access server located at a POP location. To use PPTP, the client computer must support a PPP-compliant software package. The client is installed with a PPTP driver to enable it to access to a access server (e.g., NT's RAS). The access server authenticates the client using authentication schemes such as CHAP or PAP. (Organizations use a mix of RADIUS and PPTP solutions to resolve authentication and data confidentiality issues.) After a successful connection, the client establishes a PPTP tunnel with the PPTP server. This enables the client computer to communicate over a secure tunnel to the server. The PPTP takes the regular PPP communication stream, encrypts it and encapsulates the encrypted data in IP datagrams.

The PPTP server sits at the periphery of the public and private network. After establishing a secure tunnel between the client and the server all communication between the PPTP client and the server is encrypted. Once the PPTP server receives the encrypted packets, the PPTP server, extracts the IP packets (or other protocol's packets since PPTP supports other protocols as well) and routes them to the address specified in the encapsulated addresses.

PPTP uses the RSA's RC4 encryption algorithm. Typically, PPTP client and server negotiate the encryption algorithm used to encrypt data. PPTP can also restrict network access from unauthorized users using the PPTP filtering scheme. This feature enables only authorized users to connect to the PPTP server.

8.2.1.3   Remote Authentication Dial-In User Service (RADIUS)

RADIUS is a client/server protocol that enables an organization to authenticate dial-in remote users before allowing them to connect to internal servers. Livingston Enterprises, Inc. developed the RADIUS protocol to provide authentication services for users connecting to various networks. This protocol has gained wide acceptance, and IETF's RFC document 2058 specifies RADIUS details. RADIUS decouples the authentication services from the network backbone that the user intends to connect.

Figure 8.2 illustrates the RADIUS architecture. As illustrated, RADIUS operates in client/server mode. A NAS (Network Application Server) receives user connection requests from users and forwards them to a RADIUS server (authentication server) that provides user authentication services. The connection between the NAS (acting as a RADIUS client) and the RADIUS server is a UDP connection. The RADIUS server authenticates the user and sends an accept or reject message to the NAS. For accept messages, the RADIUS server also forwards user's privileges to the NAS, which limits users connection per those authorizations.

RADIUS provides several forms of authentication. These include PAP, CHAP, and token cards. RFC 1334 specifies the CHAP and PAP protocols. CHAP authenticates users by providing them a challenge. The user's systems compute a hash based upon the user's password. The user sends the result to the NAS, which performs an identical calculation. For matched responses, the server provides access to the user. PAP, on the other hand, simply requests a user ID and password, compares the password, and upon successful match lets the user onto the network.

Recent offerings of database and operating system products have started to include security features related to remote computing in their kernels. For example, newer versions of databases (e.g., Oracle8I) include RADIUS support.

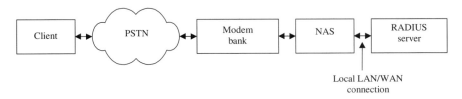

Figure 8.2 RADIUS protocol operation.

The Terminal Access Controller Access Control System (TACACS) is a similar protocol that uses TCP instead of UDP to communicate with the access control server. TAACS+ separates the authentication and authorization steps, whereas RADIUS combines the two steps.

8.2.1.4   Firewalls

Firewalls are a combination of several components—such as routers, host computers, and proxy servers—that control access to the internal corporate networks. Some advanced and newer firewall vendors provide all of these components in one platform. Firewalls restrict users from accessing corporate networks and limit authorized users to access certain authorized applications and data. ACLs and security rules allow for the definition of such rules.

Firewalls are essentially filters that examine packet contents and route or reject the routing of packets based upon the content. Firewalls enforce access control policies between two networks. The most basic types of firewalls are packet-based filters. These firewalls examine the headers of the source or destination IP addresses, and based on the access rules reject or accept the connection request. These firewalls work by simply analyzing the IP addresses individually and providing no further analysis of the network sessions. IP spoofing is the basic problem with these firewalls. Spoofing refers to masquerading as another entity to gain unauthorized access. The primary rule in the configuration of such firewalls is that filtering should be performed based on IP addresses and not on host names, as the latter is more susceptible for spoofing than the former. The Network Address Translation (NAT) feature allows the firewall to hide internal addresses from the outside network.

This form of firewalls poses a problem in some situations, as users (from certain IP addresses) usually require access to certain services on a particular host. For example, users may require access to the SMTP/POP mail services. The firewall administrator should therefore allow access to those services while restricting their access to the server from other types of connections. In such cases, there is a need for firewalls that allow access to host computers based upon TCP/UDP numbers as certain applications communicate on preestablished TCP/UDP numbers. Therefore, to reject all Telnet connections to a host computer, the administrator will configure the firewall to reject all TCP/IP requests to a specific host on port 23 (Telnet port).

Stealthy Packet Filter (SPF) firewalls incorporate additional security measures. These firewalls are dynamic packet filters that provide state checking to accommodate protocol-specific needs. These firewalls are cognizant of specific protocol implementations. For example, such firewalls can track the source and destination TCP and UDP addresses along with TCP sequence numbers and associated flags. These firewalls can prevent attacks based on protocol weaknesses. However, these firewalls are not as robust and do not perform security checking as well as the application gateways, which analyze data at the appropriate level.

Application gateways are proxy servers that run on host computers that act as firewalls. Proxy servers are software applications that act as servers for applications that request connections from within the protected network and act as clients to the outside servers. There is thus no direct connection between the internal computers and external computers. This proxy feature on the firewall allows the firewall to act as an intermediary between user requests at application, transport, or network levels. Proxy applications completely audit and filter all traffic that go through them. Proxies implement security features pertinent to each application. For example, proxies (application gateways) exist for Telnet, HTTP, and rlogin. Application gateways are general-purpose code developed specifically for each type of application.

Certain gateway servers also act as firewalls. New strategies are surfacing that require the screening of malicious software at the network perimeter, as opposed to checking for them at the client tier. For example, Security-7 Software's SafeGate is a gateway server product that filters and examines ActiveX, VBScript, and similar mobile software at the network perimeter. The server also decompresses ZIP and other files to examine their content before delivering them to clients.

With the number of connections continuously on the increase, enterprises have to install multiple firewalls to balance the load of these numerous connections. There are multiple strategies that organizations can use to achieve these firewalls. One approach is to have two firewalls communicate using proprietary protocols such as HSRP by Cisco. The other method requires the use of load-balancing firewall systems that interface with various firewalls and provide load balancing of various connections.

8.2.2   Security for Various Network Topologies

The following pages elaborate on typical network topologies, inherent security issues, and their associated solutions. These network topologies include the following:

- Security solutions associated with leased lines;
- Security solutions for connecting an organization's sites through the Internet;
- Security solutions for connecting enterprise users' computers through the Internet;
- Providing secure remote connectivity to the mobile force;
- Security solutions of typical Internet browser-based and telephony applications;
- Security issues associated with wireless networks.

8.2.2.1   Site-to-Site Connectivity Through Leased Lines

Leased lines provide point-to-point connections from one site to another. This form of connection is most popular for WAN connections. Organizations lease frame relay, T1, and other lines from various NSPs to connect their offices.

A popular approach to secure such network links requires the use of link encryptors. Link encryption refers to encryption at the data link level and provides complete encryption to all data passing between two network terminals. Link encryptors can be added in front of network devices such as routers, PBX trunks, and multiplexers. Link encryptors employ a variety of popular encryption algorithms (e.g., RSA, DES). These encryptors are usually transparent to the specific network or application protocol being used. Since link encryptors support a wide range of network protocols, they should interface with a variety of hardware equipment (e.g., T1/T3 interfaces). In WAN constructions, link encryptors usually support high data speeds to remain transparent to network operations.

Figure 8.3 illustrates a typical link encryption installation. As illustrated, link encryptors are independent of the data traffic on the network within the site. However, as a data stream leaves the site and encounters a link encryptor, it is immediately encrypted. The data remains encrypted until it reaches the other link encryptor, which decrypts the data and forwards it to the appropriate server within the other site's network. Setting up link encryptors requires

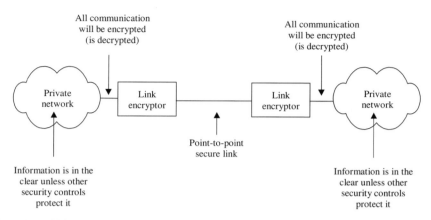

Figure 8.3   Link encryption.

configuration at both sites with common security keys that enable the encryptors to encrypt and decrypt data. Enterprises use link encryption to secure their WAN links. Various vendors provide products that secure specific types of links. For example, IRE's SafeNet/Frame provides link encryption for an enterprise's frame relay networks. Sophisticated link encryption packages support encryption for SONET, ATM, frame relay, and SMDS networks. Cylink's CIDEC-VHS encryptor is one such product.

Link encryption provides one of the safest methods of connectivity, as it is almost impossible to eavesdrop on the network connection or masquerade as another user. However, it is expensive to set up and maintain. The high cost of such connections stems from leasing lines from NSPs and buying encryption products. If the two sites belong to two different organizations, both have to agree to install the encryptors on both ends.

8.2.2.2   Site-to-Site Connectivity Through the Internet

The Internet has enabled many enterprises to start shifting from leasing lines to using the Internet network for site connectivity. An organization, for example, can establish routers and route data traffic from its site to another site. Link encryptors discussed earlier cannot protect these network links because multiple nodes may exist between the two sites. As link encryptors encrypt all information (including address information), they can only work between point-to-point connections. Also, those intermediate nodes may belong to various service providers where it may not be possible to install link encryptors.

A solution for protecting the network link between two sites over the Internet is to use a VPN. Doing so involves deploying two encrypting routers at both sites. Both routers (also referred to as peering routers) employ public

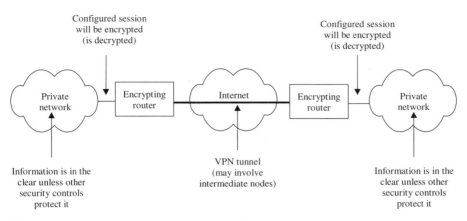

Figure 8.4   VPNs using encrypting routers.

key cryptography to authenticate each other and exchange a session key that encrypts all data between the two routers. This enables all traffic to be encrypted at one peering router. The encrypted data routes through various network hops and routers, while the intermediate hops ignore the data, as they are not part of the security association established between the two peering routers. The other peering router finally decrypts the data stream.

This solution is similar to link encryption in that the routers operate independent of the data traffic that flows within the networks of the two sites. Data is encrypted at one router and decrypted when it reaches the other router at the other site. Thus this solution does not provide end-to-end authentication or encryption of two computers situated at the two sites. Figure 8.4 illustrates the encrypting routers operation.

Certain encrypting routers support selective encryption of sessions. This implies that routers can selectively encrypt certain sessions while leaving others unencrypted.

8.2.2.3   Local User-to-Site Connectivity Through the Internet

In this scenario, the user within a site seeks a protected connection to another host that may or may not be within the same site. The earlier solutions do not apply to this scenario because they secure the network from one site's periphery to another site's periphery, and the connection between the user to the periphery encryptor is in clear text.

Figure 8.5 illustrates a VPN solution for this type of scenario. As illustrated, special software at the user's terminal (client computer) establishes a secure data tunnel to the host computer to which the client computer wishes to communicate securely. The setup of this connection requires the establishment

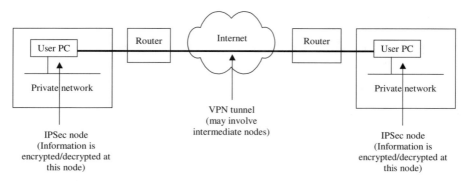

Figure 8.5   VPN implementation between user's computers.

of a security association between the client computer and host computer. IPSec software enables this type of solution. In typical IPSec implementations, the IPSec client connects to an IPSec server and both mutually authenticate each other. The client and server then establish a security association to negotiate the security controls (authentication/encryption algorithms and keys). The IKE protocol is typically used for key exchange and security negotiation. IKE ensures that the key exchange is done with the authenticated party.

Most VPN products offer client and server components. The client component is installed on the user's desktop computer or mobile computer (e.g., a notebook computer or a handheld Windows CE computer). The client component communicates with the server by first requiring the user to authenticate itself. The authentication schemes could vary and include token-based authentication, smart card, or one of the other authentication mechanisms discussed earlier. After successful authentication, the client computer establishes a secure session with the VPN server. Thus all information over the Internet through that session stays private and authenticated. SmartGate VPN from V-One Corporation is a VPN solution that includes SmartPass Client and SmartGate server components, which enable the establishment of a VPN tunnel through the Internet between remote and local users. VPNet Technologies offer other VPN products that enable users to establish a VPN connection using IPSec protocols that use varying schemes for authentication and encryption.

8.2.2.4   Internet Applications Security

Inherent weaknesses in Internet security triggered the emergence of specific security solutions for Internet applications. SSL provides security controls for numerous Internet applications and works at the network transport layer. With the emergence of Internet telephony and its potential strategic use in the enterprise, industry is actively seeking more comprehensive security solutions.

- *Secure Sockets Layer (SSL):* Netscape developed the SSL protocol for securing Internet applications. SSL is independent of the application and can secure most popular Internet applications (e.g., HTTP, Telnet, FTP). The SSL protocol encompasses both server and client authentication and provides a tunnel at the transport layer level to provide data confidentiality for two applications over the Internet. In server authentication, the client authenticates the server by first requesting the server's public-key certificate. The client generates a challenge (key), encrypts it with the server's public key and then sends it to the server. The server in turn retrieves the challenge by decrypting the message using its private key and then returns a message encrypted with the challenge that the client retrieves by decrypting it with the challenge that it had sent to the user. Successful decryption thus authenticates the server to the client. The server can in turn authenticate the client by requesting the client's public-key certificate.

- *Internet telephony security:* Internet telephony (H.323 protocol) suffers from numerous security risks. These include eavesdropping risks on the Internet and interoperability issues of the H.323 protocol with firewalls. Use of SSL to protect Internet telephony is not viable due to bandwidth requirements and because SSL does not suit media-streaming protocols. Organizations should not rely on Internet telephony to communicate sensitive information. For example, using the channel to trigger bank transfers is risky, whereas using it to discuss nonconfidential information is acceptable.

  Internet telephony also suffers from interoperability issues with firewalls. The use of Internet telephony (H.323 protocol) requires opening numerous UDP ports on the firewall to access a telephony gateway inside the network. This exposes the network to numerous attacks from the Internet. If the organization requires installation of the telephony gateway inside the firewall, then an organization has only two options to secure the network. An organization can either install a H.323-compliant firewall (installed with an H.323 proxy), or employ a security protocol such as H.235 for securing H.323 traffic. The challenge stems from the fact that very few firewalls include the H.323 proxy, and the H.235 protocol has not yet been included in the telephony gateways.

8.2.2.5   Wireless Network Security

Wireless network security primarily deals with security issues related to GSM and pager devices. As the use of GSM and pager networks became popular after the Internet, their security solutions are less mature. The following describes security issues and solutions related to the types of networks.

- *GSM security:* One advantage of GSM wireless networks over their analog counterparts is that GSM networks incorporate security functionality at both the data and voice levels. The security controls in GSM networks ensure the appropriate authentication of the subscribers to the network and encrypt voice communication over the air link. The GSM security algorithms for authentication and encryption are proprietary. The GSM association has developed these algorithms (referred to as A2, A3, and A5) to provide security functionality over the wireless links. A disadvantage of these proprietary algorithms, however, is that, unlike other security algorithms (DES, RC2, etc.), the industry has not openly tested these algorithms in the public domain. Although no known reports related to their weaknesses have surfaced, organizations planning to leverage GSM networks for sensitive transactions should be wary of the proprietary nature of these algorithms.

- *Pager security:* Pager networks do not have any inherent security controls (e.g., authentication and confidentiality). Incidents related to interception of pager messages are quite common. For example, in the past few years certain incidents involving the interception of pager messages between high-ranking government officials occurred. Pager messages cannot be relied upon to transmit sensitive information. However, certain security solutions have surfaced that work in conjunction with specific pager NSPs that extend these features to their subscribers. V-ONE's Air SmartGate is one such solution that works with SkyTel's two-way messaging network. The solution provides authentication and data encryption to mobile two-way pagers. This solution protects wireless messages between two way pagers using RSA's RC4 data encryption standard and provides PIN authentication. This solution ensures that pager messages are never stored in clear text on any node during the end-to-end transmission of the pager messages.

8.2.2.6 Remote User-to-Host Connectivity Over the Internet

In this scenario, a remote user wishes to connect to a host located at the enterprise network over the Internet. This is a two-step process. First, the user should authenticate to the network. Next, the communicating computers should establish a VPN tunnel that enables the user to communicate securely over the Internet to the enterprise host. RADIUS and PPTP enable this form of connection.

8.3 Platform Security

Platform security refers to the security of the information appliance or computing server. Platform security has two primary objectives. First, it ensures that the data on the platform is secure from unauthorized access. Second, it protects platform services against unauthorized access from other platforms and users on the network.

Figure 8.6 illustrates a typical model for platform security. As illustrated, security at a platform level requires enforcement at two different levels. The first level authorizes users to use platform's system level resources (operating system and database), whereas the second level is concerned with access control for applications that run on that platform. In a server or mainframe environment where these platforms are placed within secure data centers, various administrators have access to the operating system or database level of the platform. Security of the system-level software thus drives the access control of various users. An application's proprietary security controls, on the other hand, influence the business applications' security.

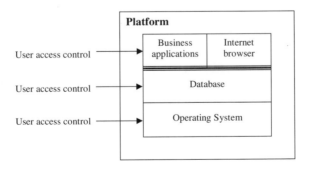

Figure 8.6 Platform security model.

8.3.1　Establishing Controls for Platform Access

Users require physical access to platforms to install and configure software and hardware. Traditional mainframe systems excelled in providing adequate system-level access control. This enabled organizations to institute adequate operational security controls. Organizations could divide security and system activities among different roles of users. One administrator had limited control to perform such sensitive operational activities on the system. However, the subsequent generation of open systems lacked such security controls. Operating systems in particular have been more notorious for their weak access control functions than database products. For example, early versions (and some current versions) of UNIX provide very rudimentary levels of access controls. Organizations use UNIX-flavored systems (e.g., Solaris, HP-UX) as the primary platform for the deployment of e-commerce systems. For mission-critical production UNIX deployments, organizations struggle to segregate security administration and system administration activities to prevent empowering one administrator with enhanced privileges, which is unacceptable for most security-conscious organizations. For example, the root account is a superuser account in UNIX that is required to perform various security- and system-level administration functions. Most versions of UNIX do not allow division of root functionality among multiple users, thus empowering one user to perform critical administration functions.

To rectify UNIX access control features, various vendors offer security products that compensate UNIX security deficiencies. For example, Platinum Technology offers SEOS, an access control product for UNIX that provides adequate controls for UNIX's root account. SEOS enables locking of the root account and definition of users that share divided responsibilities of the root account.

The newer generations of operating systems offer better access control features. For example, Trusted Solaris includes enhanced entitlement features that allows for segregation of roles thus controlling root activities. Special versions of HP's HP-UX operating systems (e.g., 10.09, 10.16) provide enhanced security features. Organizations use such operating systems for deploying sensitive applications (e.g., deployment of financial e-commerce applications). Similarly, organizations are widely using Windows NT servers to deploy e-commerce systems. NT provides numerous built-in data-center security features that include adequate access control functions and robust file permission levels.

Access control for client platforms is vital for enterprise applications, as users sign on to these platforms to access applications and services. For example, certain organizations restrict access to mission-critical applications through specific client computers due to the sensitivity of the applications. For example, certain organizations may limit the use of law enforcement applications to a specific group of users located in specific offices. One reason to restrict such applications to certain client computers is that the client computers may store sensitive data that client computers download from servers. Several PC encryption packages enable encryption of data on client computers. For example, RSA's SecurePC enables the encryption of data on users' PCs that enterprises can rely on for the storage of sensitive information. Encrypted information in certain cases is not easily transportable from one PC to another, nor can an authorized user easily access that information in a networked configuration. Another way to secure client platform data is to use security products such as LapJack. LapJack is a data security product that ensures that users cannot boot their notebooks until they authenticate to the platform.

The popularity of the Windows operating system has enabled it to be the client platform of choice for enterprises. However, Windows 95/98 provides weak security relative to Windows NT. Security-sensitive organizations thus encourage the use of Windows NT as a client platform as well (in addition to using NT as a server). Windows 2000 includes file encryption that will thus prove to be a better client platform for security-conscious users.

Databases systems (e.g., Oracle, Sybase, Informix) incorporate a base set of security features that makes them adequate for production deployment of mission-critical applications. Most RDBMSes, for example, provide segregation of security and systems administration duties.

8.3.2 Proper Configuration of the Computing Platform

Almost all operating systems and databases have their security features turned off at installation. Secure configuration of those systems is therefore vital for ensuring adequate security. For example, an NT operating system that has been certified as a secure system by the U.S. government (C2 level rating) fails all levels of security if it is installed with the default configuration. Another example is that most operating systems and databases at installation offer default passwords for sensitive user IDs (e.g., "administrator" in Windows NT). Organizations suffer numerous security breaches because of such user IDs if they do not reset those passwords. Similarly, operating systems come preloaded with various utilities that organizations do not require for their specific installations which, if left on the system, provide malicious users the opportunity to compromise system integrity. For example, the snoop utility on UNIX enables users to read data traffic, including user passwords. Organizations should remove such utilities from production systems.

Operating system vendors periodically release system patches to rectify security holes, which continue to emerge. Installation of such patches is of paramount importance to preserve the integrity of systems. For example, a security breach identified in Microsoft's IIS Web server enables hackers to control the Web server and gain illegal entry into enterprise networks. Microsoft released a patch to rectify this issue. Numerous enterprises that failed to install the patches in time suffered such attacks. Continual configuration of platforms with appropriate security settings and patches thus prevent security attacks.

8.3.3   Protecting the Integrity of Platform Applications and Data

Protecting a computing platform's integrity refers to the prevention of unauthorized modifications to data and installation of applications. Unauthorized modifications to the database are primarily the result of inadequate access controls on the system. Unauthorized users gain access to the system and either modify the database or delete database contents. Similarly, users can install malicious software on the system (e.g., viruses, trojan horses) that can compromise the integrity of the entire system. An increase in external network connections has also increased the probability of receiving malicious software. Users, for example, can receive malicious Java applets and/or ActiveX controls through the Internet, which can wreak havoc on the system and the network in general.

As mentioned earlier, instituting adequate access controls can protect system integrity. Organizations can institute security measures by installing appropriate security products that take a snapshot of specific system and data files (as established by the administrator) and periodically compare system contents against the recorded baseline. Administrators receive alerts if the product detects any modifications relative to the baseline. Various products are available in the market that provide such features.

The spread of viruses has increased manifold with the proliferation of Internet computing, and enterprises must install adequate safeguards to protect their platforms from such viruses. For example, a virus attack in late 1998 at MCI WorldCom infected numerous systems with a Windows NT virus that spread itself to other nodes as users launched infected applications. Protecting client and server platforms from such malicious software requires the installation of antivirus software. Antivirus software is also available for handheld computers. For example, iRiS Software provides iRis AntiVirus software for Windows CE systems. Considering the continual onslaught of viruses, enterprises should institute automatic or manual processes that ensure the continuous updating of antivirus files.

Mobile software such as ActiveX and Java applets pose a danger similar to viruses. As users browse various sites from within the organization, mobile code could transparently download on their systems. Malicious ActiveX controls or Java applets, for example, can destroy all files on the system or connect to other nodes on the network and cause other damage. Three countermeasures address the threat of mobile code. The first option involves blocking all such code at the enterprise firewall. Various vendors provide firewalls that provide such functionality. For example, Network Associates provides the Gauntlet Internet firewall that blocks all Java applets and ActiveX controls from entering the enterprise network. The second option involves configuring popular Internet browsers so that they do not accept any mobile code. Both Netscape's Navigator and Microsoft's Explorer provide these capabilities to users. However, configuration of the browser is difficult to enforce for every user, especially if the enterprise has many users.

The last option to block malicious mobile code from the Internet involves accepting only signed code. This means that users download mobile code from known sources that have signed their mobile code with their private keys, thus authenticating the identity of the sender. This technique does not guarantee that the downloaded code will be free of any malicious functions. The only protection this provides to the users is that users can download mobile code from well-known sources. Microsoft's Authenticode, Microsoft Office 2000 and VBA, Channel Signing for Marimba Castanet, and Netscape's Object Signing technologies provide such a mechanism.

8.4 The Certificate Authority (CA) Infrastructure

As mentioned earlier, the Internet is not a secure medium for communication or for transacting business services. The shared and distributed nature of Internet provides neither confidentiality nor authentication, the cornerstone of secure commercial transactions. Public-key cryptography facilitates secure e-commerce operations and transactions among customers, merchants, financial institutions, supplies, partners, and others. PKI services offer authentication, nonrepudiation, and confidentiality services to their internal and external customers. Due to its inherent strength of providing encryption and authentication functions and increased popularity within the past few years, public-key cryptography is becoming the backbone of all vital e-commerce applications.

Applying public-key cryptographic principles requires an appropriate certificate management infrastructure. Certificates are digital documents that bind a public-key with an individual. The use of digital certificates provides confidentiality between users involved in a transaction, mutually authenticate the two parties to each other, preserve the integrity of the communication and/or transaction, and provide nonrepudiation of transactions. Use of certificates provides the assurance that a publicized public-key does belong to the individual who claims to be its owner. A trusted authority, referred to as a certificate authority (CA), issues the certificate to the user after verifying the person's identity. The underlying philosophy behind certificates is the issue of trust. To accept a certificate from any entity, the certificate should be signed by a CA. The CA should also have either a certificate that is signed by another trusted entity or be a known trusted entity (e.g., the U.S. government). This chain of trust is essential to the validation of certificates. Thus at the end of this chain there should be a CA that is mutually trusted by the parties involved in the transaction. For example, Thawte is a trusted name in the CA industry and offers a Chained CA program to enable other, less well-known CAs to issue certificates to their customers that in turn are signed by Thawte's private-key. During verification, therefore, users track the chain of certificates to the end-signing authority (Thawte) and establish the trust required to carry out the transaction. This service is usually used by organizations that implement their internal CA infrastructure and require their root CA certificate to be signed by a recognized CA such as Thawte.

The use of public-key cryptography demands a technological and procedural infrastructure that manages various aspects of the use of public-key cryptography. A PKI provides the foundation to fulfill those requirements. A PKI consists of the entire system of certificates, associated processes, and the technology infrastructure to enable the seamless usage of public-key services. A CA usually provides such PKI services.

Before delving into the discussion of CAs, it is vital to understand the concept of a certificate in further detail. A certificate (also referred to as a digital ID) binds the user's identity with a particular key pair and confirms this validation to other parties. A certificate is similar to a passport that includes the user name, photograph, and other information and is issued by a high-level governmental agency. Issuance of the passport by a common trusted governmental entity confirms that the photograph belongs to the person whose name and other associated information is included in the passport. A certificate provides a similar assurance that the public-key belongs to the person or entity whose name appears in the certificate. In the PKI world, the CA is the third-party organization that provides PKI services and attests to the validity of the certificate. A CA providing PKI services thus provides a technological and administrative infrastructure to facilitate public-key cryptographic applications. It is therefore vital that a CA be a valid entity, as the integrity of a certificate is directly dependent upon the integrity of the CA that issues the certificate. Various CAs have surfaced to facilitate secure e-commerce. These CAs include Verisign, Thawte, and many others. Some banks are contending to become CAs as well. Alternately, an organization can establish its own PKI infrastructure.

As discussed in earlier sections, public-key cryptography is the cornerstone behind numerous e-commerce applications. These include the following applications:

- The SSL protocol on the Web uses public-key cryptography. For example, customers ordering from Web sites that use SSL place orders through secure tunnels that protect their privacy.

- In a typical VPN solution, firewalls use public-key cryptography to secure the session between them. This ensures that all data traversing through the Internet is protected point-to-point between the firewalls that establish the session.

- Users can send secure e-mail by employing S/MIME that uses public-key cryptography.

- Web-based e-commerce applications in business-to-business scenarios between partners and suppliers use a public-key–based trust.

- Physicians in the heath care industry use public-key cryptography when they need to exchange vital information about patients.

- The proliferation of mobile code increases security risks. Corporations are employing various code-signing mechanisms to distribute software securely. Code signing requires the usage of certificates.

In supporting a PKI, a CA provides various services and operations related to certificate management. The following provides an overview of some of the primary operations. An organization contending to build its own PKI services will be required to provide these functions.

- *Certificate issuance:* Certificate generation is the most vital element of a PKI service. The CA requires necessary information on the person or organization's identity before issuing a certificate. The level of information that a CA requires for issuing a certificate depends on the level or class of certificate. For example, the CA will usually issue a higher class of certificate for organizations that require the certificate for sensitive transactions (e.g., use of SSL sessions for financial transactions). In such cases, the CA will require enough information (e.g., articles of incorporation) that will authenticate the organization's identity to the

CA. On the other hand, the CA may require rudimentary information about a user that requires a certificate for sending personal secure e-mail.

- *Support for multiple applications:* The PKI may support the issuance of certificates for various e-commerce requirements and applications. Following are some of the popular applications that require certificates:

  — Web servers that require certificates for SSL sessions;

  — Certificates for signing mobile code such as ActiveX controls, Java applets, and Microsoft Office 2000 macros;

  — Secure e-mail certificates;

  — IPSec applications support—IPSec certificates identify the hardware device that uses the IPSec protocol;

  — Web client certificates for authenticating client identities through a Web session (traditional SSL sessions only authenticate the servers to the users);

  — SET certificates for SET-based credit card payments;

  — EDI application security.

- *Certificate revocation processes:* Each certificate has a validity period during which a certificate-using application can use the certificate. Certificate revocation refers to revoking a user's certificate before the expiration of the validity period so that no application can use the certificate. A CA maintains a certificate revocation list (CRL) that includes the list of all revoked certificates and periodically distributes this list to the PKI population. The CA adds a certificate to the CRL if the certificate needs to be prematurely revoked. Reasons for revoking certificates include termination of an employee, theft, or unauthorized disclosure of the private key, or violation of any mutual agreement clauses. A CA's PKI usually provides a flexible infrastructure that allows an immediate issuance of a CRL to inform other users of the status of the revoked certificate. The CA usually signs the CRL with its private key before distributing it to the subscriber population to enable it to validate CRL contents.

- *Certificate renewal:* As mentioned, each issued certificate has a validity period for which the certificate remains valid. Not renewing the certificate causes revocation of the certificate and the CA then places the certificate on the CRL. The certificate renewal process renews the certificate for continued usage. The CA should provide configurable expiration dates to match the organization's policies. Certain CAs offer products and services that automatically renew subscriber certificates based on certain attribute checks.

- *Key backup and recovery:* Key backup and recovery is a vital component of a PKI service, especially when enterprises encrypt and store business information on various media. The availability of this service enables enterprises to recover keys for data that requires recovery through decryption. Keys can be lost through various means. For example, it is not unusual for employees forget their passwords to unlock their private keys or lose devices that store their private keys. Employees may also permanently leave the organization (along with their private keys) thus forcing the enterprises to recover data encrypted through those keys. A key backup and recovery service implemented through a PKI application and augmented by various procedural controls enables enterprises to recover critical data. Key recovery however requires numerous levels of approvals to justify legitimate causes.

- *Registration authority services:* A registration authority provides certain PKI services on behalf of a CA. For example, an organization that requires a CA's services may find it cumbersome to have all its departments and employees contacting the CA individually for various public-key services. In such cases, a registration authority acts as a centralized entity within the organization and handles basic services locally. These services may include authentication of entities on the CA's behalf, distribution of certificates to users, controlling the use of keys within the organization for licensing purposes, etc. Various public CAs (e.g., VeriSign's OnSite program) offer such registration authorities to their customers.

- *Secure storage of certificates repository:* The CA should provide the infrastructure to secure the storage of certificates.

Organizations have multiple options to use PKI services. In essence, there are three primary methods. The first option is to lease PKI services from external trusted parties. Examples include Verisign and GTE CyberTrust. The second option is to act as a registration authority (RA) and provide a subset of PKI services to the local organization or corporation. The third option is to build PKI services in-house. The following sections explore these three models.

8.4.1 Public CA Services

Public CA services provide a full breadth of certificate management services. These CAs invest heavily in establishing concrete guidelines and in building a sound and secure technological infrastructure for the issuance of certificates. This is because most of these organizations provide insurance against economic theft or other liabilities when issuing certificates for secure e-commerce transactions. Various CAs have surfaced in the past few years that provide a multitude of certificate-related services for both individuals and large organizations that require certificates to conduct secure e-commerce services. Some of the more popular CAs are the following:

- VeriSign, Inc. (www.verisign.com);
- Thawte Consulting (www.thawte.com);
- GTE CyberTrust (www.gte.com/cybertrust);
- BelSign NV/SA (www.belsign.com);
- ABAecom (subsidiary of the American Bankers Association, www.abaecom.com);
- Equifax Secure CA (www.equifaxsecure.com);
- E-Certify Corporation (www.e-certify.com).

Since certificates are the cornerstones of trust in e-commerce transactions, CAs generally employ sound security and legal practices for the issuance and the entire life cycle management of certificates. Within the PKI world, such practices are referred to as certification practice statement (CPS). The Electronic Commerce and Information Technology Division of the American Bar Association defines CPS as "a statement of the practices which a certification authority employs in issuing certificates." A CPS usually includes some of the following topics:

- Verification processes for issuing certificates;
- Security procedures;
- Operating guidelines and procedures;
- CA governing laws;
- Dispute resolution processes.

In leasing CA services, the customer should explore the relevant topics of the CPS and associated technological and operational details that include:

- Disaster recovery procedures;
- Insurance for use of certificates against theft or misuse;

- Maintenance procedures of data centers housing PKI hardware and software infrastructure;
- Customer enrollment applications and processes;
- Reporting, tracking, and various administrative processes;
- Customer support center and associated processes;
- Breadth of certificate life cycle management services;
- Mechanisms for storage of private keys (e.g., hardware storage);
- Database encryption;
- Operational controls (e.g., backup procedures, continuity of business [COB] plans);
- Service unavailability thresholds.

8.4.2 Registration Authorities (RAs)

RAs enables an enterprise to operate CA services on behalf of a leading CA. This means that the primary CA provides the customer organization with the technology and process infrastructure to operate a CA, while retaining critical functions (e.g., certificate issuance functions). RAs are suitable for organizations who require an extensive suite of PKI services for e-commerce–based transactions, but do not wish to invest in the development of full-scale certificate life management services. RAs typically suit organizations that have limited groups of intranets and extranets that serve a single group. For example, an organization may establish an RA-based CA service for transacting e-commerce services with its suppliers and partners. RAs leverage the infrastructure capabilities of a public and professional CA and enable organizations to share the risks, as the primary CA still holds the responsibility for issuing certificates.

Typically, an organization authorizes a department or group of individuals to perform RA functions. Under VeriSign's OnSite program, these individuals are called local registration authority administrators (LRAA). The department caters for certificate requests for all employees within the organization by authenticating the employees based on certain criteria (e.g., Company ID number, Social Security Number, etc.). The department may use different policies for different types of certificates. For example, it may require multiple authorizations to request a server certificate. On authenticating a user's identity, the registration authority approves the requests and forwards them to the primary CA, who does not perform any other checks since the request is forwarded by an RA. The CA then issues the appropriate certificate and forwards it to the RA, who then distributes the certificate to the individual.

Establishing a RA service within the organization has numerous advantages, primary of which is low startup costs. An organization does not have to invest heavily in the technology infrastructure and processes that are usually required for establishing a CA service. The customer organization merely leases the necessary technology infrastructure from the public CA who also assists the organization to establish the RA function within the organization. This also provides the organization with the necessary experience and exposure to the intricacies of issues inherent in certificate management. The primary disadvantage of this approach is that the leasing costs can be costly when viewed over the long term. These costs are associated with annual maintenance and licensing issues.

Various public CAs offer RA functions. Among them, VeriSign's OnSite service is quite popular. VeriSign is one of the first few public CAs to emerge that provide full certificate life cycle services. Through their OnSite service, VeriSign provides most certificate life cycle services and necessary administration support to the RA. Thawte's Enterprise PKI program is another example of a public CA that offers the services of RA to other enterprises. Other CA programs include GTE's CyberTrust CA Hosting Services and Equifax's Equifax Secure services.

8.4.3   Building Internal PKI Services

Building an effective PKI entity requires the establishment of a reliable technology infrastructure coupled with sound business practices. These two elements are the basis for providing the trust that parties require from a trusted CA. An organization that does not perform adequate checks to issue certificates to its users or organizations will lose credibility because customers of those organizations will not be able to trust its e-commerce sessions. Sound business practices go hand in hand with a sound technological infrastructure that provides public key services. Technology is vital as the CA organization requires the use of sophisticated mechanisms to generate key pairs and appropriate hardware to store and protect that data and other related information.

Organizations that build their own internal PKI services therefore require the appropriate processes and procedures and a sound technological infrastructure to support PKI operations. However, an organization that intends to manage certificate issuance locally does not have to deploy the full-fledged PKI services mentioned earlier. Various vendors provide PKI products that offer different functionality, and enterprises can select the functionality that they intend to use and deploy. Following is a list of some PKI related development products:

- *Entrust/CommerceCA:* This product enables organizations (e.g., financial institutions) to issue SET certificates to appropriate parties in a SET transaction (e.g., credit card holders, merchants, transaction processors).

- *Entrust/PKI:* This product provides full certificate management services (e.g., certificate issuance, revocation, key backup and recovery).

- *Entrust/AutoRA:* This product automates most of the RA functions and enables users to automatically request certificates through a Web-based application. The application incorporates most business requirements to authenticate customers.

- *Netscape's Certificate Management System (CMS):* CMS includes multiple functions that enable an organization to internally deploy a managed PKI service. However, the current version of CMS does not support SET certificates.

- *CyberTrust Enterprise CA:* This product enables organizations to deploy full certificate management services in-house. This product works in conjunction with other complementary products. For example, CyberTrust SafeKeyper is a hardware device that provides organizations extra security for generation and storage of certificates.

- *E-Certify's Enterprise OnLine application:* This application consists of a directory for storage of certificates, RA functions, and an application toolkit that allows enterprises to establish internal PKI services.

8.5 E-Commerce Systems Security—Cost and Risk Assessment

Instituting appropriate information security controls in e-commerce systems has a cost that is both financial and architectural. Striving to gain knowledge of the various cost factors is useful regardless of the levels of security required. Organizations that require a high level of security can see the various ways in which the costs of security can add up and influence their budgets, schedules, and the performance of their systems. Similarly, organizations that institute selective security controls can assess whether instituting those controls is worth the cost and the effort. Organizations should balance these costs against the risks faced by the e-commerce systems. A detailed risk assessment should therefore drive the selection of the appropriate security controls for the e-commerce system. The sections below delineate the various costs involved in instituting information security followed by a discussion on risk assessment.

8.5.1   Information-Security–Related Costs

Instituting information-security–related controls have certain direct (financial) and indirect (system and business performance) costs. Comprehending these costs is vital in deciding which types of security controls an enterprise requires.

8.5.1.1   Financial Costs

Instituting security can result in the acquisition of numerous devices. These devices include firewalls and encryptors. Acquisition of these devices to secure an e-commerce system can be quite costly. Consider the case where a bank decides to offer video banking services to its customers from remotely placed video kiosks that connect to live agents located in a call center over a leased ISDN network. Encrypting the ISDN channel for each connection could require a minimum of two encryptors (encryption of audio and video signals) at the video kiosk alone. This could be quite expensive, especially if the bank intends to roll a system out to many geographical locations.

Direct financial costs related to instituting security in e-commerce applications and in the organization culture generally relate to the following factors:

- Security devices, including firewalls, encrypting devices (e.g., encrypting routers, link encryptors), network monitoring devices, and Internet scanners.

- Leasing of VPN services from NSPs.

- Establishing an information security organization. This refers to a department that defines the security policies of the organization and oversees their compliance.

- Development software for incorporating security in applications. This includes software such as toolkits that allow organizations to security enable their applications. For example, the industry offers numerous framework solutions and application interfaces that enable organizations to incorporate security services in applications. Examples are KeyWorks Toolkit Framework by IBM, CryptoAPI by Microsoft, Bio-API, Intel's CDSA, Entrust toolkits, and RSA's CryptoJ.

- Authentication tokens. These include smart cards and dynamic password tokens.

- Licensing costs. Use of certain software and hardware incurs annual licensing costs. For example, organizations that invest in CA development toolkits and products pay annual fees for the use of digital certificates. Similarly, usage of certain cryptography services results in tremendous licensing costs driven by the number of client computers that use those applications.

- Secure operating systems. Numerous operating systems have special security-hardened versions for security conscious organizations. For example, HP's HP-UX 10.x line of systems is security hardened and costs more than the traditional operating systems.

- Wireless security services. Organizations that intend to secure their wireless communication—such as pager communications—have to lease special services from organizations that offer such security services. These services cost more than leasing regular paging services from those organizations.

- Security audits. Security-conscious organizations such as financial institutions and governmental agencies use the services of security audit organizations to audit their security practices and procedures and that help those organizations assess their security vulnerabilities. E-commerce and Internet-related infrastructure has increased the potential of these vulnerabilities. These security audits expose any security holes and vulnerabilities in the Internet infrastructure of the organization. Some organizations also analyze business processes and identify security issues inherent in steps of those business processes. Depending upon the breadth of such audits, these can prove to be very expensive.

8.5.1.2 Performance Costs

These indirect costs attribute to both the system's technical performance and business process performance. The following highlights some primary areas that fall into this category:

- *Web connection performance:* Using SSL to connect to Web servers to establish secure sessions affects the performance of each Web session. This degradation in performance may be tolerable for a Web site that handles a small number of Web connections. However, for a popular shopping Web site that offers secure Web ordering and services millions of concurrent Web connections, such degradation may impact customer service levels. The Web site therefore may need to consider the use of special crypto-hardware modules to boost the performance of SSL connections.

- *Security administration processes:* With the proliferation of extranets and other Internet-based transaction Web sites, organization customers, merchants, and employees require IDs and other authentication tokens to transact e-commerce services. Depending on organizational security policies, the processes to receive such information may be quite bureaucratic and burdensome, thus leaving users frustrated and unable to perform business activities. The design of user applications should therefore minimize security administration bottlenecks and balance them against required security policies.

- *Storage of encrypted data:* Some organizations require archival of critical data in encrypted form. During contingency, decryption of these large databases lengthens recovery time.

- *Platform performance:* Certain security products installed on operating systems affect the entire system's performance. This is especially true for those products that install at the operating-system kernel level to augment kernel security. For example, as mentioned earlier, UNIX systems have inherently weak file-level controls. Products such as AutoSecure from Platinum Technology, Inc. install at the operating-system kernel level and inspect all commands before passing them to the operating system. Any violations at the security product level are not allowed to occur at the operating system kernel. These types of security products thus effect the platform's performance by a great percentage. This degradation can range from 3% to 10% and depends upon both the product and installation configuration.

- *Application performance:* An application that incorporates security controls such as encryption has to pay the performance penalty as well. For example, an electronic wallet application that stores customer's credentials (credit cards and other profile information) in the database may store it in encrypted format due to the sensitivity of the information. This implies that in accessing each user wallet's record through the Internet, the application will have to decrypt the data before using it for customer purchases. For numerous user connections, the encryption requirement may result in the purchase of additional servers for application load balancing purposes.

- *Business process performance:* The various tiers involved in an e-commerce system may increase the authentication steps for a user, and this may impact system usability. Typical corporate users usually have to authenticate to multiple applications, including signing on to a LAN, to e-mail applications, firewalls, intranet sites, and other applications. This proves to be quite burdensome for users and in some cases has direct financial impacts. For example, in call-center applications, where time is a primary metric, multiple sign-ons can restrict an agent to a limited number of calls. An organization may therefore need to invest in single sign-on (SSO) packages to counter such concerns. SSO packages authenticate users against the SSO security product database and, upon successful authentication, enable the user to use the various applications by transparently signing on to other systems.

- *IT software process performance:* An organization that develops software may have to institute appropriate checks and balances at all levels of its software development life cycle. This can increase development cycle times due to the potential need to incorporate security software in application modules, and it can influence application-testing times as well.

8.5.2 Information Security Risk Assessment

As IT has penetrated all facets of an enterprise and its processes, it is vital to understand the business risks associated with information security violations. Assessment of these business risks serves two purposes. First, it helps an enterprise assess its vulnerabilities. Second, risk assessment helps the organization determine the costs of securing the business process. Instituting security controls without understanding their value can prove very costly. The business should assess the nature of security controls and their value. To cite a simple example, an SSL session that sends a credit card number provides protection against eavesdropping, but it does not provide the merchant any guarantees about the identity of the user using the credit card number. For such cases, the SET protocol (explained in Chapter 7) should be used. In addition, for a news organization to deliver news, an SSL session may not be useful. Thus, all risk assessments in performing a security evaluation of an e-commerce system should be valued against these direct risks to the business.

Instituting information security risk management is a higher-level activity that addresses information security issues at the strategic levels of business process design. Strategic risk management highlights information security issues that may deem activities in the business process to be impractical, challenging, or very costly to address at later stages. This practice is analogous to that of IT systems implementation, in which vital details left out in the early stages of planning could greatly influence the activities in later stages of the system implementation life cycle.

An organization can resolve information security risks through any of the numerous risk management methodologies. The essence of a risk methodology involves three primary steps, which are as follows:

- Risk identification.
- Risk analysis, involving the analysis of business risks in light of financial, customer service, and business continuity impact.
- Risk mitigation, including the establishment of relevant technological security controls and operational processes and procedures for risk abatement.

8.5.2.1   Risk Identification

Risk identification involves delineating all information security-related risks inherent in the design and operational plan of the e-commerce system. The objective of this step is to identify those risks that can impact the financial, customer service, and business-continuity performance indicators. Risk identification encompasses three primary areas, which are described next.

Considering Technical Source of Risks

This refers to weakness in the technical design of the system, the exploitation of which can affect business performance. Various security protocols and controls have potential weaknesses that vary in their impact depending upon specific business requirements. Technical deficiencies in some security controls are deemed acceptable for various business processes due to the limited business process impact, while other business processes are vulnerable to the same technical weaknesses. For example, a few years ago the IT industry considered the breaking of DES to be computationally difficult. However, an increase in the computing power within the past few years has facilitated the means to carry out such attacks, as witnessed by numerous examples that have surfaced in the news. The use of DES is therefore no longer an option for critical business processes (e.g., financial and military processes), whereas it is still useful for noncritical business functions.

Considering Procedural Source of Risks

Numerous security weaknesses exist due to the inherent weak procedural controls in the operational aspects of an e-commerce system. Procedural security controls lie at the heart of information security initiatives. For an operational e-commerce system or any IT system, procedural controls apply especially to the security administration processes and data/network center access procedures. Regardless of the technological sophistication, the strength of the information security controls is directly proportional to these security procedural controls. For example, an e-commerce system that employs sophisticated authentication controls has little value if associated security administration processes for the system are weak. If users can obtain pertinent authentication tokens (e.g., passwords, smart cards) without due checks and controls from the security administrator, they can easily circumvent system security. These procedural controls have greater relevance for such systems due to the common trends of collocating e-commerce systems at external data/network operation centers.

Probability of Intrusion

Organizations should consider the probability of intrusion when identifying potential risks. This refers to the practical possibility of the occurrence of the security violation. As the purpose of the risk identification stage is to identify risks for which organizations can devise appropriate risk mitigation steps, identifying risks that have a low probability of occurrence will have little value. For example, numerous security conscious organizations strictly regulate the use of network sniffers in their data centers. Thus, identifying a risk that mentions the use of network sniffers within the organization's data center has little value. To mention another example, Lucent Technologies recently detected a flaw in the SSL protocol. By exploiting this flaw an intruder can break the encryption key of an SSL session by generating and studying one million sessions on the server. As the generation and storage of such a high volume of messages is impractical and easily detected by an administrator, the probability of intrusion in this case would be considered quite low. However, the probability of an intrusion of a non-SSL message is relatively very high.

8.5.2.2 Risk Analysis

After risk identification, the organization should analyze the business impact of those risks to the organization in cases of a security violation. Understanding the impact of information security violations to a specific business process is vital in assessing the investment needed for a specific e-commerce project. As mentioned earlier, e-commerce security has high associated costs and investment in information security initiatives should be directly proportional to the negative impact that a particular intrusion may cause to the business.

Information security risk analysis involves the analysis of three forms of business risks.

Financial Impact

This refers to the financial loss that an organization incurs due to an information security violation. With the proliferation of electronic payment systems and networks, security of those networks and transactions is of paramount importance. Individuals can tap public or private networks to spoof illegal transactions that may result in money transfers and thus cause financial impact to the organization. In other cases, intruders can use an organization's facilities or systems to steal other types of services. For example, hackers can misuse an organization's misconfigured PBX to launch fraudulent calls that may result in a substantial financial impact to the organization. Some types of customer service impacts cause an organization to lose customers, thus affecting financial numbers. Consider, for example, a financial institution that offers a server-side electronic wallet application to its customers. As explained in Chapter 7, server-side wallets store user credentials in the organization's data centers. A security breach involving theft of user-wallet records will drive customers away from using that institution's electronic wallet application, thus affecting the organization's revenues.

Customer Service Impact

Information security violations in e-commerce systems present numerous threats to customer service levels. The primary reason for this is the fact that e-commerce systems interface with a large volume of external customers. A hacked Web site, for example, can result in a denial of service that inhibits customers from accessing services on that site. This can cause a customer-service impact to the organization. A customer, for example, who intends to shop on the Internet will not hesitate to switch to another shopping Web site. Similarly, the unavailability of a call center, established to service customer calls through the Internet and other electronic channels, will leave calling customers frustrated. Theft of customer information, as mentioned earlier, can result in a potentially serious customer service impact as well.

Business Continuity Impact

Certain security intrusions can impact the continuity of a business process without resulting in a direct financial or customer service loss. For example, if an organization has implemented a procurement solution over the Internet, a lack of appropriate security controls may disrupt that process without causing any financial or customer services impact. These intrusions thus impact business process performance levels and their criticality varies for each business process. In a complex supply-chain process, the unavailability of one business process may disrupt the entire supply-chain, which may not be acceptable to an organization.

The next section elaborates upon the primary risk mitigation strategies.

8.6 E-Commerce Systems Security Controls Enforcement

This section discusses processes and practices that an organization can institute to mitigate and control information security-related threats due to the continual deployment of various IT applications. Three strategies can help mitigate such risks.

8.6.1 Formulating an Enterprise's Information Security Policy

An information security policy is a set of rules that govern the use, handling, and processing of information technology assets. The information technology assets constitute both information and information systems. The policy delineates enforcement of various forms of controls for the protection of IT assets. It facilitates a process by which IT-related risks—particularly those pertaining to information security—are minimized in a cost-effective manner. The information security policy of an organization specifies rules and attributes that are more specific to the organization's businesses.

The following delineates some of the characteristics of an information security policy. The information security policy:

- Is not very detailed;
- Is enforced throughout the enterprise and is directed toward the entire staff of the organization and not toward a specific group of staff or individuals;
- Is supported by a well-defined process that derives and maintains the information security policy;
- Is documented in intelligible natural language statements;
- Has a defined structure addressing all security controls mentioned in the earlier sections of this chapter;
- Is upheld by the highest level of management;
- Is realistic for implementation in the appropriate enterprise.

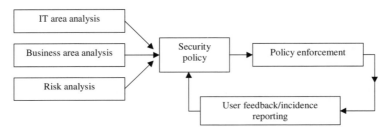

Figure 8.7   Information security policy derivation model.

Figure 8.7 illustrates an information security policy derivation model. As illustrated, three factors drive the derivation of the security policy. A detailed understanding of the IT initiatives planned in the enterprise (IT architecture) is the first factor. Uncontrolled injection of new IT initiatives is a very common cause for failure of IT initiatives within organizations. Inherent security risks contribute to the failure of those initiatives. For example, many enterprises that embraced the flexibility of mobile code technology (ActiveX and Java applets) for mission-critical business applications were unaware of the inherent security risks in those technologies. An analysis of new initiatives and their payback must be factored into the formulation of the security policy. For example, during the early days of ActiveX controls, numerous organizations adopted the policy of prohibiting their use until the technology was mature enough to handle the security issues inherent in their deployment and use.

The nature of the business processes (analysis of the business area) is the second critical factor in the derivation of an information security policy. E-commerce initiatives applied ad hoc to certain business processes could prove dangerous to an enterprise. This applies especially to financial institutions and governmental agencies, where breaches in information security through the Internet can prove to be very expensive financially and pose a threat to national security.

A high-level risk analysis based on the nature of the business processes and IT initiatives within the enterprise is the third driver for deriving the security policy. The nature of the risks should control the thresholds of various security controls defined in the security policy. For example, a certain enterprise may require the use of 128-bit browsers for certain types of transactions, whereas the use of 40-bit browsers may be adequate for other enterprises. Similarly, the use of traditional DES may be appropriate for certain businesses, whereas other enterprises may require the use of Triple DES for all its transactions.

An organization should consider the following subject areas for inclusion in a security policy:

- Regulating the use of sniffers and data scopes within production environments and networks.

- Strictly regulating the use of live information assets for testing or other purposes. For example, in testing a Web site containing live information about customers, organizations should never allow the use of live customer data for development and testing purposes.

- Organizations should use data classification strategies for selective security enforcement. For example, an organization may require that all data classified as TOP SECRET be fully encrypted while it is stored in a database or transmitted over a network, whereas data classified as SECRET may not need encryption while in storage.

- An organization should enforce appropriate policies and procedures for reviewing audit trail information as a means to not only investigate incidents but also deter potential future incidents. For e-commerce systems, where numerous platforms are involved, the challenge is to identify incidents that require a vigilant reviewing approach. Numerous tools have surfaced in the market that allow automated alerting mechanisms, thus alleviating administrators from the burden of manually reviewing audit trails. Some tools, for example, monitor a Web site to detect illegal hacking attempts and send alerts to the administrators.

- No system is free of security flaws. Security organizations such as CERT and others continually post numerous warnings related to various e-commerce systems platforms. An organization has to be diligent to respond to these warnings and react accordingly. However, lack of an effective organization structure and processes may inhibit enforcement of these security controls on numerous systems spread throughout the organization.

8.6.2 Controlling Software Processes

Effective procedural controls instituted in an organization's software development processes ensure a planned and methodical approach to building security controls in e-commerce applications. Conversely, ad hoc software development processes could have potentially devastating effects. Malicious code inserted into software by the development staff could intentionally or accidentally result in serious security breaches. Likewise, failure to test the incorporated security functionality in the application can have serious repercussions when the organization deploys the application in operations. It is imperative to execute appropriate checks and balances in the software-development life cycle, as they have a direct bearing on the information security environment of the organization. The following outlines critical milestones in the software development processes when an e-commerce systems project should assess the need to institute appropriate security controls:

- *Requirements management and project planning:* Planning for information security requirements can begin with the requirements stage of the software life cycle. Requirements management establishes a common understanding of the requirements between the e-commerce system's customer and the development and implementation team. Intergroup coordination practices will ensure the setting of appropriate expectations for the effect of security controls on budgetary requirements, the system's technical performance, customer enrollment processes, required security administration processes, and business performance in general.

  Certain information security requirements entail extensive training for the development staff. For example, as mentioned in earlier sections, implementation of an in-house PKI infrastructure requires expertise in PKI technology and administration processes. Such activities require early planning.

- *Software development infrastructure:* An organization should institute relevant security controls in the software development infrastructure of the enterprise. This refers especially to those development machines that the organization uses to transfer software in production. A compromise of software work products and their integrity (inserting code without anyone's knowledge) could result in the injecting of malicious code during production.

- *Outsourcing e-commerce systems hosting:* As increasing numbers of organizations outsource many facets of IT operations and services, it is vital that an organization assess the security controls of the external service provider. Some of the questions that the organization should ask are:

  — Does the service provider use firewalls or router filtering to ward off external connections?

  — What are the service provider's policies for transmiting information between various points?

— What is the security of data in storage?

— What services does the service provider employ for the storage of backup tapes containing critical information?

— How mature are the external service provider's security administration processes?

— What are the physical security controls of computing platforms?

— What are the physical security controls of the data center where the e-commerce systems are collocated?

— What type of background checks does the service provider perform in hiring security administrators?

— What are the change control procedures relevant to installing new releases in production?

• *Penetration testing:* Penetration testing should be part of the organization's software testing phase. Penetration testing provides assurances for the security of an automated system. Popular methods of information security breaches for each of the specific IT systems are executed against the system to ensure it is secured against any potential future external intrusions. The scope of penetration testing spans numerous areas such as the security configuration of the system, operational considerations, system architecture, and hardware. Penetration testing uses next-generation network utilities (e.g., SATAN, ISS) to take a snapshot of the network and uses predefined patterns to penetrate networks through the Internet and intranets. This enables the detection of security holes and misconfigurations before the final release of the application to users on the Internet.

8.6.3 Formulating Security Administration Processes

Security administration functions deal with the assignment of computing privileges and resources to users. As e-commerce systems include user and computing resources spread across multiple organizations, the security administration functions have grown more challenging. Security administration functions span beyond simple ID management to comprise other critical functions as well, some of which are the following:

• Enforcement of consistent security policy rules for external entities (e.g., extranets).

- Responsibility for the incidence response function, including the merging of these processes with external service providers' incidence response processes. For example, the organization's security administration function should be appropriately notified of any security breaches in external service provider's data centers.

- Responsibility for triggering remedial measures for information security-related incidents.

- Responsibility for triggering remedial measures in operational environment set-ups related to security issues.

- Assigning users and resources for networks.

- Setting up of secure networking connections (e.g., setting firewall ACLs, establishing keys in encrypting routers for VPN connections).

- E-mail access control.

- Tracking employee, Internet customers, and external partners' movements.

- Responsibility for audit trail management. This activity focuses on intrusion detection and escalation measures and is considered to be an ongoing proactive risk control activity. Relevant processes include activities such as audit trail review processes for internally- and externally-hosted systems, detection of security incidents, escalation of detected incidents, and initiation of corrective actions.

Notes and Web Sites

[1] csrc.nist.gov/encryption/aes/aes_home.htm

[2] www.rsasecurity.com/news/pr/990924.html

9

Managing E-Commerce Systems Implementation Risks

The success of an enterprise's e-commerce initiatives depends upon three primary factors. First, enterprises should institute appropriate processes to ensure the delivery of high-quality e-commerce systems. This requires businesses and the IT industry together work to understand customer requirements and translate them into high-quality systems. Traditional organizational cultures may have been tolerant of defects found in live operational environments that included system traps, incomplete functionality, and other problems, but such defects in a production environment in an e-commerce paradigm directly affect the bottom line, as external customers are less tolerant of such defects. Customers can switch organizations in the e-commerce arena with a few keystrokes or mouse clicks. The delivery of high-quality systems is thus of paramount importance in retaining the integrity and stability of a business.

Second, organizations require a stable and high-quality operations environment to operate their e-commerce systems. But high-quality e-commerce systems alone are not enough to ensure the stability of an operations environment. A stable operations environment requires the proper configuration of systems, network security controls, internal processes and procedures, and other parameters discussed in earlier chapters of this book. The organization should ensure the building of a stable operations environment that delivers high-quality service levels. If an organization chooses to outsource such operations, appropriate service level agreements should provide organizations with firm commitments about the quality of services that the organization receives from external service providers.

The third issue has to do with legal and governmental regulations. Doing business on the Internet raises numerous legal issues. An enterprise should therefore ensure that its e-commerce initiatives address all legal and other regulatory issues, as the violation of certain laws can lead to litigation and may ultimately have a tremendous impact on bottom line results.

The remainder of this chapter further elaborates upon these issues and recommends strategies that enable an organization to reap value from its e-commerce systems.

9.1 Business Process Alignment

An essential indicator driving the success of e-commerce systems in particular and IT systems in general is business-IT alignment. A misalignment or "disconnect" between IT and business management causes a domino effect that inhibits IT management from setting up an effective infrastructure and instituting processes that deliver solutions to the business. Charles Wang, CEO of Computer Associates, defines disconnect as "a conflict—pervasive yet unnatural—that has misaligned the objectives of executive managers and technologists and that impairs or prevents organizations from obtaining a cost-effective return from their investments in information technology" [1]. Organizations that fail to address this issue appropriately will carry this phenomenon forward in their initiatives to implement e-commerce systems as well. Such misaligned e-commerce initiatives will either be ineffective to resolve appropriate business cases or will target the wrong business processes for reengineering. It is, therefore, vital for an enterprise to adopt strategies that will effectively inculcate management's vision and objectives in all of its facets to leverage e-commerce trends appropriately for its benefits.

Among the many strategies organizations can employ to align their business and IT facets is the institution of new polices and their enforcement throughout all layers of the organization. Secondly, organizations need to continuously monitor their performance and institute appropriate improvement indicators. The following sections discuss these indicators in further detail.

9.1.1 E-Commerce–Relevant Organizational Policies

Policies translate an organization's vision and mission into actionable statements. For example, an organization whose mission includes providing excellent customer service will reflect this in its business policies. These policies will in turn drive the engineering of business processes and supporting systems that will focus on building an appropriate customer service function. Similarly, an organization that commits to securing its customers' transactions and information formulates an appropriate information security policy that addresses issues relevant to instituting an appropriate information security culture within the organization. Policies thus help organizations uphold a set of values and standards.

Corporate policies also define the boundaries within which an enterprise carries out its operations. Policies provide an organization's businesses with direction and a set of choices for executing specific business functions. In this environment of rapid technological and regulatory change, lack of such direction will result in business implementation inconsistencies. These inconsistencies inhibit the portrayal of standard values required to build an organization's image or brand for customers and the external world in general.

An organization's vision statement, industry trends, public policy, governmental regulations, and internal and external risks guide the formulation of an organization's policies. An organization's vision statement reflects an organization's mission of existence and its priorities in achieving that vision. Public policy and governmental regulations are as vital to adhere to as local laws and regulations. For example, a government regulation requiring e-commerce sites to protect customer's privacy can hold organizations accountable if they do not take appropriate measures (through formulating polices and implementing those policies) to protect consumer privacy. Factoring internal and external risks into its policy formulation helps the organization limit its risks. For example, a financial institution may require the use of certain security controls in transmitting financial information over external networks.

Organizations formulate numerous policies to tackle different subject matters relevant to the existence and operation of the organization. Some of the common policies include:

- Technology policies
- Information security policies
- Business risk management policies
- Technology risk management policies
- Human resource policies.

The emergence of e-commerce and the Internet, while providing organizations with numerous opportunities for advancement and growth, also poses challenges that can seriously inhibit an organization's ability to deliver value-added products and services to customers. First, the Internet provides organizations with innumerable opportunities and options to engineer their business processes. This, coupled with rapid technological advancements, requires a contained approach that will help organizations embark upon e-commerce initiatives in a controlled fashion. This will prevent organizations from wasting their investments. Another challenge is the threat of information security. Serious violations related to the enterprise's information assets as well as its customer information can threaten an enterprise's existence. Such challenges and threats therefore necessitate the updating of all organizational policies that are in one way or another affected by the Internet and e-commerce.

Following are some of the policies that an organization should consider instituting when embarking upon Internet and e-commerce initiatives:

- *Internet usage policies:* An organization should clearly stipulate acceptable practices for Internet usage within the organization's facilities. An organization, for example, may prohibit Internet access for personal use, commercial activities, or engagement in any behavior that would jeopardize an organization's image.

- *Monitoring employee behavior on the Web:* Non-work-related Internet usage affects an organization's network bandwidth as well as employee productivity. Organizations can adopt varying policies to regulate such use. For example, an organization may simply use statements to prohibit such usage, while others may deploy tools to track employees' usage of the Internet. On the other hand, organizations may opt limiting Internet usage during specific periods (e.g., restricting personal access to lunch hours).

- *Regulation of bulk commercial e-mail for marketing purposes:* Bulk commercial e-mail is also referred to as *spam.* There are numerous government regulations against spamming. An organization in most cases is obliged to formulate antispamming policies, the lack of which can hold an organization fully responsible for any such activities from its employees or business units.

- *Allowable policies for the usage of e-mail:* Corporate e-mail systems connected to the Internet can cause employees to send e-mail that can jeopardize an organization's credibility. Organizations usually establish policies directing employees to use corporate e-mail systems only for work-related activities and prohibit its use to harass employees, send chain letters, or engage in any malicious activities. An organization may also control the use of e-mail for transmitting sensitive information (depending upon the security of its e-mail system),

- *Using external ISPs and NSPs for hosting services:* Another good business practice is the establishment of criteria in the selection of ISPs for hosting or other business services. The Internet policy should prohibit an organization's business units from selecting ISPs or NSPs that do not conform to those guidelines. Chapter 3 outlines some ISP selection criteria.

- *Internet information security policies:* As explained in an earlier chapter, information security-related policies govern the flow of information through the enterprise's networks and external networks connected to the enterprise as well as the criteria for securing such information. The policy may require the use of certain encryption algorithms or encryption key lengths for securing specific classes of information.

- *Penetration testing policies:* Also referred to as ethical hacking, these policies ensure that the deployed site institutes appropriate security controls to prevent security threats against the Web site. The organization's Internet policy should require businesses to undergo Internet penetration testing before opening the site to the general public.

- *Regulating Internet usage for business reengineering initiatives:* An organization should institute policies for engineering high-liability systems. For example, public certificates issued by certain certificate authorities provide liability coverage up to certain limits. Organizations that plan to deploy e-commerce systems with associated higher liability and who depend upon those certificates to protect the integrity of its systems should either consider alternatives or institute compensating controls to limit liability.

- *Privacy policies:* An organization should require all Web pages deployed over the Internet to display appropriate privacy statements and/or include certain privacy seals from service providers that provide appropriate trust to consumers (e.g., usage of TRUSTe and BBBOnline seals of trust). This chapter further elaborates upon the privacy issues in later sections.

- *Business's responsibilities for deploying Internet-based systems:* An organization's policy may require all businesses to scrutinize all systems deployed on the Internet to ensure that they enforce all organizational standards and procedures related to Internet deployment. Review of these systems ensures the addressing of various legal and other concerns.

9.1.2 Measuring Business Performance

An organization's success depends upon the stability and efficiency of its strategic business processes. These business processes are comprised of both revenue generation and support processes. The ability of an organization to continuously monitor the health of these processes and institute appropriate improvement initiatives drives the organization toward achieving the required efficiency and stability.

Instituting strategic measurement programs across the various organizational processes provides management with the necessary knowledge and insight into the overall performance of the organization. An organization can institute these measurement programs to measure both process and product quality. Measurement of product and service quality can help differentiate the organization in its offerings in the marketplace. As for processes, measuring both revenue-generation and support processes is vital. Measuring revenue-generation processes provides organizations with insight into opportunities that lead to higher revenues. Measuring support processes provides organizations with an opportunity to reduce costs.

Trends and initiatives for taking measurements for organizational improvement have existed for a number of years. Their specific use in the IT discipline surfaced only a few years ago. Initial use of such measurement programs did not result in appropriate ROI due to two primary challenges. First, a lack of defined and streamlined IT processes posed difficulties in measuring various parameters associated with those processes. It is well recognized that the quality of products greatly depends upon the quality of processes that produce those products. On the other hand, organizations that measure the quality of IT products faced difficulty in influencing changes due to the ad hoc nature of the processes that produced them.

The second challenge that organizations face in reaping value from measurements is misalignment between the IT and business sectors. Although the industry in general recognizes IT's enabling power to engineer an organization's business processes, the misalignment does not yield any fruitful results for the enterprise as a whole, even in cases where IT departments successfully institute measurements.

An organization can take measurements at various levels of the organization. These measurements in some cases have to be very granular to accurately depict the quality of the product or process being measured. The primary challenge in instituting any measurement program is determining the right mix of measurements that can directly benefit the enterprise. Another challenge is selecting an appropriate measurement model that translates those granular measurements into higher-level indicators that can in turn directly depict the state of an enterprise's various facets.

The Balanced Score Card (BSC) approach is one such framework that has proven quite successful [2]. The BSC approach proposes to measure an organization from four primary facets that determine an organization's short term and long term health. As enterprises embark upon e-commerce initiatives, the BSC approach rightly applied can assist an organization measuring the success of its e-commerce initiatives.

BSC is a performance measurement system that provides a mechanism to measure various aspects of business performance. BSC measures business performance on four different fronts. The performance is not merely financial, as BSC formulators contend that financial indicators indicate only the present state of the organization and do not indicate performance of the general direction of the business. Measuring on the other three fronts ensures future success and helps an organization to adjust its course toward its mission and objectives. The framework proposes that indicators be tied to the specific strategic organizational objectives that they measure. The objectives in turn relate to the organization's overall mission and vision. Measuring those indicators will thus determine an organization's progress in marching towards its mission.

The BSC framework, initially proposed by Robert Kaplan and David Norton, proposes measurement of the business from the following indicators:

- *Customer performance:* These indicators determine the overall satisfaction of an organization's customers and in general assess customers' views about the organizations. Organizations can use various criteria to measure this indicator including customer enrollment, customer retention rates to gauge customer satisfaction, and customer satisfaction measured through surveys.

- *Internal process efficiency:* These indicators measure an organization's operational effectiveness and efficiencies in delivering products and services to its customers. An organization can institute indicators such as cycle times, lead times, and other such measures to assess the health of its processes.

- *Learning and growth:* This determines an organization's ability to continuously learn and grow its employees to meet future challenges. An organization can use the number of training days or number of courses offered to measure this indicator.

- *Financials:* Financial indicators measure an organization's financial performance. An organization uses the revenues of various business units periodically to assess the financial health of the organization.

An organization can apply the BSC framework to various aspects of a business. While the BSC framework should be instituted at a higher level to assess the well-being of the organization, various businesses within the organization can institute such measures to measure their internal performance. An IT department, for example, can apply the BSC framework to assess its performance and the value that the department provides to the business.

While performance measurement systems such as the BSC can provide numerous benefits to an organization, care should be taken to ensure that an organization does not start measuring non-value-added parameters or tries to measure too many parameters. The organization should be careful in selecting parameters that directly relate to its objective.

An organization can use the BSC framework to measure some of the following critical indicators for the success its e-commerce initiatives:

- *IT process effectiveness:* As mentioned earlier, the quality of products (e.g., an organization's e-commerce systems) depends upon the quality of processes (e.g., an organization's IT processes) that produce them. The processes should be effective enough to build e-commerce products that align with business requirements, efficient enough to rapidly respond to business needs in delivering appropriate products, and stable enough to enable an organization to predict the performance of its IT processes.

- *Capturing prospects:* One of the first steps in e-commerce initiatives, especially in the business-to-consumer paradigm, is the capturing of prospects (subjects that have the potential to become customers). This enables an organization to engage in one-to-one marketing initiatives to turn prospects into customers. Knowing the number of prospects can help organizations determine the effectiveness of its marketing and advertising strategies.

- *Customer retention:* An e-commerce system's ultimate success is in its retention of cutomers. An organization can employ intelligent techniques and good customer service to retain such customers. The number of loyal customers can help an organization determine the value of content that it offers through electronic channels and various aspects of its Web navigation features.

- *Revenues:* The ultimate success of an organization's e-commerce initiative is its ability to measure revenues through its e-commerce systems.

- *Reengineered products:* An organization spends a great fortune on its business processes. Organizations can leverage the power of the Internet (cost, ubiquity) to reengineer its processes to reduce costs. Total cost reduction because of such reengineering can measure the effectiveness of such initiatives.

- *Legal stability:* E-commerce initiatives entail substantial legal issues. A minimum number of legal issues may indicate the organization's due diligence in addressing those issues.

Subsequent sections elaborate upon some of these indicators in more detail.

9.2   IT Process Alignment

As discussed earlier, one of the primary issues in the IT arena has been the failure of enterprises to reap appropriate return on IT initiatives. IT solutions suffer from long cycle times, insufficient definition of requirements, numerous defects, missed commitments, and many other issues. These issues have driven enterprises to question their effectiveness in reaping appropriate ROI. With e-commerce systems demanding higher investments, enterprises are struggling to optimize their IT processes to maximize the value of their investments.

The industry in general has been tackling these issues for a number of years. Numerous methodologies (e.g., information engineering and related computer-aided software engineering tools) surfaced to partially address the problem. Such tools and methodologies focused on the engineering aspects of systems. However, the primary issues lurked at a higher level and were attributed to the defective and ad hoc processes that organizations followed to deliver software solutions to their businesses.

Software Engineering Institute's (SEI) Capability Maturity Model (CMM) emerged a few years ago to provide a comprehensive process framework to address the quality of software solutions and certain other parameters relevant to the success of the IT projects. The underlying philosophy behind the CMM is that the quality of software products depends upon the quality of process(es) that produce the software solution. CMM therefore focuses on the process aspects of the issues and encompasses critical subject areas required for the production of software work products. These subject areas in CMM terminology are referred to as Key Process Areas (KPAs).

The CMM defines five levels of process maturity for an organization's software processes. Each maturity stage constitutes a set of KPAs that an organization institutes in its culture to produce software products. Level 1 (initial level) indicates the state of instability within the organization for developing software products. Organizations fitting this maturity level do not have any defined software processes or practices that will enable the organization to adequately predict the outcome of a particular project's efforts. At this level, the organization usually depends on individual skills and heroics to deliver software projects.

In Level 2 (repeatable level), the organization has instituted basic project management processes and procedures to effectively track and control the delivery of software products. In Level 3 (defined level), an organization has defined a software process for the organization and delineated process-tailoring guidelines to address requirements of various software products. A Level 3 organization establishes a software engineering process group that oversees the organization's software process definition aspects.

A Level 4 (managed level) organization establishes quantitative goals to measure the quality of software products. The organization uses various measurement indicators (e.g., size of system, cycle time for delivery, defects identified) to assess the quality of its software products. In Level 5 (optimization level) the organization uses measurement data to continuously improve its processes.

Table 9.1 highlights all the KPAs associated with the CMM levels. Delivery efficiency and quality of software products corresponding to IT solutions in general and e-commerce solutions in particular depend upon the organization's maturity levels and the KPAs that an organization institutes within its infrastructure. Although all KPAs directly influence the quality of e-commerce solutions, this section focuses on the issues of project management, intergroup coordination, and software product engineering, as they have a direct bearing on the quality of e-commerce systems.

Table 9.1
SEI's CMM and Key Process Areas

Maturity Level	Key Process Area for the Level	Brief Description
2	Requirements management	Deals with establishing and controlling software product requirements between the customer of the software product and the software engineering organization.
2	Software project planning	Deals with establishing project plans that highlight the deliverables and commitments of the project. The project plan clearly delineates resources, schedules, and work products delivery timetable for the duration of the project.

Table 9.1 *(continued)*

SEI's CMM and Key Process Areas

Maturity Level	Key Process Area for the Level	Brief Description
2	Software project tracking and oversight	Deals with tracking the progress of the software project against the preestablished software project plan. This practice involves the tracking of various software parameters such as delivery schedules, size of work products, resources, etc.
2	Software subcontract management	Deals with managing various activities of the subcontractors including their delivery schedule, committed software products, and delivery schedule.
2	Software quality assurance	Deals with establishing software quality criteria plans for the project and providing management with visibility into the process followed to deliver software products.
2	Software configuration management	Deals with controlling changes of software products throughout their life cycle, from development phases to maintenance. This practice requires the establishment of baselines to track changes to software products.
3	Organization process focus	Deals with establishing a dedicated group to oversee, assess, and continuously improve an organization's software processes.
3	Organization process definition	Deals with developing software processes of the organization and associated work products (e.g., methodologies for the development of various types of software, tailoring guidelines for the organization's software process to suit different project requirements).
3	Training program	Deals with identifying the training needs of the organization in general and specifically for individuals and projects and developing plans to address those needs.
3	Integrated software management	Deals with defining a software process tailored to a project's characteristics. The software process is derived from the organization's overall defined software process. The software process or project includes all engineering and management activities required to manage the project.
3	Software product engineering	Deals with performing the required engineering activities to deliver the software product of the appropriate project. These activities include establishing technical architectures, and developing code.

Table 9.1 *(continued)*
SEI's CMM and Key Process Areas

Maturity Level	Key Process Area for the Level	Brief Description
3	Intergroup coordination	Deals with establishing the mechanisms to effectively deal with all the necessary groups in order to enable successful delivery of the software product and prevent bottlenecks. A software project may need to deal with organizations such as data centers, the SCM organization, users, and testers.
3	Peer reviews	Deals with reviewing software products during the development life cycle to remove defects from software products.
4	Quantitative process management	Deals with measuring performance of the software process that organization or project uses for developing software products. Measured parameters include defects identified in products, efficiency of the testing process, cycle times to deliver products, etc.
4	Software quality management	Deals with preestablishing quality goals for a project's software products and managing the software project to ensure that the project plans result in products of desired quality. Software plans and other parameters of the process are continuously adjusted to ensure the delivery of software products of appropriate quality.
5	Defect prevention	Deals with the identification of defects during the life cycle of software product development and operations, identifying the causes that lead to the injection of those defects, and making all necessary changes (related to engineering, process, or other causes) to prevent their future recurrence.
5	Technology change management	Deals with instituting a systematic approach to introduce new technologies within the organization. This may involve activities such as performing evaluations and analyzing the impact of new technology on the IT and business environment in general.
5	Process change management	Deals with ensuring that the organization's software process is continuously improved. Measurements are used to track performance and institute remedial and improvement changes.

9.2.1   E-Commerce Project Management

Project management refers to the formulation of work activities required to achieve a project's objectives and the monitoring of those activities until project completion. Formulation of a project plan considers the parameters of time, cost, and resources for the execution and control of the associated work activities. As erecting a full-fledged e-commerce system may require the involvement of numerous external entities and their associated resources, managing an e-commerce project involves detailed planning, coordination, and execution of various work activities. The following depicts some of the key parameters relevant for managing e-commerce projects.

- *Project scope:* One of the primary activities associated with project management is determining the scope of the project. This requires a clear definition of the project's desired functionality and aligning that with the project's customer. Lack of thorough delineation of work activities results in confusion and hides complex aspects of designing appropriate software modules. For example, it is easy to confuse an electronic bill payment system with a full-fledged EBPP system. An EBPP system is a composition of two distinct functions, both of which entail individual complexities. The first component is the bill-presentment function, which includes numerous work activities within the organization. The other function is that of electronic bill payment, which requires interfacing with various payment networks and financial institutions that reconcile payments across various entities that include customers, customers' banks, billers, billers' banks, and intermediate payment networks. An organization rushing to offer e-commerce-related services to customers should therefore clearly define its scope and the underlying activities that will enable it to assess the magnitude of the effort.

- *Identification of software work products:* An e-commerce system usually consists of multiple software modules distributed across systems that work in conjunction with each other to fulfill the system's various requests. The software components may vary in size and functionality. Identification of all software components required for development or for modification to achieve interoperability between systems is of paramount importance to prevent delays toward the end of the software development life cycle. Because e-commerce systems may span a multitude of organizations, identification of such software products and components requires mature intergroup practices and mechanisms. Furthermore, identification of all software components is vital to ensure that all parties involved in making those changes can plan the necessary testing resulting from the modification of those software components.

- *Allocation of resources:* Organizing a project within an enterprise entails fewer complexities due to a relatively better control on internally assigned resources. If an e-commerce project engulfs multiple organizations, managing resources across external entities complicates the project management process. Subcontracting whole functions and holding subcontractors to their commitments is one method to control resource allocation. This is because the subcontractor masks such details from the customer organization and manages that function internally. The subcontractor coordinates only the deliverable aspects of the project with the customer.

9.2.2   Software Product Engineering

Software product engineering refers to using the appropriate tools and methods to engineer the desired IT system. Various tools and methods exist for engineering software products. The e-commerce trend has triggered the emergence of numerous products. These products facilitate the development of Internet-based applications, provide universal access to a wide range of data repositories, and enable rapid development and deployment of applications to cope with a business's changing requirements.

Software product development tools for developing e-commerce projects incorporate a suite of functionality such as the following:

- Facilitate easier integration of various software components required in e-commerce systems such as browsers, middleware, and backend databases.

- Provide an integrated development environment that includes wizards to simplify various aspects of software development such as debugging and testing.

- Facilitate the development of thin client applications that interoperate with HTML-, Dynamic HTML- and XML-based data.

- Include support for ActiveX Data Objects and OLEDB technologies that enable the integration of client applications with relational and nonrelational data.

- Support mobile appliance applications.

- Support the popular and emerging Linux operating system.

- Facilitate development of transaction based applications based on CORBA, COM, and MTS.

- Incorporate diverse software components in software libraries that facilitate the rapid development of Internet-based applications.

- Include rapid application development (RAD) features that allow for the rapid development and integration of e-commerce software modules across servers and systems. RAD techniques focus on prototyping, the reuse of software, and reduced software development that in turn enable software development in short cycle times. For example, certain RAD tools enable database access without the need for extensive code development.

Popular development tools that selectively incorporate the above features and more include Inprise's Delphi, TopSpeed's Clarion, Magic Software's Magic, and Microsoft's Visual Basic application development tools.

Testing is another vital phase of engineering software products. The new generation of tools include a multitude of testing features to support testing of Internet-based applications, some of which are as follows:

- Include features that facilitate the testing of CORBA- and Java-based applications;

- Support features for testing embedded applications;

- Support end-to-end testing of Internet-based transaction management applications;

- Facilitate testing at various levels such as functionality testing, stress testing by simulating Internet-based transactions, and testing for browser and GUI interfaces;

- Support regression testing for multicomponent software modules and systems;

- Support load testing of applications distributed across servers and systems;

- Support testing at the API level for non-GUI software modules.

SunTest is a Sun Microsystems unit that provides a host of Java testing tools. These tools include JavaSpec (facilitating testing at API level), JavaStar (testing Java applications and applets in a windows environment), JavaLoad (allows for distributed load testing), and others. Similarly, Segue Software, Inc. provides a host of automated testing solutions for building e-commerce solutions. Their products include SilkTest (for automated functional and regression testing), SilkPerformer (for load and performance testing), SilkPilot (for unit-testing CORBA objects), SilkObserver (for end-to-end testing of CORBA solutions), SilkRealizer (for scenario-based testing), and others.

9.2.3   Intergroup Coordination

Intergroup coordination calls for instituting intelligent synergies between various groups that are involved in the delivery of IT systems. This KPA is especially relevant for e-commerce system implementation, which involves numerous organizations for successful implementation and operation. The CMM defines three primary goals for intergroup coordination. The first requires that the customer requirements for the project are agreed upon when coordinating with various groups for engineering a system. The second goal is to ensure that various groups involved in the delivery of the system mutually agree to their commitments. The third goal is to ensure that the groups track and resolve all intergroup issues. The project management function thus has to ensure the coordination of all such functions. This can be very challenging in building e-commerce systems that require interfaces to multiple organizations. For example, an e-commerce system involving the establishment of an EBPP extranet requires interface to some of the following organizations:

- ISPs and NSPs to gain access to the Internet;
- An organization that will manage the operations of the EBPP extranet;
- Network interface to multiple billers that will link to the extranet to upload bills for bill presentment;
- Interface to payment processors (e.g., ACH, VISA, or others) for processing payments;
- Subcontracting EBPP application development;
- Establishing a centralized security administration that will perform or oversee the security administration activities of the extranet;
- Establishing a call center to handle customer requests by either outsourcing to call-center organizations or self-erecting the entire infrastructure.

9.3 Capturing and Retaining Customers

A business's customers dictate its ultimate success. Establishing an online business requires innovative strategies and approaches for capturing and retaining customers. Figure 9.1 illustrates the high-level steps involved in customer acquisition and retention. The objective is to build awareness and generate the appropriate response. Capturing customers involves marketing campaigns that include advertising to customers. This involves marketing through traditional channels such as television and radio, and new techniques of advertising over the Internet. Once such campaigns draw users toward the site, these users become prospects and the next step involves leading the customer to make a sale. The success of this step depends upon content offered for sale, value of contents, and other differentiating services that a site offers to generate interest. After generating a sale, the next challenging step is to retain those customers. An organization can offer a multitude of services to retain those customers. The following sections elaborate upon the technical aspects of capturing customers through Internet advertising and retaining those customers through various means.

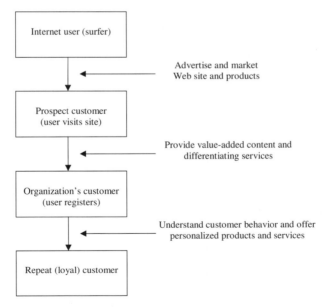

Figure 9.1 Capturing and retaining customers.

9.3.1   Internet Advertising

Advertising builds awareness and generates response. Organizations have relied on various channels for advertising their products and services as well as their brand names. Traditional channels include radio, billboards, television, newspapers, and other means. Doing business on the Internet provides organizations with another opportunity to reach customers. The primary advantage of the Internet is that it enables one-to-one marketing. Using customer-profile databases and database marketing techniques, organizations can target their ads to customers' specific needs. Organizations use Internet advertising to increase sales and revenue, build customer relationships, and promote brand recognition.

Early use of Internet for advertising was frowned upon by a group of Internet users who wanted to use the Internet solely for research and educational purposes. However, Internet-based advertising and the online revenue that it generates has picked up tremendously in the past few years. Most of the public portal sites such as Yahoo! and Excite earn most of their revenue through Internet advertising. Advertisers rent spaces on publishers' Web sites to target their ads and campaigns. Popular means of Internet advertising include banner ads, interstitials, shockwave plug-ins, mobile code (e.g., applets and ActiveX controls), and others. These are described later in the chapter.

Typical advertising steps include the following activities:

- *Scheduling an ad campaign:* Scheduling an ad campaign first requires defining clear goals, the target audience, and the indicators that an organization will use to measure success. The planning phase includes deciding on the site(s), frequency of ad appearance, and date(s) and time(s) of appearance. Other parameters include number of ads delivered, delivery of multiple ads per page, and so on.

- *Deciding on the media used for advertising:* The next step is to decide on the content of the ad and the media that will be used to deliver the ad. Potential media include animated GIF and JPEG files, HTML content, Shockwave, and so on.

- *Targeting ads:* Targeting refers to delivering ads to the right Internet audience. Targeting ensures that advertisers deliver ads to sites whose visitors will have enough interest to follow the ad campaign through. Advertisers can target ads based upon criteria such as zip code, telephone area code, country, income level, search-based targeting (public portal sites use this to display ads corresponding to the entered search strings), times of day, and other parameters. Organizations can maintain user profile databases and target ads based upon customer demographics and other stored information. Organizations need to determine the breadth and depth of profile databases used for advertising to customers. Some organizations provide rich profile databases, while others use internal data to target ads. Engage Technology's Engage Knowledge is a product and service that maintains a rich database of anonymous user profiles and uses numerical identifiers to identify customers.

- *Ad delivery mechanism:* This involves the delivery of ad files to the Web pages that should display those ad files. Various techniques allow the placement of ad files onto the Web pages. One technique requires the inclusion of special tags in publishers' Web pages. When a user requests a Web page, the tags reference the ad delivery server, which in turn serves the appropriate ad to the requested Web page. Another method delivers the ad directly to the user browser, where it displays in a predetermined position.

- *Tracking ad effectiveness:* This involves creating and producing value-added reports that can help advertisers assess the effectiveness of their ad campaigns. Sophisticated ad tracking services and products track customer's reaction to an ad campaign by tracking user's actions till the end. For example, such a technique would assess whether the user clicked on an ad and opted not to continue browsing or whether the user continued through the Web site to place an order through that campaign. Various ad products and services produce intelligent reports depicting various trends of ad campaigns and provide users with deep insights into the effectiveness of their campaigns.

- *Building an end-to-end system for managing ads:* End-to-end ad management involves multiple activities that have to be thought through properly and requires the establishment of a workflow to generate, deliver, and track effectiveness of ads. Certain organizations deploy internal ad server solutions, while others rely on external agencies to deliver an end-to-end solution. Centralized ad management organizations provide services that enable advertisers to reach out to the right customers through the right electronic channels. These organizations also enable ad publishing organizations to control the process of placing ads and delivery schedules. AdForce is an example of an organization that provides end-to-end ad management services and matches Web-advertisement publishers with advertisers based on their individual profiles and requirements. AdForce maintains a high-performance data center to serve numerous ads to its customers.

9.3.1.1　Internet Advertising Techniques

Multiple techniques exist for advertising on the Internet. Some of them include the following:

- *Banner Ads:* These are varied-sized rectangular areas on a Web page that display ads. The Internet Advertising Bureau (IAB) has designated various sizes for these areas to suit various needs. The IAB is a non-profit organization that focuses on research, standards, and associated initiatives to maximize the effect of Internet advertising. This organization defines standards for the various facets of Internet advertising, including advertising terminology, ad-size formats, and the counting of ad hits. The IAB has stipulated the following sizes [3]:
  — 480×60 pixels (full banner)
  — 392×72 pixels (full banner with vertical navigation bar)
  — 234×60 pixels (half banner)
  — 120×240 pixels (vertical banner)
  — 120×90 pixels (button 1)
  — 125×125 pixels (square button)
  — 88×31 pixels (microbutton)
- *Interstitials:* Interstitial means something in between. In Web terminology, an interstitial is a page that is inserted in between the normal Web page flows. A user, for example, while browsing to a URL may see an intermediate page that displays an ad. The intermediate page or interstitial automatically moves the user to the desired page after a brief pause.
- *Animated graphics:* This refers to displaying 3D images or animated graphics to customers.
- *Ad sequencing:* This type of ad campaign appears in a sequence and has the effect of telling a story to customers. The challenge in delivering these types of ads is delivering the portions of the ad sequence to the appropriate sites, thus enabling customers to properly perceive the desired sequence.
- *Listing in Internet yellow pages:* This form of advertising is similar to the traditional yellow pages directory in which businesses list their business names by category. Customers browse through the yellow pages to look up a business in the appropriate category. National Direct Internet Yellow Pages (NDIYP) at www.ndiyp.com is one such service pro-

vider that lists businesses by category and locality. Most portal sites also offer their own yellow pages. Following are yellow page services offered by the popular portal companies.

— AltaVista's (www.altavista.com) Zip2.com (avyellowpages.zip2.com);

— Yahoo!'s (www.yahoo.com) Yahoo! Yellow Pages (yp.yahoo.com);

— Go Network's (www.go.com) Go Yellow Pages (uses BigYellow, a service of Bell Atlantic Electronic Commerce Services, Inc.).

- *E-mail-based advertising:* Certain organizations can send products and services updates to customers if the customer specifically asks the organization to send such information. For example, customers may register with the organization when they buy a certain product and ask the vendor to keep them informed about new product releases or updates.

- *Listing in portal sites' search engines:* This allows advertisers to list their domain names with popular public portals' search engines. An organization's listing appears when visitors type in specific categories for searching the Web. Various advertising agencies exist that provide such search engine registration services. Costs for submissions to search engines vary depending on the number of search engines. There are more than 100 search engines on the Web. The challenge for organizations is using the appropriate descriptions and keywords that will enable the search engine to index the Web site's name and list it higher up when returning search results.

- *Traditional channels:* Traditional means of advertising on the radio and TV continue to be popular. The availability of these channels through the Internet has further fueled interest in these forms of advertising.

- *Interactive advertising:* This refers to the use of multimedia software, applications, and demonstration software to explain an organization's offerings. Multimedia tools captivate customers' attention while providing the user with a better understanding of the organization's products and services. Internet-connected video kiosks enable organizations to offer richer graphics and more captivating presentations.

Various vendors offer end-to-end advertising products. The advertisement process consists of planning campaigns, targeting, ad serving (using sophisticated and captivating technologies), and measuring the effectiveness of those strategies. Organizations have the option to implement an in-house ad-management system or outsource it to organizations like AdKnowledge that provide end-to-end systems. AdKnowledge is a very popular ad-serving service on the Internet. It provides a suite of tools and services that enable an organization to direct Web-based advertising and measure the effectiveness of campaigns. Organizations can then use that data to enhance and adjust their strategies and campaigns. The system is comprised of various modules. The Planner module helps an organization identify the target audience and Web sites suitable for launching campaigns by providing them with access to a searchable database indexed by several criteria. Another module called the SmartBanner automates the delivery of ads and advertising campaigns to the right Web sites. Organizations can control the direction of the ad campaigns by frequency, platform type (browser, operating system), and other criteria. The system allows same day changes so that organizations can react to special events and other business drivers.

ROI Advisor is another component that enables organizations to assess customers' actions after they respond to the ads (e.g., by clicking on the ad at a particular Web site). This provides better data for decision-making analysis, as opposed to just assessing clicks on a Web site. The Reporter component allows the generation of detailed reports that provide various aspects of the ad campaigns.

Accipiter's AdManager is another product that automates the ad management process and provides multiple reporting features. The product works with Engage's Precision Profiling system to target ads based on specific user profiles. Organizations that wish to outsource Ad management solutions can chose to use Accipiter's AdBureau service, which enables organizations to completely outsource all technical aspects of ad management to Accipiter. The service builds upon the AdManager solution.

Other solutions include NetGravity's AdServer, Real Media's Open AdStream, and W3.com's Ad Optimizer.

9.3.2   Customer Retention

Advertising can help organizations attract prospects and may even help toward their first sale. However, pleasing customers enough to get them to return to the organization's Web site requires incorporating numerous quality points into their Web sites and into the services that organizations offer through those Web sites. The following are some considerations to retain customers on an organization's Web site.

- *Understanding customer requirements:* Understanding customer requirements is more vital for the design of e-commerce systems than it was for traditional systems. One reason is that e-commerce customers are external entities who have a direct bearing upon an organization's financial portfolio. It is vital to factor acceptance criteria into the design of the application before embarking on any e-commerce systems initiative. For example, an organization may require the usage of certain payment schemes (e.g., offering electronic cash payments) on its e-commerce site that may not be popular with customers in that market niche. Similarly, an organization may wrongly assume that all of its customers in a certain market segment have multimedia PCs and are knowledgeable enough to engage with a business in a particular fashion. This may prove to be a very costly mistake.

- *Analyzing customer behavior through browsing habits:* Organizations can analyze customers' interactions with its Web site and provide better offers, customer service, and other services to keep customers interested in its business. An organization can analyze customers' buying patterns and offer services and products that suit those patterns. For example, if a customer orders gifts through a Web site during certain times of the year, the organization can remind customers in the future (e.g., through e-mail). Similarly, based on customer interests identified through their browsing behavior, an organization can cross-sell and up-sell relevant products and services. For example, an organization can offer the customer an automobile security system and other amenities when the customer buys an automobile. Andromedia, Inc.'s Like-Minds is an example of a product that records user behaviors and analyzes that behavior to enable organizations to efficiently cross-sell and up-sell products and services to customers.

- *Fulfillment steps and cycle times:* The organization should offer customers streamlined request fulfillment steps to keep them interested in their site. Request fulfillment includes all steps required to process a customer's request. For example, an e-commerce site that enables customers to browse through an organization's Web site for products but requires them to call an 800-number to enter orders will deter customers.

- *Updating content:* An organization should continually update content on its Web site. Technological breakthroughs enable an organization to provide dynamic updating of content. Certain site visits offer the impression of outdatedness. Technological breakthroughs (e.g., interactive media) facilitate the means to present content that portrays a lively image of the site to customers. The organization may also display the latest press releases to provide an up-to-date impression.

- *Including search capabilities:* A site that offers a multitude of content should include search capabilities in its Web site. Investment in sophisticated search engines prevents the dumping of information onto customers.

- *Respect for privacy:* Many customers are sensitive about privacy issues as they visit a Web site. Aside from addressing the legal ramifications, the organization should be forthright in openly addressing their concerns. If the organization assesses privacy to be a major concern for its particular business, then it should address it accordingly by making the policy clearly visible and not include the privacy statement in a small font at the bottom of the Web page.

- *Customization and personalization:* Public portals effectively use these techniques to retain customers. Using customization and personalization, customers can tailor a Web site's various views with different contents. This technique enables organizations to build better customer relationships. ART Technology Group, Inc.'s Dynamo suite of products give Web merchants and service providers the ability to offer these features to customers. Dynamo's Personalization Server provides organizations with the ability to target profile-based content to users. The product offers adapters that integrate diverse data sources and sends targeted e-mail and other messages based upon user profiles. By recording user behavior on the Web, the data can be used to cross-sell and up-sell products and services. Besides, the product suite also enables business users to request personalized content to perform various business functions.

- *Site performance:* An organization with inadequate site performance will frustrate users and drive customers away. Organizations can deploy numerous tools to measure a Web site's performance. For example, I/PRO provides tools for Web-based measurement and auditing. I/PRO's NetLine product provides detailed analysis and reports on a Web site's traffic patterns. The product enables organizations to measure a Web site's statistics for a variety of architectures and configurations. For example, certain Web sites are implemented on a single server, while others are spread over multiple servers and include subhierarchies of sites. The product enables central viewing of critical data. Similarly, WebCriteria provides a measurement product called SiteProfile that enables organizations to assess Web performance based on certain criteria. These criteria consist of four primary indicators of load times (time it takes to load a Web page), accessibility (time that it takes to navigate through various Web pages), freshness (content age on the site), and composition (media types included in a Web site, such as text, audio, and video).

- *Security:* Organizations that do not facilitate secure order entry are less likely to attract customers. SSL and other security technologies enable organizations to offer secure sessions through which customers can enter private personal and payment information. Organizations that attract customers sensitive about entering payment information on the Internet should alleviate customers' concerns accordingly by providing a plain description of the security features of the transactions.

- *Measuring a Web site's success:* An organization should continuously monitor the success of its online business to improve its brand image and revenue indicators. During the infancy of the Internet and e-commerce applications, organizations gauged customer success by measuring the number of customers who visited a Web site in a given period of time. Organizations that offered customers the ability to register on their Web site measured their success in user registrations. Such a metric was adequate during the Internet's infancy, but it can no longer be considered a reliable indicator of success. Many customers who register on certain Web sites rarely revisit those sites. Although capturing a customer's attention to join a Web site's registered users list does indicate some measure of success, only repeat visits reflect a site's true popularity. Organizations can use some of the following to measure a Web site's success:

  — Number of repeat visits;

  — Number of visits measured over time (e.g., which day of the week or day of the year has the maximum number of visitors);

  — Sales volumes;

  — Studying customer demographics (refers to specific customers who visit an organization's Web site);

  — Content access frequency (refers to the visits per page, file, or other content available on the site);

  — Access paths (refers to the manner in which customers access the Web site; e.g., direct access by typing the site's address or referral through another site such as a portal);

  — Number of site hits (customers visiting the Web site);

  — Average time that a customer stays on the site;

— Referring links that bring customers to this Web site;

— Tracking Web problems that customers face during their visits;

— Ad campaign results.

9.4   Using SLAs to Manage Outsourced Services

Service Level Agreements (SLAs) are an important element of the IT Service Management (ITSM) strategy. These documented agreements specify performance thresholds for various outsourced IT services. In the traditional IT infrastructure, SLAs primarily concentrated on internal "data center" services. However, their role in the e-commerce paradigm has been transformed to encompass external entities engaged in delivering e-commerce services to the enterprise. Negotiated SLAs hold external service providers accountable for disruption to any of the performance indicators.

An SLA is a contract of services between the service provider and customer. The customer may receive such services from entities within the organization or external to the organization. Typical examples include businesses hosting IT services in data centers or an organization hosting IT services at external NSPs. In such scenarios, the data center or NSP is the service provider and the organization is the customer. SLAs have to be established between these entities in order to ensure that the service provider provides the various computing services per established SLAs.

SLAs stipulate the services that the service provider will provide to the customer organization and respective performance levels. The service provider is expected to adhere to those levels in delivering those services to customers. The granular definition of those service levels facilitates the monitoring and execution of the agreement.

Formulating SLAs provides customer organizations with two primary benefits. First, SLAs enable organizations to limit risks as well as share risks with the service providers. By establishing and agreeing upon various threshold levels for various services, an organization can expect to consistently receive those services from the service providers. Organizations can demand various forms of compensation should the service provider fail to meet those levels.

Second, SLAs provide organizations with confidence in outsourcing services, thus lowering TCO. Delivering services at consistent quality levels requires enormous investment in technology and process infrastructure. The outsourcing of services backed by SLAs gives organizations the advantage of focusing on its core competencies while expecting desired services.

Since businesses rely on IT for the operation of their businesses, SLAs are required at all levels of an organization that ultimately influence a business performance. Typical service providing organizations include the following:

- Internal data centers
- External data centers
- Internal software development organization
- External software development organization
- Network service providers
- Security administration functions
- Call centers.

An SLA has several components, some of which are as follows:

- Parties covered under the SLA contract (name and address of the service provider and customer organization);
- Listing and description of functions and services covered by the SLA;
- Clear definitions of all legal and technical terms used in the SLA (e.g., site unavailability, system unavailability);
- Methods used to calculate or measure SLA thresholds (e.g., latency, network outage time, network delays, stipulating the customer demarcation point to the service provider demarcation point);
- Thresholds and levels associated with those functions (e.g., latency of 100 milliseconds, transaction fulfillment time of 3 seconds);
- Validity period of the SLA along with a discussion of renewal conditions and cancellations;
- Clear definition of processes in executing SLAs between the service provider and the customer, problem resolution processes;
- Actions that the service provider will take when not meeting certain SLA clauses:
- Refunds associated with various services;
- Refund of liquidated damages;
- Termination of the contract;
- Contact names and address of individuals to be contacted for escalation or trouble reporting.

The concept of system ownership is vital for understanding and formulating SLAs. Identification of an IT system's owners ensures the effective and efficient operation of the business's (system owner's) applications and systems. Usually, the business is the system owner of systems that execute in a production environment, such as a data center that provides services to the system owner. The data center acts as the custodian of those systems and maintains those systems according to the terms and conditions agreed upon with the service provider. A system owner has (or should have) a comprehensive understanding of the IT systems, including applications, supporting platforms, system level software, configuration requirements, and performance requirements.

During the mainframe era, the concept of system ownership existed quite loosely. Most of the organization's businesses relied on the organization data center(s) to squeeze an organization's entire applications portfolio onto one or more large mainframes. The data center assumed an active role in capacity planning and other such activities, to ensure upkeep of the system(s). The data center acted as the primary hub to provide other services as well, including network services.

In the e-commerce era, organizations' applications run on a multitude of data centers or network operations centers (NOCs). Furthermore, as highlighted in earlier chapters, organizations rely on multiple service providers for their business operations. In fact, an organization will find it quite challenging to transact e-commerce services with out relying on the services of service providers, which minimally include ISPs and NSPs.

Explicit ownership of a business's systems can help alleviate numerous issues. Taking explicit ownership of one's systems can be beneficial, as it drives the system owner to define end-to-end operational requirements of systems and assess future requirements. Most businesses (system owners) assume that an organization's data center (or other external service providers) will define and tackle operational issues for the system owner (e.g., capacity planning). While this is an adequate scheme, in the midst of the technology diversity and rapid innovations in e-commerce, it is only prudent that the system owner identify its operational requirements. The system owner can then communicate these requirements to the service provider (e.g., any data center) and match the data center's existing operational practices and procedures with the new system's requirements. Lack of such planning results in delays in deployment (e.g., a data center's primary enterprise-level UNIX platform may not be adequate to handle the new application's memory requirements) and other similar issues. Furthermore, the application or system may not be adequate (from a security, performance, or other perspective) to be shared on one platform with other applications that have other system owners. For example, system owners sharing one platform for their applications may have conflicting requirements about the operations of that platform that may effect each other's operations.

System ownership also benefits the service providers. A data center (or other service center) that services numerous businesses and their applications and systems can set appropriate expectation levels of their owners by defining adequate SLAs. By receiving defined requirements from the system owners, the service provider can cater to the special operational requirements and other needs of the system owner.

Service providers and centers may pre-assess the common needs of the system owners and commit to a set of defined service levels. This model is very popular among service centers that serve like businesses (e.g., a call center established to service financial institutions that have similar needs). NSPs or ISPs that collocate other business's systems may have to accommodate varying requirements and define different SLAs to provide services.

9.4.1 ISP/NSP/Data Center SLAs

The following delineates some common services that IT operational centers provide for their customers.

- *System installations:* A data center usually receives defined installation requirements from the system owner. The SLA may require the service provider to perform installations (e.g., installing new releases, maintenance releases) within a certain time, or per schedule as defined in the installation requirements. The SLA may stipulate that the service provider should inform the customer in advance if the installation cannot proceed due to certain reasons.

- *Change management:* This SLA may require the provider to institute adequate change management procedures to ensure that the system can revert to the original configuration if the new release malfunctions.

- *System troubleshooting:* For malfunctions, the SLA may require the provider to troubleshoot the system within certain predefined limits rather than triggering the next level of support immediately.

- *System backups:* This SLA parameter requires the provider to follow regular backup procedures. For example, the system owner may stipulate special physical security of backup tapes that store a portal's customer profiles. This may cause the provider to build physical floor space and vaults to provide such services.

- *Network availability:* This refers to the percentage of time that the network is available during the expected network uptime. Therefore, this parameter does not include maintenance times and other customer requested downtimes. A customer may stipulate in the SLA that the service provider will credit the customer for any unavailability during this period, based on predefined criteria (one hour of unavailability results in an entire day's credit, for example).

- *Network performance guarantees:* The customer and service provider may agree upon specific thresholds for network performance. Examples of network performance indicators are as follows:

  — Latency guarantees (e.g., 100 milliseconds for average roundtrip latency);

  — Committed information rates (CIR) in frame-relay circuits;

  — IP QoS indicators (response times at packet levels);

  — Delivering network statistics and reports to customers on the network usage and performance;

  — Access link availability (e.g., 24 hours a day or 99%);

  — Throughput of networks to support multimedia applications.

- *System response times:* This parameter refers to application performance. For business-to-consumer Web sites, end-to-end performance is a critical indicator that affects customer satisfaction levels. The customer may specify thresholds for each application. For example, a shopping Web site may require the service provider to not exceed a certain time limit (e.g., 3 seconds) for an end-to-end transaction fulfillment time. Similarly, a customer that outsources e-mail services may require the service provider to monitor end-to-end usage of e-mails and commit to certain delivery service levels (e.g., mail will be delivered within 5 minutes).

- *Business continuity plan:* This refers to the business plan that a service provider executes in the event of a service disruption to ensure continued delivery of business services. The customer can require the service provider to submit a detailed plan that will guarantee service continuation in case of a business disruption. Depending on the business requirements, the customer may agree to a basic set of services until the service provider resumes full operation. Some of the pertinent service-level indicators that customers can define with their service providers are the following:

  — Recovery time to restore backup (contingency) operations;

- — Recovery time to restore full operations;
- — Service restoration times;
- — Contingency sites and their locations;
- — Data backups and recovery mechanisms;
- — Identification of a recovery team.

- *Reporting:* Customers can request a specific list of periodic reports from the service provider. These reports can be performance-related, that the service provider will deliver to the customer periodically.

- *Problem management:* This refers to the problem resolution process that the service provider institutes in providing services. The problem management process can be a basic manual process or the customer may require the service provider to install a full-fledged ticket-resolution-based system to handle and resolve system issues. The problem management process should include the following:

  - — Notification channels (people and their contact information);
  - — Escalation of problems in an outage situation;
  - — Service response time (e.g., 24-hour help desk, 4 hours to send technician on site).

- *Capacity planning:* Operational systems continuously require more computing resources to support business processes. For example, an organization enrolling customers over the Internet requires more disk space to store their profiles. Similarly, new applications may require additional memory. A system that reaches its resource limits may result in business unavailability. Capacity planning tools ensure continual monitoring of systems, predict such resource demands, and alert administrators. An organization that expects a continual increase in customers should include specific capacity planning requirements in the SLA.

9.4.2 Software Development SLAs

Organizations approach software development issues in numerous ways. One way relies on internal software development departments. Others outsource software development to external organizations. In either case, the business depends upon the development organizations to deliver quality systems on a committed schedule. A business may define SLAs with software development service providers and request certain levels of service. The following entails the definition of certain SLAs for a software development organization.

- *Scope of application development:* This parameter stipulates the specific functions and services that the development organization will fulfill in delivering systems to its customers. The scope may include only application development and unit testing or may include full lifecycle development.

- *Adopted IT processes:* The application development service provider should clearly state the processes that it will follow to deliver systems to its customer. The customer can audit the application service provider against those standards. For example, the customer can require the service provider to maintain a certain CMM maturity level. The customer may also require the service provider to meet certain quality thresholds for delivered software products such as defect levels, cycle times, and meeting of commitments.

- *Continuity of business plans:* Unavailability of the application development and testing environment could prevent developers from delivering applications to the business in due time. In the e-commerce arena, where businesses rely on the application development organization for various business initiatives (e.g., introducing promotions, rectifying a security defect), appropriate business continuity plans are mandatory. Unavailability of the development environment can hamper a business from responding effectively to a customer's requirements. The service provider should demonstrate plans that will provide the customer the assurance of business continuity to cope with various disasters. The customer should set those thresholds and business continuity commitments in accordance with business requirements.

- *Configuration management infrastructure:* A configuration management infrastructure controls the various software work products that comprise an e-commerce or IT system. As a complex e-commerce system can include numerous software components, the tracking of changes and versions associated with those components becomes increasingly essential. The wrong version of a single component can result in the unavailability of the entire system. An organization can thus require the service provider to guarantee adequate configuration management controls.

- *Software peer-review practices:* Peer review allows software developed by one individual to be reviewed by another to ensure adherence to certain standards and to detect any other injected defects. Lack of code review practices increase the threat of malicious software (e.g., time bombs, viruses), which may disrupt operations of a mission-critical system. Expecting application service providers to institute such practices thus provides the assurance of reasonably defect-free software. For cases where the application provider offers a turnkey solution, the service provider should assure the customer (through legal contracts) that the delivered code is free of viruses and other malicious code that may impact business availability or loss.

- *Protection of intellectual property:* If the organization uses an external service provider for application development, it is bound to share critical business requirements with the application developer. The service provider should ensure that all sensitive information related to business propositions and requirements are not disclosed.

9.5 Understanding E-Commerce Legal and Regulatory Issues

Engaging in e-commerce initiatives over the Internet raises numerous privacy issues related to customer information. Misuse or mishandling of customer information can result in litigation and can ruin an organization's image. Internet-based computing also poses numerous software piracy risks and endangers software publishers' bottom line figures. The following sections further elaborate on these two issues.

9.5.1 E-Commerce Privacy Issues

Organizations with Web sites hold the potential to collect various types of information from the customer. An organization has ethical and legal obligations to prevent misuse of this information, protect it from unauthorized disclosure, and refuse to share it with other organizations without the customer's knowledge. The organization should declare its practices regarding usage of such information. These privacy statements should minimally include the following:

- *Notification to customers about specific information collected about them:* An organization can collect information about customers by directly asking them for certain information (e.g., name, addresses), or by collecting information without the customer's direct knowledge. This includes information such as IP addresses of customers and navigation behavior.
- *Notification about the usage of data:* Some organizations use customer-related data for marketing purposes and self analysis. Others use this

information to provide customized content or user interfaces to their customers. For example, information collected by portal sites (e.g., NetCenter) enables the portal site to offer a customized view of the portal to their customers. Certain information collected from customers could be very sensitive. This especially applies to sites that offer numerous calculators or tools for their customers (e.g., health and financial assessment tools). Organizations that collect this type of information have to specifically notify their customers about the organization's intent to use that data and about the internal safeguards and security controls that protect that data from unauthorized access.

- *Notification about the distribution and sharing of customer information with other entities:* Organizations usually do not share customer information directly with other organizations without notification to the customers. Some portal organizations maintain customer profile information and respond to advertisers' requests by targeting ads to specific user profiles (e.g., certain age group, gender). Web sites should notify customers about their intent to share customer information with other entities.

- *Security controls incorporated into the site to protect their visits and transactions:* Sites should notify customers on the various security mechanisms they employ to accept financial transactions over the Web and how they protect that data internally.

- *Provide the opportunity to opt out:* An organization's privacy statement should provide the customer the option to turn off information collection. These steps may direct the customer either to use certain browser's features to turn off such information collection, or may offer an application that does it for the customer.

Various organizations exist to assist customers in the building of appropriate privacy statements. The following organizations provide such services and provide seals of trust to attest to the acceptable privacy practices followed by organizations.

9.5.1.1   BBBOnLine

Better Business Bureaus are nonprofit organizations that provide various reports to consumers on the business practices of their enrolled businesses. Various reports from the BBB inform consumers about the age of the business and their record on customer complaints. The reports also indicate whether the business has been subject to any governmental actions. BBBOnLine is a fully owned subsidiary of the BBB that extends the same principles of trust to the Internet and online community.

BBBOnLine has undertaken various initiatives to offer services that indicate whether businesses comply with online privacy principles. The BBBOnLine provides various seal programs (BBBOnline Reliability seal and BBBOnline's Privacy seal) that require participants to meet stringent standards for the usage and treatment of personal information in cyberspace.

The BBBOnline Reliability seal provides customers the assurance that the site meets certain criteria and follows required business practices. Some of the criteria that the BBBOnline requires from sites requesting such a seal are the following:

- The organization has been in business for a minimum of one year;
- The organization addresses customer complaints by responding to them effectively;
- The organization follows rules for advertising guidelines and agrees to withdraw ads if found in violation of certain practices and standards;
- The organization is a member of the local BBB chapter.

BBBOnline's Privacy seal ensures that the organization has taken the appropriate technical and other security measures to secure customer information. Organizations, through privacy statements are also required to share with their customers their practices about sharing and distributing this information to other entities. Any business that displays this seal is required to notify BBBOnline of any modifications to their privacy policy. BBBOnline's Kid's Privacy seal includes the same requirements as the BBBOnline's Privacy seal and additionally ensures that sites take effective measures to control children's communications and do not collect any information unless by parental consent.

9.5.1.2 TRUSTe

TRUSTe provides privacy-statement–building wizards for their customers. TRUSTe awards a seal (referred to as *trustmark*) to businesses that meet the standards and follow the practices stipulated by TRUSTe. Clicking on this seal on any site takes the user to the privacy statement of that site. TRUSTe continually monitors the holders of their seal to ensure adherence to their standards and practices. TRUSTe enables users to look up businesses that are awarded the TRUSTe seal.

9.5.1.3 VeriHost

VeriHost is an organization that certifies Internet Web-hosting organizations per certain quality standards [4]. VeriHost uses a multitude of criteria to certify Web-hosting organizations. Upon appropriate certification, a Web-hosting organization receives a seal of trust that it can display on its Web site. Upon clicking on the seal, customers receive authentication from VeriHost about the Web site that is certified by VeriHost. VeriHost considers financial, technical, and other information about the Web-hosting organization—such as its security practices—before providing its seal of approval.

9.5.2 E-Commerce Piracy Issues

Piracy is the unlawful copying and/or use of software. Electronic piracy controls protect publishers' investments in content. Illegal use of software includes copying original software for use without the publisher's permissions and exceeding usage of a software product over the authorized license limits. Various governmental regulations exist to curb piracy activities. For example, those who violate piracy laws within the United States can be fined thousands of dollars or serve a jail term. Other countries vary in their enforcement of antipiracy initiatives and penalties.

Electronic piracy costs software organizations enormous amounts of money. A recent study by the Software and Information Industry Association revealed that the top ten metropolitan areas in the U.S. were responsible for a total loss of more than a billion dollars (the worldwide figure was 11 billion) [5].

Numerous antipiracy schemes have surfaced, including the following:

- *Dongle protection:* This method requires the use of a dongle that attaches to a parallel port of the computer. The program reads the dongle for a special serial number and upon detection and successful comparison allows execution of the program.

- *Embedding 3D holograms on CDs:* Many organizations use this technology to communicate the authenticity of the media that they buy to copy software. Microsoft has actively used this technology.

- *Watermarking:* Watermarking enables digital content publishers to embed watermarks, or special codes, into content to combat piracy. This prevents illegal copying and also in certain cases enables publishers of digital content to track the usage of their published material. Various watermarking technologies exist to combat content piracy.

- *Software metering:* This technology allows organizations to track software usage from a centralized location. Software licenses act as binding contracts between software publishers and users. Organizations can deploy software-metering products to track software licenses throughout the organization. This is especially very useful for organizations that buy large volumes of licenses because tracking licenses through manual means is daunting. Popular products that facilitate software metering include Tally Systems' CentaMeter and Microsoft's Systems Management Server (SMS).

Numerous technological and procedural initiatives have surfaced to counter piracy and related issues. Technology-based piracy controls prevent the use of existing technological methods (e.g., using cut and paste for copying) to violate piracy controls. Procedural initiatives rely on ethical policies, administrative controls, and governmental enforcement to thwart piracy practices. The following elaborates upon some of the many initiatives that exist to prevent IT-related piracy activities.

9.5.2.1 Digital Millennium Copyright Act (DMCA) of 1998

The DMCA (S. 2037) is an amendment to the United States' original copyright law. Copyright law is a form of intellectual property law that protects original works of authorship [6]. The law extends to include technological measures that may be used or circumvented to infringe upon an entity's copyright laws. This law forbids illegal duplication of technologies and forbids the use or invention of technologies to break security technologies (e.g., encryption). This act, however, frees ISPs and other service providers from liabilities if they are unaware of the use of their services for piracy-related activities and, if discovered, they take due measures to curb such activities.

9.5.2.2 AudioTrack Secure Watermark

AudioTrack places a special code in audio files (e.g., MP3, WAV, AC3) to prevent piracy. This code is referred to as the AudioTrack Secure Watermark. The code stays permanently in actual content regardless of the distribution media and data manipulations (e.g., encryption, compression). The company offers software that allows the embedding of these codes into sound files without increasing the original file size. Publishers can use this technology to track copyright violations. Sophisticated audio analysis systems can read these codes to identify the copyright.

9.5.2.3   Business Software Alliance

Business Software Alliance is a group of software publishing organizations that undertake initiatives to fight software piracy by influencing government policies and educating end user communities on the issues of software piracy. BSA members include Microsoft, Adobe, Visio, Novell, and other software publishing organizations. BSA encourages reporting of all forms of software piracy.

9.5.2.4   Digimarc's Image(s) Piracy Protection

Images are a vital component of most Internet and Web publishing activities. However, cutting and pasting technologies facilitate piracy opportunities. Digimarc, Inc. has invented an antipiracy technology that controls the use of proprietary images. Using Digimarc's technology, an organization can publish copyrighted images and protect them from tampering. Digimarc's technology embeds a special ID into the pixels of the image that survives all copying, distribution through electronic channels, and basic editing operations. This ID is unique for every organizational entity that registers with Digimarc and hence acts as a permanent signature that attaches to the image, thus enabling organizations to track image usage. Digimarc facilitates this through their MarcSpider service, which scours the Web for such copyrighted images and reports on their usage. This facilitates the detection of any copyright violations. Numerous image editing applications such as those from Adobe, Corel, and Micrografx use this technology.

Digimarc claims that their technology is not foolproof, but it provides publishers with an adequate measure to prevent the piracy of images.

9.5.2.5   Secure Digital Music Initiative (SDMI)

SDMI is an initiative sponsored by organizations from the multimedia and IT industries to prevent against music piracy. SDMI is not a music format; rather the initiative provides piracy protection for digital musical content and thus protects artists from copyright infringement resulting from copying of digital music. The ultimate objective of the standard is for compliant devices to detect pirated music. This standard would thus regulate and control the distribution of music through the Internet and other electronic channels. Since the current base of users do not own SDMI-format content, the framework's current specification supports both current and secure formats. Manufacturers can adopt the standard to build SDMI-compliant portable devices to enable playing of protected and unprotected music. SDMI-compliant devices can also play MP3 music.

Notes and Web Sites

[1]   Wang, Charles B. *Techno Vision*. New York: McGraw-Hill, Inc. 1994.

[2]   Kaplan, R. and Norton, D. *The Balanced Scorecard: Translating a Strategy into Action*. Cambridge, MA: Harvard Business School Press, 1996.

[3]   www.iab.net/iab_banner_standards/bannersource.html

[4]   verihost.com/press.html

[5]   www.siia.net/news/releases/piracy/6.8.99-Piracy-Release.htm

[6]   lcWeb.loc.gov/copyright/faq.html#q1

Index